Professional Grooming & Care of the Racehorse

ISBN 0-935842-10-1

P.O. Box 535547 Grand Prairie, Texas 75053
(U.S. & Canada) 1 (800) 848-0225
(Other Countries) 1 (214) 660-3897

Author
T.A. Landers

Veterinary Editor
Chris King, BVSc, MACVSc, MVetClinStud

Editing, Appendix
research staff of Equine Research, Inc.

Editor/Publisher
Don Wagoner

Dedication

In loving memory of Mary Haff, Eileen Landers, and Grace Sheehan.

Acknowledgments

The author gratefully acknowledges the many people who contributed their time and effort to make the production of this text possible. Together they have made my dream come true.

My wife, Jeriann, for her moral support and patience. My daughters, Kristin, Jennifer, and Meghan, who remained patient and amiable during the writing of this book.

My friend, Joel Silva, who provided most of the expert photography. David and Ann Gribbons, owners of Knoll Farm, Brentwood, New York, for their valued assistance in providing the facility, horses, and equipment needed for the photographs throughout the text. Ms. Suzanne R. Morrissey-Ambrosio and Willow Keep Farm for their valued assistance with some of the photographs appearing in the text. Mr. Joseph Allocco and Ms. Janet Janifer, for their original art work.

Ms. Catherine Sheehan, for her professional typing of the manuscripts. JoAnn Cardillo, for her typing contribution. Mrs. Diane Brintzenhofe for her efforts in editing the rough drafts of the text.

For their help with the Standardbred portions of this text I would like to thank the following people: Ms. Dolores Zanieski and Hugh “Slugger” McIntosh of Sound Turnouts, Jamesport, New York; Tom Corelli, Driver/Trainer; Dawn Fillips, Groom; Bob and Paulette Greenfield, Owners and Trainers; Louis and Kathy Grisolia, Grooms; W. Budd Benner, Farrier; G & B Farrier Supplies, Westbury, New York; Jim Lowe, Farrier; Dan Cervat and Jim Knox, Nassau Sulky Corp., Babylon, New York; Gary Lewis, Trainer and Author.

The Thoro’Bred Racing Plate Co., Inc. of Anaheim, California for providing information and materials on horseshoes for the book.

Finally, to all the expert horsemen whom I have had the privilege of meeting over the years during the production of this book. Especially, Mr. Frank Whiteley Jr., David Whiteley, C.R. McGaughey III, William Gibford, and fellow teacher and friend, the late William Ardito.

Disclaimer

Every effort has been made in the writing of this book to present quality information based on the best available and most reliable sources. Neither the author nor the publisher assumes any responsibility for, nor makes any warranty with respect to, results that may be obtained from the procedures described herein. Neither the publisher nor the author shall be liable to anyone for damages resulting from reliance on any information contained in this book whether with respect to feeding, care, treatment procedures, drug usages, or by reason of any misstatement or inadvertent error contained herein.

Also, it must be remembered that neither the publisher nor the author manufactures any of the drugs, feeds, or training products discussed in this book. Accordingly, neither the publisher nor the author offers any guarantees of any kind on such items—nor will they be held responsible for the results that may be obtained from the use of any of those items.

The reader is encouraged to read and follow the directions published by the manufacturer of each drug, feed, or product which may be mentioned herein. And, if there is a conflict with any information in this book, the instructions of the manufacturer—or of the reader's veterinarian—should, of course, be followed.

To ensure the reader's understanding of some technical descriptions offered in this book, brand names have been used as examples of particular substances or equipment. However, the use of a particular trademark or brand name is not intended to imply an endorsement of that particular product, nor to suggest that similar products offered by others under different names may be inferior. Also, nothing contained in this book is to be construed as a suggestion to violate any trademark laws.

Preface

For as long as I can remember, I have heard racetrack trainers and farm managers complaining about how hard it is to find good grooms. Today, more than ever, racehorses are a business and represent major investments. And like most major investments, they require a professional staff to develop their potential and maximize their abilities. While the trainer may devote several hours each day to these horses, their well-being is primarily in the hands of their grooms.

As there are so many young people who love horses and are drawn to these and other horse-related industries, the problem must not lie in the availability of groom candidates, but in the availability of knowledgeable, professional grooms. As such, the purpose of this book is to define the responsibilities of the groom and to describe the various skills necessary to provide *top* care for a racehorse. And since few horses are pampered like racehorses, this book can also be a vital asset to owners, trainers, and caretakers of show horses, polo horses, eventers, hunter/jumpers, western pleasure horses, backyard horses—any horse at all.

Understanding that a groom's job emphasizes skills more than book knowledge, I have provided many step-by-step instructions, "action photographs," and detailed illustrations in an effort to make the learning process less painful. I also made an effort to arrange topics in the order that a groom might encounter them, to make the book into a handy reference.

Horseracing is an exciting sport, perhaps more so for the groom than for anyone else. There is nothing quite like seeing the result of your hard work leading down the homestretch.

Contents

Chapter 1

Chapter 2

Chapter 3

Chapter 4

Chapter 5

Chapter 6

Chapter 7

Chapter 8

Chapter 9

Chapter 10

Chapter 11

Chapter 12

Chapter 13

Chapter 14

Chapter 15

Chapter 16

Chapter
1

Introduction to the Backstretch

For those who have never seen it, the backstretch is the area behind the racetrack's back-stretch where the race-horses are stabled and where all the work and training takes place. Anyone who lives or works there can tell you the backstretch is a world of its own and always full of excitement. In the morning, the area is buzzing with trainers, grooms, and riders who are working hard to prepare racehorses for that all-important end re-sult—the winner's circle. In the afternoon, the action

Fig. 1–1.

shifts to the "front side," where the racetrack determines the winners and losers for that day.

This chapter introduces the newcomer to some common elements and events of the backstretch as well as provides the new groom with valuable insights on how to find a job here.

BARNS

The racetrack barns are designed to house from 20 to 50 horses and in some cases, even more. Each barn consists of a shedrow, box stalls, a tack room, and a storage loft.

Fig. 1–2. The shedrow is between the box stalls and the outside wall of the barn.

Shedrow

This is the area between the stalls and the outside wall of the barn. The floor of the shedrow consists of soft dirt. The width of shedrows varies, but at most racetracks the shedrow is about 10 feet wide.

Horses are sometimes walked inside the shedrow. When you walk into a shedrow while horses are being walked, always be on the left side of the horses to avoid being kicked. In other words, when you enter a barn, you should walk next to the box stalls and not next to the outer wall of the barn. If the barns are built in a square, as they are at some racetracks, you will be walking in a counterclockwise direction. Likewise, when you are walking a horse inside the shedrow, always place yourself between the horses in the stalls and the horse you are walking. This practice discourages the horses in the stalls from trying to bite at the passing horse.

Box Stalls

A box stall (approximately 12 feet x 12 feet) is a racehorse's living quarters. The doorway is approximately 8 feet high and 4 feet wide, and the stall door is usually the Dutch door type, which means the top and bottom halves can be opened separately. Each stall has a window (usually no lower than 8 feet above the floor for safety) to allow enough light and ventilation. The stall floor usually consists of clay, which is easier on the horse's legs than cement.

Tack Room

This room serves primarily as a storage area for tack and equipment. However, it is sometimes used as an office by the trainer. Occasionally there will be a specially designed tack room, but more often the tack room is just an extra stall that has been converted.

Storage Loft

The storage loft is located above the stalls. Stall bedding material (straw or wood shavings) and hay are usually stored here. Take care when working in this area as excessive noise above the horses may upset them. *You should always alert any horses or people standing below before dropping items from the storage loft into the shedrow.*

SECURITY

Racetrack security is important—especially when one realizes the enormous monetary values placed on racehorses today. For this reason, the backstretch is restricted to owners, trainers, grooms, veterinarians, and other people who work with the horses; it is not open to the general public. All persons affiliated with the backstretch are licensed by the racetrack security force, which patrols the racetrack grounds 24 hours per day. Backstretch employees must wear identification badges at all times, and their vehicles must display special

Fig. 1–3. The goal of racetrack security is to protect valuable racehorses.

stickers when entering the backstretch. (The backstretch is one place where the automobile must yield the right of way to the horse.)

Security Office

This office is the headquarters for the staff in charge of providing security for the entire racetrack. Any visitor wishing to gain access to the backstretch stable area must apply for a visitor's pass at the security office. It is also here that employees must file any complaints concerning safety while on the backstretch.

This office is also responsible for keeping track of horse vans that are entering and leaving the racetrack. The security office must verify the paperwork (such as registration papers and veterinary certificates) on all incoming and outgoing horses.

BACKSTRETCH AMENITIES

Farrier's Workshop

Professional racetrack farriers perform most of their work out of their vehicles, driving around to the particular stable where the horse is located. However, many racetracks have a permanent workshop on the grounds for times when a farrier must create a special shoe or brace for a specific foot problem.

First Aid Station

Most racetracks provide a first aid station that is staffed with a registered nurse. The first aid station is strictly for non-emergency health problems of the backstretch employees. Any serious medical conditions are usually referred to a local hospital by the nurse or another staff member.

Maintenance Office

This office is responsible for the upkeep of the entire racing facility, including barns, fences, and dormitories. For example, if there is a broken wallboard in a stall, the trainer contacts the maintenance department, and someone will be sent out to repair it.

State Racing Commission's Testing Barn

This barn, also known as the "spit box," is where horses that have officially finished first, second, or third in a race must report after a race. In addition, any horse in any race on any given day may be "spot tested" regardless of its finishing position in a race. By law, each of these horses must also undergo blood and urine tests administered by state racing officials. These samples are then sent to a state-appointed laboratory to test for the presence of illegal drugs.

Typically, a racing commission employee escorts the horse, trainer, groom, and/or hotwalker to the testing barn. The horses are washed, walked, and cooled out inside the testing barn. It is up to the trainer to decide when the horse is cool enough to be tested, at which point the racing commission official will collect the sample.

No person is allowed in this barn unless he or she is affiliated with one of the horses being tested. Only after the samples have been taken are the horses permitted to return to their stables.

Fig. 1–4. A racing official escorts the winner to the testing barn.

Training Track and Clocker's Stand

Some of the major racetracks throughout the United States and Canada not only provide a main track, but a training track as well for its resident training stables. The distance of this training track may vary from ⅝ mile to 1 mile. Somewhere along the track there is usually a clocker's stand, which is typically a shed (with a glass front) set on a raised platform. It is usually from this stand that track officials watch horses and clock morning workout times for publication in the *Daily Racing Form*. Trainers can also watch their horses work and time their workouts from the clocker's stand.

Veterinary Hospital

A fully-equipped veterinary hospital can be found at most racetracks. It is a comfort to trainers and owners to know that such a facility is available to perform emergency surgery if necessary. The hospital may also serve as an autopsy facility in the event that a horse must be put down.

THE PROFESSIONAL GROOM

Fig. 1–5. A good groom possesses a genuine love for horses.

The professional racetrack groom plays a very important role, and is, therefore, a major contributor to a successful racing stable. The groom as caretaker is directly responsible for the care of the horse under the supervision of the trainer. The basic, fundamental quality of a good groom is his or her love for horses. This fondness for horses is what makes the required hard work and long hours tolerable, even enjoyable. Without this love and compassion for the animals, a groom quickly becomes unproductive, careless, and irritable with the horses. Trainers do not want an employee who does not care what happens to the horses and does the minimum amount of work necessary to continue drawing a paycheck. They would much rather hire a person who respects and appreciates the horses.

A good groom should exhibit the following characteristics:

- possess a genuine love for horses
- take pride in the work as well as in the overall appearance and condition of the horses
- be punctual, trustworthy, and maintain a neat, clean appearance during working hours
- never smoke or drink alcoholic beverages around the stable during working hours
- communicate with both horses and fellow workers in a quiet, pleasant voice
- refrain from using offensive language around the stable during working hours
- be willing to work long hours if circumstances require it
- keep the trainer informed of any changes in the well-being of the horses

A person who possesses all of these characteristics will do much to promote the physical condition and attitude of a horse, making the trainer's (and the groom's) job easier and more rewarding. It is highly unlikely that the trainer will experience much success with a horse that is constantly nervous and irritable due to its handlers. However, if a horse is healthy and content, a trainer can expect that horse to perform at its peak on the racetrack.

Fig. 1–6. A healthy and content horse is more likely to perform at its peak on the racetrack.

Accommodations for Grooms

Dormitories

Most racetracks in the United States provide co-ed living quarters for backstretch employees. In the past, some backstretch housing facilities were neglected by racetrack management, whose primary concerns seemed to be the comfort and convenience of the patrons on the front side. Today, it is enlightening to see that the newly constructed racetracks are providing above standard housing for their backstretch employees. More and more racetracks are now budgeting for the creation of better housing as well as arranging housing repairs for older backstretch facilities.

Cafeteria

The cafeteria is a vital part of the backstretch, as most of the backstretch employees rely on it for their meals. The cafeteria is also an area for socializing, as well as a gathering place for the racetrack unemployed. (It is not uncommon for a trainer to enter the cafeteria seeking a desperately needed, licensed, new employee.)

Recreation Room

The recreation room is usually in the same building as the cafeteria, or located nearby. It is here that the backstretch employees relax by playing cards or billiards, watching TV, or reading. In some cases, the recreation room may serve as a house of worship for backstretch employees. Various offices in the recreation building may also offer substance abuse programs, housing assistance, educational material, or legal advice to backstretch employees.

FINDING A JOB

Getting a job on the backstretch is not a difficult task. However, there are some qualifications a potential groom must have to be employable at the racetrack.

First, you must have some practical experience working around horses. Trainers will not trust an absolute beginner to care for horses worth hundreds of thousands of dollars. If you do not have a background in horse care, the first thing you must do is find a way to gain practical grooming experience, perhaps at a local farm.

Second, you must gain access to the racetrack backstretch, which, as mentioned earlier, is off limits to the public. If you happen to know a trainer or other backstretch employees, tell them you wish to work on the backstretch. They can help you obtain a visitor's pass from the security office and gain access to the stable area, where you may approach each trainer and seek employment. (If you are under 18 years old, you may be required to obtain working papers from the state Department of Labor before seeking employment. Regulations vary from state to state.)

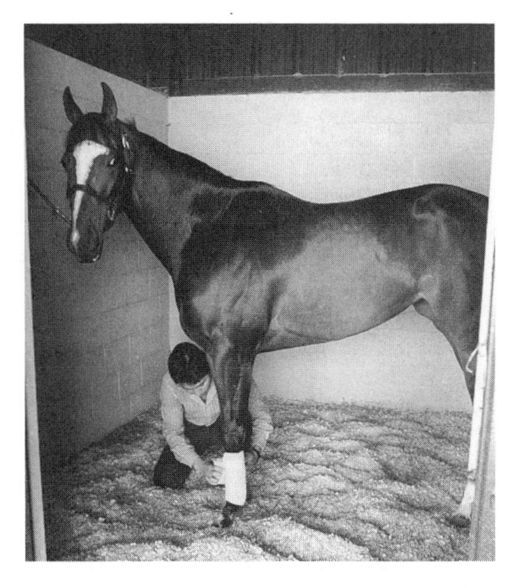

Fig. 1–7. Some practical experience with horses is required for a potential groom.

If you do not have any connections, one alternative is to go to the security office and request a temporary pass to seek employment on the backstretch. You may also consult the classified ad section of a horseracing magazine or newspaper for available jobs. Check your local telephone directory for local horse farms and training centers that cater to horses. Then make a list of potential employers and plan to contact each of them in person. You may also obtain permission to post an index card announcing your availability for employment (with your name and phone number) at tack shops, laundromats, veterinary offices, and feed stores near a racetrack.

Personal Interviews

Before you arrange any personal interviews or visits to potential employers, it would be wise to prepare a complete portfolio of your qualifications, which should include the following materials:

- personal résumé
- list of professional and character references
- copies of degrees or diplomas
- copies of awards or certificates of achievement
- articles reflecting your equine accomplishments
- list of memberships to horse-related organizations
- copies of any official licenses pertaining to the position you seek

Fig. 1–8. A groom should project a neat, professional appearance.

Be prepared to have several copies of the above materials to leave with any potential employers for their review. Make sure all those people who are listed as references are aware that you are actively seeking employment, and that they may be contacted by employers in the near future.

When approaching any prospective employers, you should arrive early and project a clean, professional appearance. Dress the part (for example, clean boots and jeans, not a business suit) and be polite. Unfortunately, most trainers and stable employees are very busy between the hours of 6:00 A.M. and 11:00 A.M. and may not welcome an interruption. Yet, for the job seeker, this is the ideal time to be seen and get a feel for potential employers. You might even ask several grooms if they like working for a particular trainer.

It is possible that you may be hired on the spot as a hotwalker, then given a stable pony as your first charge to look after. Eventually, you can expect to care for two or even three horses, depending on the total number of horses and employees in the stable.

Fig. 1–9. Many professional grooms began as hotwalkers.

Negotiating Pay

Before accepting a permanent position, you should discuss salary and bonuses with the trainer. Racehorse trainers generally provide a set salary each week. Some trainers issue bonuses or "stake money" in addition to a base salary, depending on the performance of the horses in your care or in the entire stable. For example, if a stable has two winners and a third place during a one-week period, all of the stable employees may receive extra pay at the end of the week. Some trainers award 1% of the purse money on the spot if your horse is the winner of a stake or handicap race. Other trainers prefer to hold all bonus or stakes money for the employees until the end of the year. If you decide to leave the trainer's employ before the end of the year, all accumulated bonuses and stakes money are forfeited.

It is customary, but by no means required, for the owner of the winning horse to give the groom an additional bonus for the win.

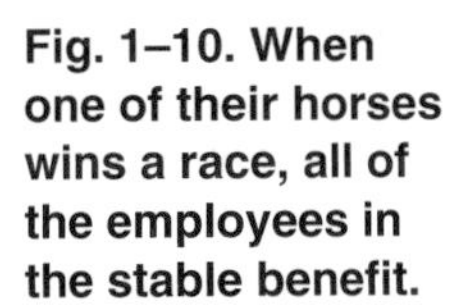

Fig. 1–10. When one of their horses wins a race, all of the employees in the stable benefit.

Insurance

As an employee of a racehorse owner, you may be covered by worker's compensation if you are hurt while working. For further protection, both the owner and trainer may have liability insurance. As for health insurance, the Horsemen's Benevolent and Protective Association (HBPA) uses its resources to provide licensed, duly employed grooms with health insurance. For instance, in New York, the

HBPA takes a portion of the purse money to fund health coverage for backstretch employees.

Getting a License

It should be understood that *you do not have to be licensed as a groom before you apply for a grooming position.* After you are employed, the trainer places you on the stable badge list, which requires you to appear and apply for a groom's license at the backstretch security office. Most racetracks require an applicant to submit a signed, completed application form which asks for personal data as well as employment history.

You will then be photographed and fingerprinted, for which a processing fee may be required. Once your application has been approved by the State Racing Commission, you will be issued an official license and photo badge. The license is usually good for one year. The photo badge must be worn on your outer clothing and be visible at all times while working on the backstretch. If you own a car, you will be issued a bumper sticker which allows you to enter the backstretch area with the vehicle.

RACING INDUSTRY OVERVIEW

A professional racetrack groom will be working with one of the two main types of racehorses: running horses and harness horses.

Editor's Note: When we refer to running horses, we are primarily talking about Thoroughbreds and Quarter Horses, as they make up the bulk of the running horse industry. However, we realize that there are also running Arabians, Paints, and Appaloosas. Of course, it would be cumbersome to list the names of all the breeds each time we refer to them. Therefore, we will use the term running horses as often as possible. When it is not possible, we hope the Arabian, Paint, and Appaloosa owners, trainers, and grooms will forgive us. (In the same way, Standardbred trotters and pacers are often referred to as harness horses.)

The following sections briefly describe the racing industry as it pertains to the three major breeds of racehorses: Thoroughbreds, Quarter Horses, and Standardbreds.

(For the definitions of many common racing terms, refer to the Glossary on page 439.)

Thoroughbred

Thoroughbred horses are bred for speed and endurance. Their races begin from a standing start in a starting gate and average from 4 furlongs (½ mile) to 1½ miles, although some races are even longer. Thoroughbreds are usually trained to pace themselves and maintain their speed over a long distance. (If a Thoroughbred ran all-out in the first ¼ mile of a 1-mile race, it would have no energy left for a strong finish.)

Fig. 1–11. Thoroughbred horses are bred for speed over a long distance.

The shorter races for Thoroughbreds are usually either for 2-year-olds or "sprinters" (horses that have great speed, but only for a short distance). Most classic stakes races are a mile or longer and are designed for "stayers," or horses that have speed over a long distance.

Thoroughbred racing is popular all over the United States, and grooms are a very important part of the Thoroughbred industry. Trainers depend on their grooms to keep their horses happy and healthy and to notice when something is not right. In a small Thoroughbred stable a groom might also be an exercise rider, in which case the responsibilities are even greater. Most large stables, however, employ both grooms and exercise riders.

Fig. 1–12. Quarter Horses race at top speed over short distances.

Quarter Horse

Quarter Horses are bred for speed, and race at top speed over short distances. Like Thoroughbreds, they begin from a standstill at the starting gate. However, as a Quarter Horse is trained to run all-out from start to finish, its time for a quarter mile is usually faster than a Thoroughbred's. Quarter Horses race at distances from 220 yards (⅛ mile) to 870 yards (almost ½ mile).

While Quarter Horse racing is more common in the southern and western United States, the Quarter Horse racing industry as a whole is similar to the Thoroughbred industry. Likewise, the role of the Quarter Horse groom is very similar to that of the Thoroughbred groom.

Standardbred

There are two divisions of harness racing: trotting and pacing. Thus, most Standardbred horses are usually bred to either trot or pace. The trot is a two-beat gait in which the horse moves its legs in a diagonal pattern; the left front leg and the right hind leg move forward at the same time. The pace is also a two-beat gait, but the horse moves its legs in a lateral pattern; the left front leg and the left hind leg move forward at the same time. Because the pace is a faster gait than the trot, trotters and pacers do not race against each other.

Fig. 1–13. The trotter moves its legs in a diagonal pattern.

Standardbreds start from a rolling gate. The distances they race vary from ⅝ mile to 1½ mile, but most Standardbred races are 1 mile. They often exercise and race in three "heats," or three separate races. In some cases, the over-all winner must win two out of three heats.

Standardbred racing is more prevalent in the eastern and midwestern United States. The Standardbred groom plays a vital role in the overall conditioning of the racehorse. Not only does the groom provide for the daily care of the horse, but he or she is also responsible for much of its daily exercise as well.

Fig. 1–14. The pacer moves its legs in a lateral pattern.

A GROOM'S DAILY SCHEDULE

A groom rarely has the luxury of sleeping in. Some trainers require their employees to report to work at 5:30 A.M. seven days a week, but allow them two or three afternoons off per week, depending on the workload and racing schedule. Fortunately, the majority of trainers allow the workers in their employ at least one full day off per week. But not every racing stable starts its day this early. Much depends on the region of the country you are in. Trainers have various reasons for implementing their time schedule, including season of the year, climate, location of the track, and access to the racetrack facilities.

The groom's responsibilities are discussed in detail throughout the book. But first, example days for both a running horse groom and a harness horse groom, with different times given, are outlined in the following two schedules. These schedules are just examples: do not assume that your trainer will adhere to either of them.

Running Horse Groom

5:30 A.M.

Remove the feed tubs from each stall. (The night watchperson or a designated feed person fed the horses earlier.) Rinse the feed tubs out and hang them outside the stall to be ready for the midday feeding.

Remove any stable bandages or night sheets from your horses. Then take each horse's temperature.

Grooming

Groom each horse thoroughly and pick out the feet. Check with the trainer as to how each of your horses is scheduled to be trained on this day. Some trainers post a daily exercise schedule in the barn.

Tacking Up

Then tack up the horse with all equipment prescribed by the trainer.

If you are in charge of more than one horse, you may have to prepare several horses to go out at the same time. In this case, be as efficient as possible in preparing each animal.

The first set of horses may go to the track for exercise as early as 6:00 A.M.

6:00 A.M.

If the horse is tacked up before the rider is ready, place a halter over the bridle. Walk it around the shedrow until the rider is ready. Be sure to run the stirrups up. If you cannot walk the horse, tie the horse to the stall wall.

Giving the Rider a Leg Up

Once the rider is ready, attach the lead shank to the bit ring and walk the horse under the shedrow at least one turn before giving the rider a leg up onto the saddle. When the exercise rider is mounted, lead the horse once around the shedrow until the rider gives the OK to release the shank from the horse.

Mucking

Prepare the wash buckets and washing materials for the daily bath. After the horses have left, begin mucking out the empty stalls. Afterwards, be sure to clean the water buckets with a stiff brush to remove any film, hay, or grain, and place the buckets back in the stalls with fresh, clean water. Fill the hay nets or racks with fresh hay. Then place another clean bucket of fresh water in the shedrow for the horse to drink from as it is being cooled down.

Untacking

When the horses return from the track, bring each one inside its stall and untack it. If the weather is nice, the horses may be untacked outside the barn.

Cooling Out

Thoroughly wash and cool down each horse. After each horse is dry and is returned to its stall, groom it and clean its feet out. Wash and pack the feet with mud at this time, if required.

Bandaging

Apply any necessary liniment or poultice to the legs. Then wrap any stable bandages on the legs that the trainer specifies. Apply any other treatments to the body such as ointments, fly repellent, etc.

10:00 A.M.

When all the horses in your care have completed their morning routines, remove their halters so each horse is loose in its stall. Clean the halters and shanks and hang them outside the appropriate stalls.

Cleaning

After the halters and shanks have been put up, clean the brushes and wash dirty bandages, saddle cloths, rub rags, and girth covers. After washing, hang or set everything out to dry.

Rake and sweep the shedrow in front of the stalls occupied by the horses in your care.

10:30 or 11:00 A.M.

It is again time to feed the horses. Top off the water buckets first, then make sure the horses have enough

Feeding

hay in their hay nets. Next, pour the pre-measured feed in the feed tubs (if necessary) and place the tubs inside the appropriate stalls.

Remove the feed tubs as the horses finish eating. Rinse out the tubs and hang them up outside the stalls to be ready for the evening feeding.

Break Time

If you are finished feeding and none of your horses is scheduled to race that day, you need not return to the barn until 2:00 or 3:00 P.M. Some stables have hotwalkers, and many riders return in the afternoon to take the horses out to be walked under the shedrow or to be grazed. Thus, the grooms' services are not usually needed until mid-afternoon.

2:00 or 3:00 P.M.

When you return in the afternoon (or before the evening feeding), take the temperatures of all the horses in your care.

Stall Maintenance

With each horse either outside (being walked or grazed) or tied up in its stall, sift through each stall and remove visible piles of manure and urine-soaked bedding. Clean the water buckets thoroughly and place them back in the stalls with fresh, clean water. Be sure the hay nets or racks are filled with fresh hay.

4:00 P.M.

Feeding

Cleaning

After the water and hay have been replenished, the feed tubs are filled with feed and placed in the stalls.

Rake and sweep the shedrow area. Begin rolling the clean, dry bandages and folding all laundry from the morning training session.

When the horses are finished eating, remove the feed tubs and scrub them thoroughly. After washing the feed tubs, hang them up outside the stalls to dry.

5:00 or 6:00 P.M.

Your day is usually complete.

Note: The length of each workday for the running horse groom varies with the type of training that is scheduled each day. *The sequence of events described herein does not include racing.*

If the horse is scheduled to race, the groom must accompany the horse at all times. It is not uncommon for a groom to finish all of his or her duties very late on a race night.

Harness Horse Groom

6:00 A.M.

Feeding

Feed the horses their morning meal. It is not uncommon for some stables to have one employee report to the stable earlier to feed all the horses before the grooms arrive.

After the horses have finished eating their morning meal, remove the feed tubs, leg bandages, blankets, etc. Note whether any of the horses did not eat properly and if so, report it to the trainer immediately.

Check with the trainer or the posted training schedule to determine the distance and type of training each of the horses shall receive.

Mucking

Clean the stall by removing the wet and soiled bedding and replacing it with fresh, clean bedding.

Grooming Harnessing

Groom the horse in the stall. Place the harness and bridle on the horse, and then lead it outside and hitch it to the cart.

Exercise

Hook up the overcheck rein and mount the cart. Drive the horse to the track and exercise it according to the instructions of the trainer. Or, you may turn the horse over to the trainer for exercise.

Unhitching

Upon completing the exercise session, return the horse to the barn. Dismount from the cart and unhook the overcheck rein. With the aid of a second person, unhitch the cart and back it off the horse.

Cooling Out

Take the horse to its stall and remove the harness and bridle. Place the halter on its head and allow the horse to drink a little water. Then wash the horse and walk it to properly cool it out.

Grooming

Apply water therapy to those horses that require it. Then groom the horses and clean their feet. Pack the feet with mud. After grooming, massage the legs with liniment and wrap bandages on the legs according to the trainer's instructions.

11:30 A.M. or 12:00 P.M.

Feeding

Feed the horses their second meal. Clean the grooming tools, harness, boots, and carts. Store the equipment in a secure area of the stable. Check the water buckets to be sure the horses have plenty of fresh clean drinking water. As the horses finish, remove the feed tubs from the stalls and clean the tubs.

Laundry

Rake the shedrow and wash the bandages and towels. Hang them out to dry on a washline.

Now you can take a break.

3:00 P.M.

Stall Maintenance

Return to the barn and go through the stalls with a fork to remove the wet spots and visible manure piles. Fill the hay nets or racks with hay to last the night. Clean and refill the water buckets. Roll or fold the dry laundry, and store it for the next day.

4:00 P.M.

Feeding

Feed the horses their evening meal. With the other grooms, rake and sweep the shedrow clean. When the horses are finished eating, remove the feed tubs and clean them.

5:00 P.M.

Your day is usually complete.

Note: The length of each workday for the Standardbred groom varies with the type of training that is scheduled each day. *The sequence of events described herein does not include racing.*

If the horse is scheduled to race, the groom must accompany the horse at all times. It is not uncommon for a groom to finish all of his or her duties very late on a race night.

SUMMARY

Now that you know the general outline of a groom's daily routine, we will move on to the specifics. The following chapters discuss common situations at the racetrack and describe each of the groom's responsibilities in detail. However, the first thing a new groom must do is get to know the racehorse.

Chapter 2

Proper Horse Handling

Good communication is the key to a successful relationship between the groom and the horse. Most of the problems which a caretaker encounters can be avoided with a little background knowledge about how horses think and react to their environments. Understanding *why* a horse acts a certain way will help the groom decide how to react in an appropriate and professional manner.

While this chapter provides many useful tips on understanding horses and handling them with confidence, it

Fig. 2–1.

should be understood that these few pages cannot substitute for years of practical experience.

ACQUIRING HORSE SENSE

An inexperienced horse person may make the mistake of trying to impose human behavior on horses. This misunderstanding is not only useless, it can sometimes be dangerous. Horses are noble creatures, but their behavior is not based on the same logic or instincts as human behavior.

Before you can understand the behavior of domestic horses, you must first understand how horses lived in the wild. Over the centuries, their experiences in the wild led to the development of instincts which still influence domesticated horses' behavior today.

Fig. 2–2. Flight is a horse's principle means of defense from danger.

Horses were (and are) vegetarians. They roamed around in herds, grazing for over 12 hours per day (15 – 20 hours if food was scarce) and drinking water at intervals. They were alert to every strange noise, smell, or sight. When a predator approached, the herd thundered away at great speed.

Even today, "flight or fight" is the horse's principle means of defense from danger. When confronted with a strange event or object, a horse's first instinct is to run. However, when cornered, a horse will fight using its teeth and hooves. Excitement, nervousness, confusion, and fear can all result in a horse acting defensively or trying to run away.

Intelligence

The average adult horse has the approximate intelligence of a three-year-old human child. Therefore, a good handler should be just as patient with a horse as he or she would be with a young child.

Like humans, horses vary in mentality and personality. They have ticklish spots and pet peeves. They have many moods and individual reactions to situations which they encounter every day. They are also creatures of habit and become very upset, both emotionally and physically, if their daily routines are changed.

Fig. 2–3. A good groom understands each horse's personality.

Horses have very long memories, especially for experiences of pain or fear. This allowed their ancestors to survive in the wild. However, a domesticated horse's long memory for bad experiences can be quite exasperating for a trainer because it can result in bad habits. In fact, almost every bad habit a horse has acquired can be traced to a human being. *Do not be the source of a horse's bad habits.* Always respond to a horse's behavior first with calm, gentle understanding. Rewards work better than punishments.

If you must punish a horse for bad behavior (otherwise defined as dangerous behavior), you have only three seconds to scold the horse or slap it on the shoulder with your open hand. After three seconds, the horse has forgotten what it has done and cannot link the punishment with the bad behavior. Remember, though, a horse should never be punished just for being a horse.

For example, punishing a horse for being frightened by a piece of paper blowing in the wind will not make that horse less frightened of a piece of paper in the future. In fact, punishment is likely to result in the horse becoming even more frightened of blowing objects of all kinds. Try reassurance—pet the horse gently and talk to it in a soothing manner.

Fig. 2–4. Horses are herd animals and are therefore very social.

Horses have very limited reasoning abilities. They cannot work out the solution to a problem in their minds. They stumble on the solution through the process of trial and error. The fact that they remember that solution (like how to open their stall door) for years afterward makes horses seem more logical than they really are. For the most part, horses have no sense of right and wrong. Therefore, humans must teach them by rewarding good behavior and discouraging bad behavior.

Social Behavior

A discussion of the interaction between horses can help you understand how a horse may react to humans. Horses are herd animals and are therefore highly social. They can become unhappy when they are not in contact with other animals. For that reason, companion animals such as goats, dogs, and cats are very common on the backstretch.

In the wild, a herd consisted of a stallion, several mares, and their offspring of various ages. The stallion was not the leader, but he served as the protector if necessary, and drove other stallions away. A dominant mare was usually the leader. This mare had influence over the other horses, dominating them either by actual fighting or threats of fighting. (Dominance with horses is not based on age or size, but rather on personality and strength.) Some horses may demonstrate more dominance in some situations, and less in others.

The Five Senses

Horses have the same five senses as humans: sight, smell, taste, hearing, and touch. How they react to what their senses tell them is based partly on the experiences of their wild ancestors, and partly on the careful training of their human handlers.

Sight

Unlike the human eye, each of a horse's eyes can operate independently of the other, which means the horse can see almost all around itself. One eye can be watching the handler near its head while the other watches the veterinarian near its hind end. Or, the eyes can operate together when the horse focuses them forward. To see something at a distance, the horse raises its head. To focus on something closer, it lowers its head.

Fig. 2–5. This horse raised its head to focus on something in the distance.

Horses have two blind spots where they cannot see at all. One is from their nose to about four feet in front of them. Many people are surprised to learn that, unless it makes a special effort, a horse cannot see its own feet. But you can see both the horse's feet and your own feet. Therefore, if a horse ever steps on your foot, you can be sure that the accident is your own fault for not paying attention. Remain calm, and with the palm of your hand, gently and firmly push the horse away from you. Yelling will only frighten the horse, in which case you could get even more seriously hurt.

The other blind spot that horses have is directly behind them. As you can imagine, many predators have taken advantage of this blind spot by sneaking up on wild horses from the rear. That is why horses are so sensitive to anything which comes at them unexpectedly from behind. Thus, it is best to always approach a horse from the side. If you must approach a horse from behind, get its attention by calling its name while you stand at a safe distance (at least six feet away from the horse's hind end). When the horse turns its head to look behind it, and is not frightened, it is safe to approach quietly, still talking.

It is also important to remember that most adult horses sleep standing up. (Young horses and male horses are more likely to lie down to sleep than other horses.) If you are approaching from behind, it may be difficult to tell whether the horse is sleeping. It is safest to assume that the horse is sleeping, and wake it up before approaching too closely. If you do not, you may be hurt when the horse lashes out in self-defense, or the horse may be hurt as it struggles to get away.

Although they can see well in the dark, horses' eyes do not adjust to sudden light or darkness as quickly as our own. Therefore, it is best to allow the horse a little extra time when entering a dark area.

Fig. 2–6. Allow the horse a little extra time for its eyes to adjust when entering or exiting a dark area.

Smell

The horse's sense of smell, while better than a human's, is not as advanced as some animals. However, horses are very good at identifying odors which are carried to them by the wind. In the wild, horses relied on this sense to identify other animals, either from a distance or close up. It is, therefore, important to allow a horse to smell new or unfamiliar objects: a new brush, bucket, blanket, etc.

Horses can sniff out the additives in their feed. And stallions can smell a mare in heat (or another stallion) from quite a distance.

Horses do not like the smell of smoke—most panic and become difficult to handle. In fact, some trainers leave their horses' halters on at night so the animals may be more easily led to safety in the event of a fire.

Fig. 2–7. A flehmen response is when a horse (usually a male) curls its upper lip back and raises its head.

A strong smell can cause a horse to curl its upper lip back, called a flehmen response. (A stallion often does this when he scents a mare in heat.) Although the teeth are exposed when the horse does this, it is not an aggressive act.

Taste

The sense of taste is closely related to the sense of smell. Some horses will eat things which by human standards are extremely distasteful. Others refuse to eat something for seemingly no reason at all. In general, horses prefer things which are salty or very sweet. Still, it is not advisable to give a horse sugary treats—it spoils them and makes them "nippy," or prone to bite when no treat is given.

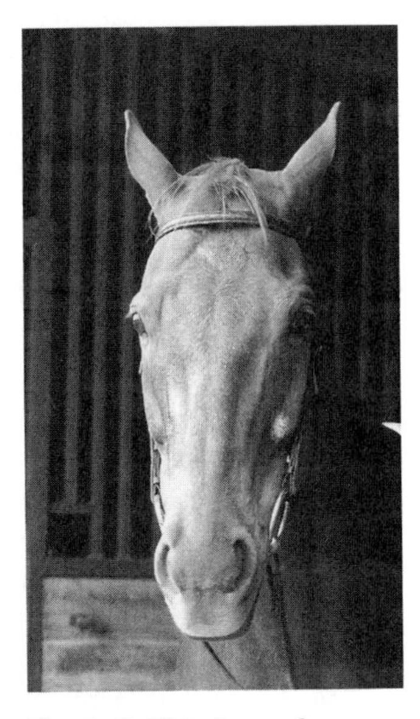

Fig. 2–8. This horse has both ears back, listening to noises behind it.

Hearing

A horse's ears are like its eyes—they can operate independently or together. An alert horse's ears are in motion, rotating up to 180° in response to various noises. By swiveling its ears in this way, a horse can pinpoint the origin of a sound quite accurately. A horse can also hear things that a human cannot, a behavior which often perplexes the horse's handlers.

A horse's acute hearing and pinpoint accuracy can be used to the groom's benefit. While working with or around a horse, any professional groom should constantly be making some kind of noise: whispering, talking, humming, or singing. This is a very useful tool because it lets the horse know exactly where the person is at all times, and soothes and relaxes the horse as well. A vocal groom is less likely to startle a horse and cause it to act out in self-defense.

Fig. 2–9. Horses are trained from an early age to stand quietly while their ears are being handled.

Because the ears were so important to a wild horse's survival, a domesticated horse is usually very shy about having its ears handled. Thoroughbreds tend to be more thin-skinned and usually have more delicate ears than Standardbreds or Quarter Horses. While some horses are trained from an early age to stand quietly while their ears are being handled, many horses still object to having their ears touched. With this in mind, the groom should be careful not to pinch or pull on the horse's ears sharply. It takes only one careless action to undo good training and make a horse headshy for a long time.

Touch

It is the horse's sense of touch which has enabled it to become so useful to humans. Properly trained and handled, horses are sensitive to the slightest leg pressure by a rider. They are trained to respond to pressures on the head from a halter and on the head and mouth from a bridle and bit. An abused horse can become insensitive to leg, mouth, or head pressures, usually requiring extra attention or retraining.

Horses like to be scratched and stroked on the withers and neck. Heavy-handed patting, though, may be interpreted as an aggressive gesture, with the result that the horse may become either frightened or resentful.

Emotions and Body Language

The parts of the horse which receive information from the outside world—eyes, nose, mouth, ears, and skin—can also transmit the horse's emotions to the observant handler. Other parts of the horse which transmit emotions are the legs and tail. Understanding the horse's emotions and responding to them appropriately is what makes a groom into a real horse person.

Alert and Interested

A healthy horse has alert, clear, prominent eyes. The ears are pricked, but not stiff, and are pointed toward whatever interests it. The nostrils are flared slightly, and the horse may blow softly. The tail is arched slightly away from the buttocks. This horse may nicker when its caretaker appears and may nudge at him or her in a bid for attention. (A sick horse may also nudge its groom to relay that it is not feeling well.)

Fig. 2–10. This horse is interested in the activity in the saddling paddock.

Irritated or Frustrated

An irritated horse puts its ears back slightly. It may toss or shake its head. It may stomp a forefoot or hind foot, or swish its tail as if to dislodge an unusually pesky fly. If the horse does not get the result it wants, namely the end to whatever is irritating it, the irritated horse may become an angry horse.

Angry

An angry horse's eyes roll, and the eyelids open a little wider, showing the white around the eyes. (If a horse is looking behind it, some white may show, but the horse is probably not angry.) An angry horse's ears are pinned back, flat against the skull. The flatter the ears, the more angry the horse is. It may lift a hoof to warn of rearing or striking, or it may turn and present its hind end to you in a similar kind of warning.

The angry horse may hold its mouth open tensely and bare its teeth, perhaps even lunge toward you in an attempt to bite. An attempt to bite is a very aggressive act and should be dealt with by scolding loudly in a deep tone of voice and slapping the horse soundly on the shoulder. *(See Chapter 14 for more information on this vice.)* A stiff tail is also an indication of anger: it swishes back and forth violently, and the horse may lift it up and slap it back down hard.

Nervous

The nervous horse's ears twitch and flick about wildly. The horse moves its head jerkily, and the eyes are opened very wide. The mouth is stiff, but the teeth are not bared. The nostrils usually flare and the horse may snort. The nervous horse will most likely pull back or bolt at the next strange sight or sound.

Fig. 2–11. The horse on the left appears nervous before a race, whereas the horse on the right appears calm.

Frightened

A frightened horse's eyes roll and show the whites, just like an angry horse's, but instead of acting aggressively, the frightened horse will probably rear up or pull away from whatever frightens it. The ears are stiff and pricked straight up toward whatever alarms it. The mouth is also stiff, but the teeth are not bared. The nostrils flare widely. The tail may be flattened against the buttocks.

Fig. 2–12. A frightened horse may pull back and rear. Note that the lead shank, which was wrapped improperly, is tightening around the handler's left hand.

Sleepy or Depressed

A horse with dull eyes may be depressed or, if the lids are half closed, merely dozing. The sleepy horse's tail is relaxed, touching the buttocks, but not clamped to them; you can lift the tail easily. A depressed horse's ears flop out sideways, like airplane wings. Ears which are merely droopy means the horse is either in pain or deeply asleep. The lips of a sleepy horse often droop also.

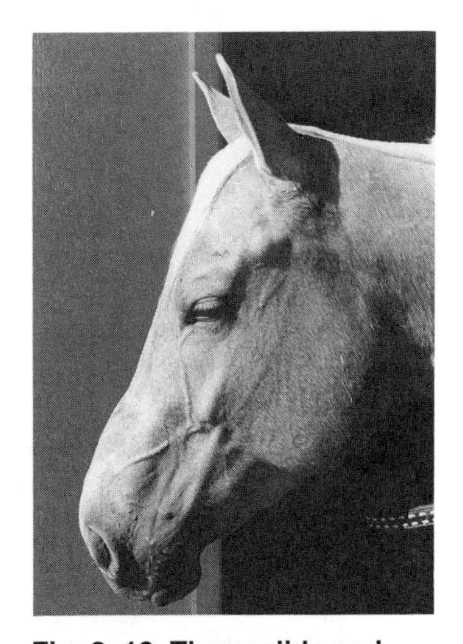

Fig. 2–13. The eyelids and sometimes the lower lip droop on a sleepy horse.

Pained

Glassy eyes may indicate a horse in pain. Closed eyes may also indicate pain if the horse is obviously awake, but unresponsive. The tail is flat against the buttocks. The ears may droop sideways. The mouth is stiff, but the teeth are not bared. The body may be held stiffly, and the horse may sweat or have muscle tremors. The horse's facial expression is one of anxiety, and its pulse and respiration rate are high. *(See Chapter 11 for more information on pulse and respiration rates.)*

Barn Etiquette

Now that you understand some of the psychology and behavior of horses, you will be better equipped to handle them. There are also a few basic rules of human conduct to which everyone in the shedrow should adhere:

- Do not make any sudden or loud noises while working around horses.
- Do not make any sudden or threatening gestures with your hands and arms.
- Do not run in the barn.
- Work slowly and methodically around the horse—do not rush.

WALKING THE RACEHORSE

A new or less experienced employee may at first be given the job of walking the racehorse, later to progress to actual grooming work. Therefore, knowledge of horse behavior and skill at walking the horse is a firm foundation for a groom.

Walking is one of the best forms of exercise for the racehorse. Because the racetrack has a limited amount of space, and horses are unable to run loose as they do on the farm, most racehorses are walked in the shedrow. At some racetracks, there are walking rings outside the barn where a horse may be walked in pleasant weather.

Since a racehorse is confined to its stall for much of the day, many trainers have a horse walked in the morning and then again in the afternoon. The afternoon walk helps to put the horse in a good frame of mind, which is essential if it is to perform at its peak.

In addition, walking is a means of "cooling out" a horse after a race or workout. A horse's body temperature rises during a race or workout. To bring the body temperature slowly back to normal (about 100°F), the horse is first washed with warm water and then walked. The length of time the horse is walked depends on the climate and the judgment of the trainer. For most racehorses, 45 minutes is sufficient time to cool out.

Haltering

Before discussing haltering, you should understand the concepts of "near" and "off" sides of the horse. The near side is the horse's left side, and the off side is the horse's right side. Some procedures, such as walking the horse, are only performed from the near side, which is why the groom should know the difference. *(A complete diagram of the horse's anatomy can be found in the Appendix.)*

The purpose of the halter is to control the horse from the ground. Most halters used on horses are made of leather or durable nylon. The straps of leather or nylon are held together by five rings and one or two fastening buckles. One ring is under the horse's jaw, called the center ring. There are also two rings on each side of the halter: the upper and lower rings of the near and off sides. Some halters have two fastening buckles—one on the near side and one on the off side. If there is only one fastening buckle, it will be on the near side of the halter.

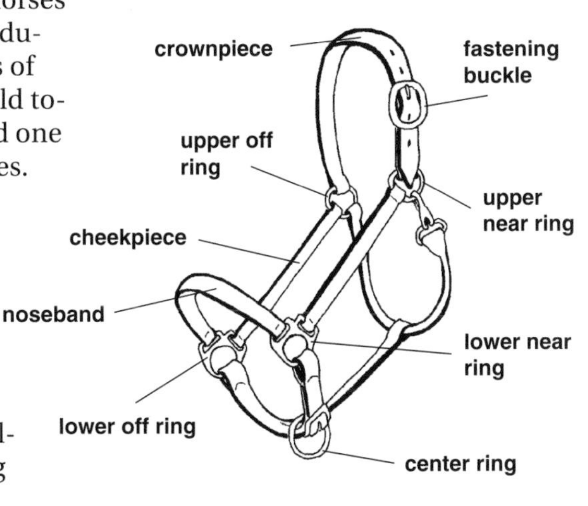

Fig. 2–14.

When trying to catch a horse to halter it, *do not walk into a stall if the horse has its rump facing you.* You are asking to be kicked if you approach the horse in this manner. Get the horse to turn around and face you by calling its name.

The procedure for haltering a horse is illustrated in Figure 2–15.

1. Stand on the near (left) side of the horse and place your right arm over the neck, holding the unbuckled crownpiece in your right hand. With your left hand, hold the fastening buckle on the near side. By using this method you have the horse "captured" so it cannot move its head away from you.
2. Slide the halter on so that the noseband rests loosely over the bridge of the horse's nose (about two inches below the protruding cheekbones).
3. Slip the end of the crownpiece through the fastening buckle

Haltering a Horse (Fig. 2–15.)

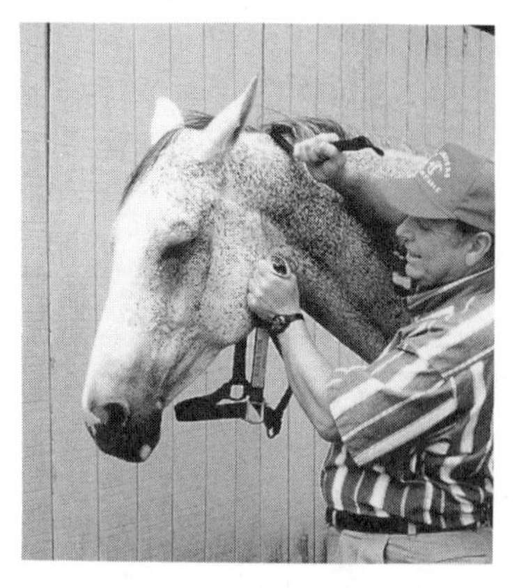

Step 1

Stand on the near side, next to the horse's neck. Place your right arm over the horse's neck behind the ears, holding the crownpiece of the halter in your right hand.

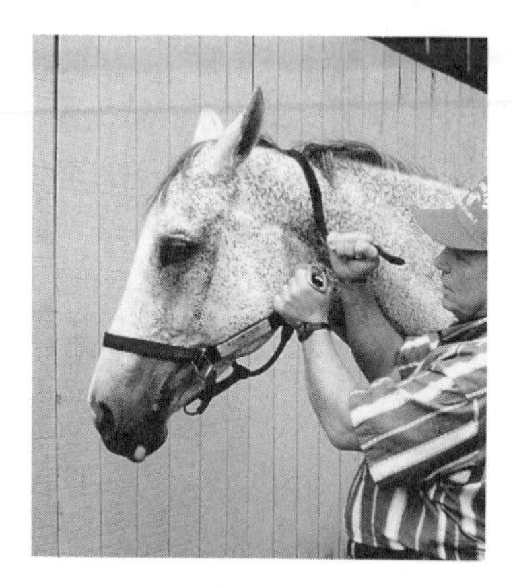

Step 2

With the buckle in your left hand, slip the halter over the horse's nose. Lift your arms, sliding the noseband up to its proper position.

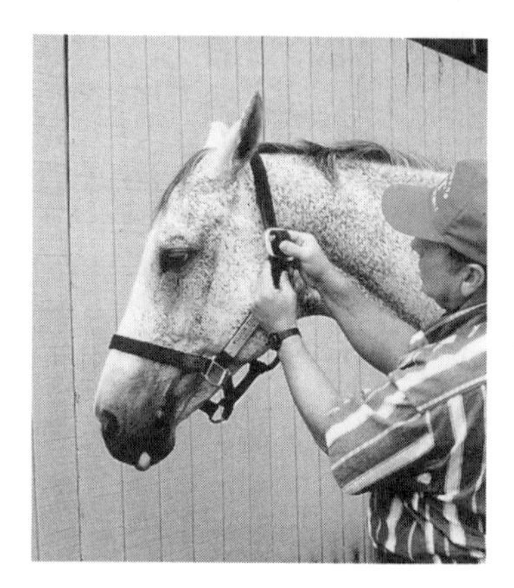

Step 3

Buckle the halter. Be sure not to buckle it too tight or too loose.

and adjust the halter to the appropriate tightness. If the halter is fitted too loose, the noseband will rest too low on the horse's nose and disrupt air flow through the nasal passages. This can cause a horse to throw its head and can lead to injuries for the horse or groom. If the halter is adjusted too tight, the noseband will leave a mark on the horse's nose when the halter is removed.

A halter is sometimes slipped on over the bridle to lead a horse to the saddling paddock before a race. Before the halter is put on this way, the reins should be pulled over the head to rest on the horse's withers. Once the halter is fastened over the bridle, it holds the reins back out of the handler's way. Keep in mind that the halter may need to be readjusted to fit properly over the bridle and reins. After the halter is adjusted, a lead shank must be attached before walking. *Never try to lead a horse with only a halter*—a lead shank is necessary for proper control and safety.

Fig. 2–16. When a halter is put on over the bridle, it holds the reins up against the neck.

Lead Shanks

The lead shank is used to lead, and when necessary, to reprimand a horse. The most common types of lead shanks are yearling shanks, rope shanks, and chain shanks.

Yearling Shank

As the name implies, the yearling shank is used primarily on yearlings on the farm. This shank is made of leather and is about 78 inches long and ¾ inches wide. A yearling shank has a metal snap on one end. The metal snap is clipped onto the center ring of the horse's halter.

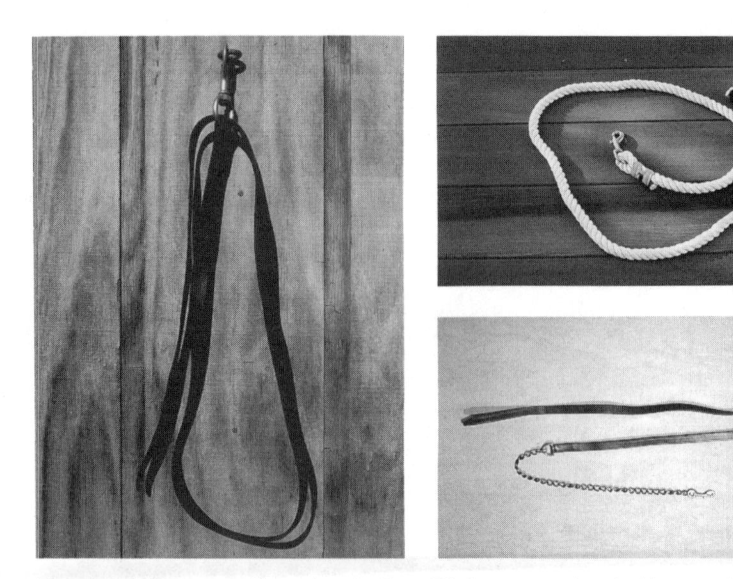

Fig. 2–17. Left: yearling shank. Top Right: rope shank. Bottom Right: chain shank.

Rope Shank

Rope shanks are made of a durable rope material such as hemp or nylon. The total length is about six feet. The shank has a metal snap at one end. The metal snap is clipped onto the center ring of the horse's halter. At the racetrack, rope shanks are used primarily on lead ponies or older horses that are quiet. Using a rope shank on young or nervous horses is not recommended because they are often unpredictable. If the horse rears or pulls suddenly, a rope shank may burn the hands of the person holding the horse.

Chain Shank

The chain shank is the most popular type of shank used to walk or restrain a horse. It consists of a metal linked chain at least 30 inches long attached to a leather or nylon lead. At the end of the chain is a metal snap, which attaches to the halter.

There are several ways to apply the chain shank:

- over the nose
- under the lower jaw
- over the nose and under the jaw
- through the mouth
- over the upper gum

The chain shank can also be used in combination with a Chifney bit, which attaches to the halter, for walking or restraining the horse. *(See Chapter 15 for more information.)*

While use of the chain shank is very common at the racetrack, an inexperienced horse person should be aware that misuse of this shank can be harmful to the horse and its handler. Because the chain shank is so severe, it requires very little force to be effective. In fact, some horses will straighten up and behave as soon as the chain shank is put on.

When using the chain shank in any of the five positions listed above, use the minimum amount of force necessary to control or restrain the horse—there is usually no need to jerk on the shank. Hold the shank with just enough tension so that the chain is kept in position. If the chain is held too loose, it may slip out of position and irritate the horse. If the chain is held too tight, most horses will object by shaking their heads, throwing their heads up, or pulling back.

Over the Nose

This commonly used method allows the handler to apply pressure from the chain over the bridge of the horse's nose. This application can be used when walking or grazing the horse.

To apply the chain shank over the horse's nose, begin by threading the chain through the lower near ring of the halter. Pull the chain over the nose and thread it through the lower off ring and up the right side of the jaw. Attach the metal snap to the upper off ring of the halter. This procedure is illustrated in Figure 2–18.

Under the Lower Jaw

This method is also common, but should not be used when taking the horse to graze. (If the chain becomes slack under the jaw, the horse could step on the chain. At this point the horse might panic because it cannot lift its head up.) To apply the chain shank under the horse's lower jaw, begin by threading the chain through the lower near ring of the halter. Pull the chain downward under the lower jaw and thread it through the lower off ring. Pull the chain upward and attach the metal snap to the upper off ring of the horse's halter. This procedure is illustrated in Figure 2–19.

Other variations are to attach the metal snap to the center ring of the halter, or to attach the metal snap to the beginning section of the chain shank. All of these techniques allow the handler to apply pressure from the chain under the horse's jaw.

Applying a Chain Shank Over the Nose (Fig. 2–18.)

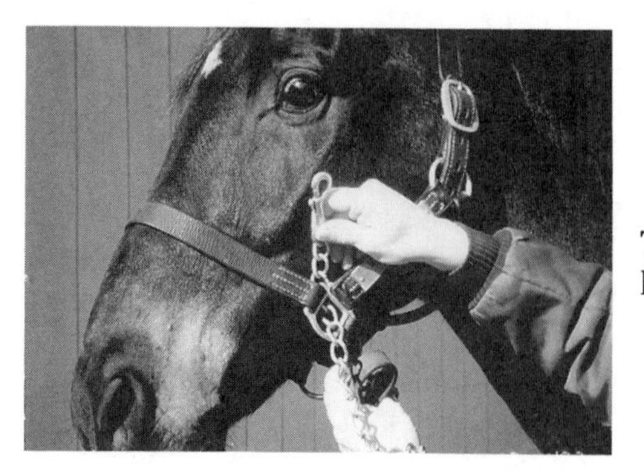

Step 1

Thread the chain through the lower near ring of the halter.

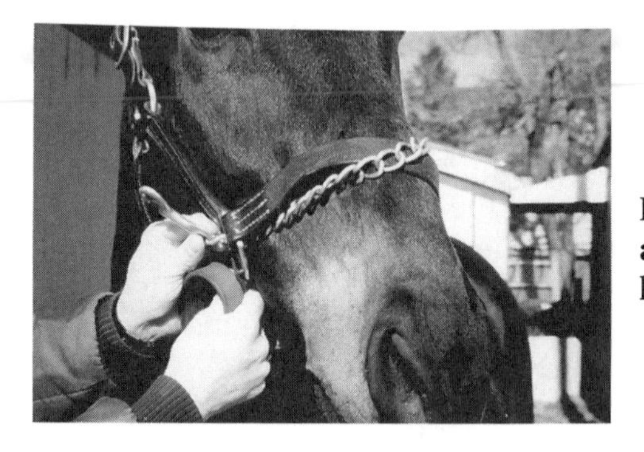

Step 2

Pull the chain over the nose and thread it through the lower off ring.

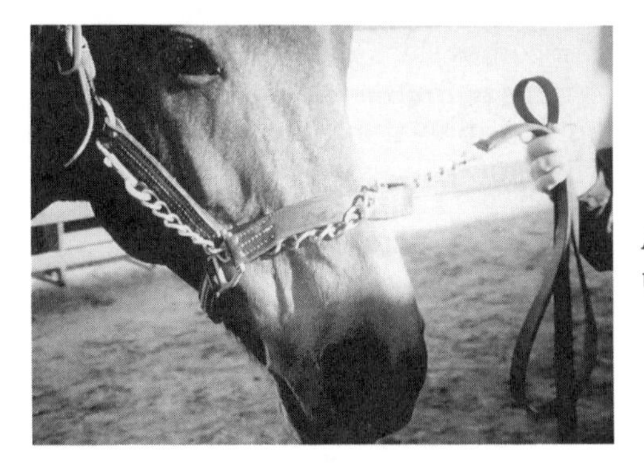

Step 3

Attach the metal snap to the upper off ring of the halter.

Applying a Chain Shank Under the Lower Jaw (Fig. 2–19.)

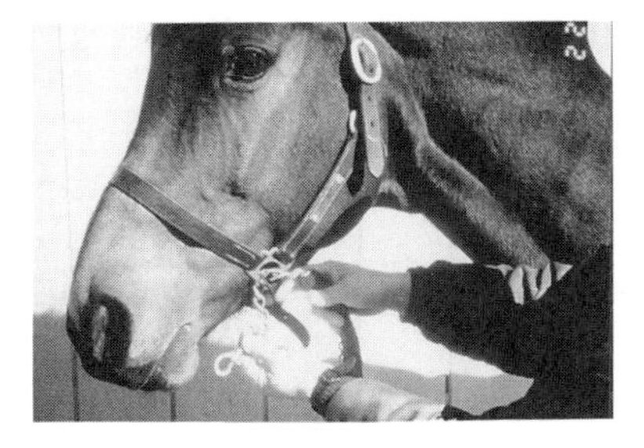

Step 1

Thread the chain through the lower near ring of the halter.

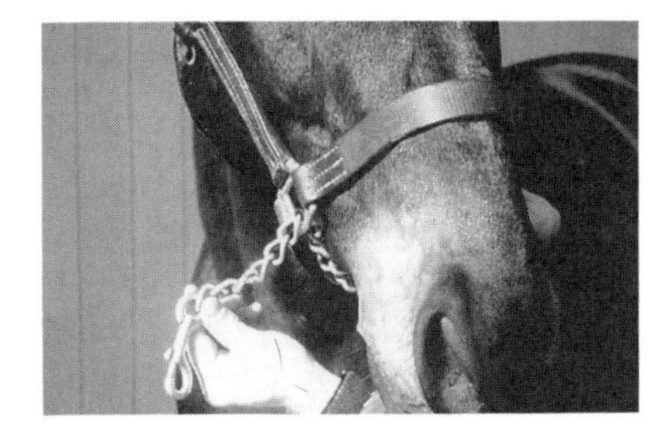

Step 2

Pull the chain under the lower jaw and thread it through the lower off ring.

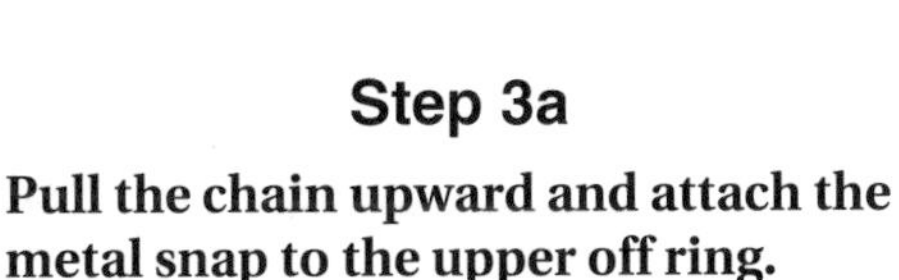

Step 3a

Pull the chain upward and attach the metal snap to the upper off ring.

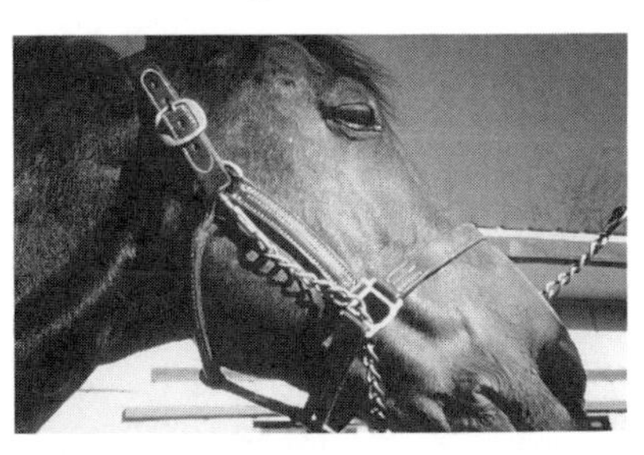

Step 3b

Or, you may attach the metal snap to the center ring of the halter.

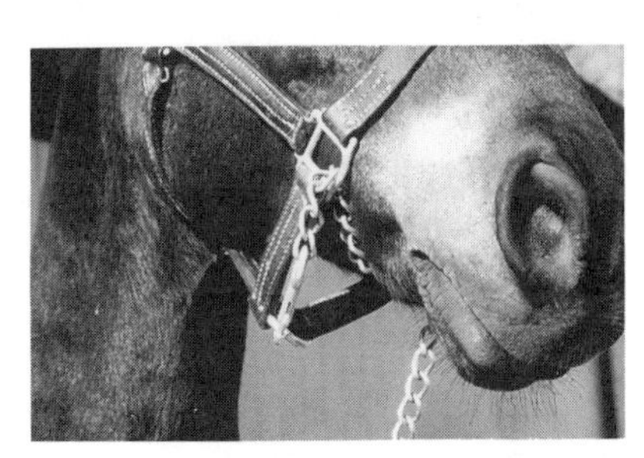

Step 3c

Or, you may attach the metal snap to the beginning section of the chain shank.

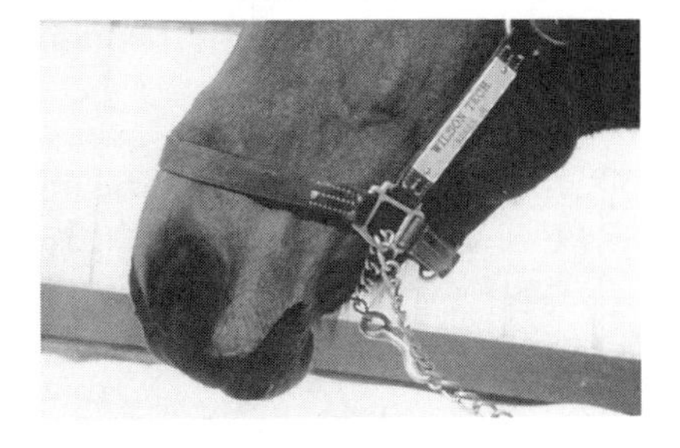

Over the Nose and Under the Jaw

This method of applying the chain shank is a combination of the previous two—it places pressure over the nose and under the jaw at the same time—and may be used for walking the horse.

Place the chain over the nose in the same manner as described earlier. However, instead of attaching the snap to the upper ring on the off side of the halter, the chain is threaded down through the lower off ring, extended under the jaw, and clipped to the upper near ring.

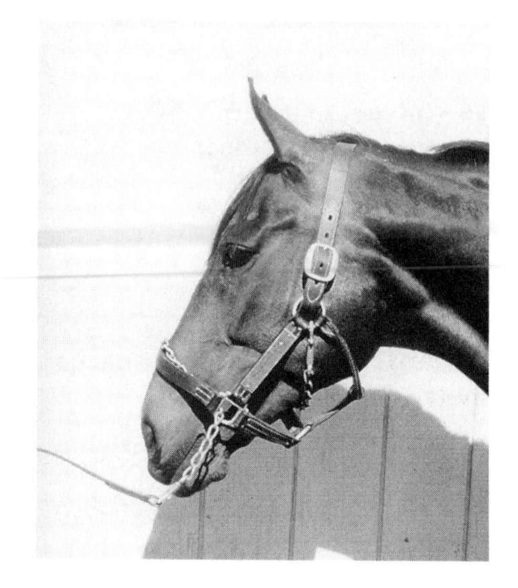

Fig. 2–20. Placing a chain over the nose and under the jaw puts pressure on both of these areas at once.

Through the Mouth

Placing the chain shank through the mouth is also a severe method that should only be used for restraint or walking a horse that is difficult to control. (Do not use this method unless the trainer recommends it.) While this technique allows the handler to apply pressure in a manner similar to that of a bit, excessive force on the chain can ruin a horse's mouth.

To place the chain through the mouth, begin by threading the chain through the lower near ring of the halter. Run the shank through the horse's mouth and over the tongue. Thread it through the lower off ring. Pull the chain upward and attach the metal snap to the upper off ring of the halter. This procedure is illustrated in Figure 2–21.

Applying a Chain Shank Through the Mouth (Fig. 2–21.)

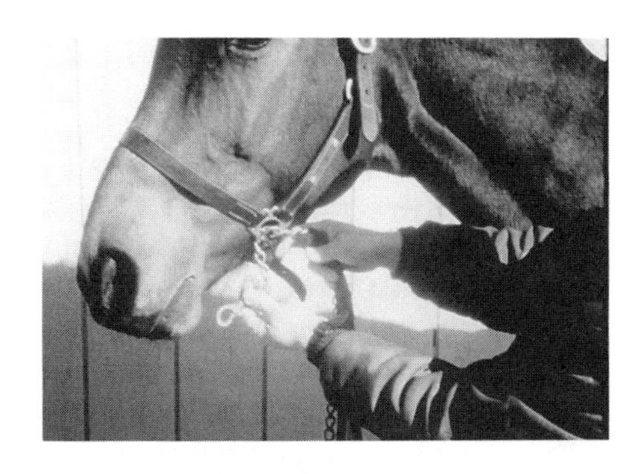

Step 1

Thread the chain through the lower near ring of the horse's halter.

Step 2

Run the shank through the mouth and over the tongue. Thread it through the lower off ring.

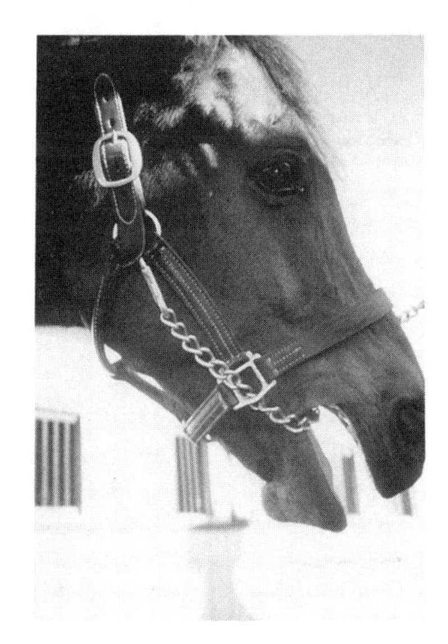

Step 3

Pull the chain upward and attach the metal snap to the upper off ring of the halter.

Over the Upper Gum

This method of applying the chain shank is more severe than the previous methods and is not recommended when walking horses. (Do not use this method unless the trainer recommends it.) It is used primarily as a restraining method when the horse is undergoing veterinary treatment, etc.

To apply a chain shank over the upper gum, begin by threading the chain through the lower near ring of the halter. Then pull the chain over the upper gum and attach the metal snap to the lower off ring.

Again, be careful not to jerk the chain when it is in the horse's mouth, as this is very painful for the horse. (Adding a leather covering to the chain before using any of the previous methods is one thing the handler can do to make the chain shank more humane.)

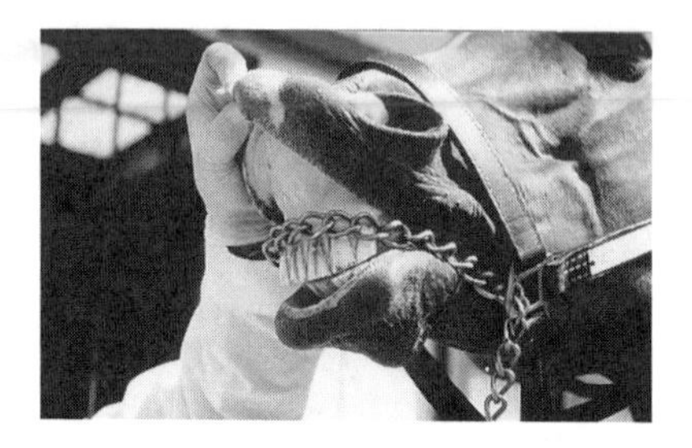

Fig. 2–22. A chain applied over the upper gum is a severe method of restraint.

Fig. 2–23. A leather covering makes the chain shank more humane.

Rules for Walking

Once the groom puts the halter and lead shank on the horse, the horse can be walked. Walking a racehorse is serious business and there are some important points to remember.

Begin by standing on the horse's near side. Your right hand should be holding the leather or nylon part of the shank, *not the chain or halter.* Your left hand should hold the excess portion of the shank. It is important not to wrap this excess portion of the shank around your left hand. If the horse pulls away or rears, the shank may tighten around your left hand and you could be seriously injured. Instead, fold the excess length of the shank back and forth within your hand, grasping the center of the folds. (It should look like a figure-eight.)

Fig. 2–24. To make the horse halt, say "Whoa," pull back on the shank, and stop walking.

Speak to the horse softly, tug on the shank gently, and begin walking forward confidently. If you are hesitant, the horse will be also. Stay at the horse's shoulder. It may seem that the horse is ahead of you, but horses are trained to lead this way. (Remember, it can see you with its left eye.) If you turn to face the horse, it will not move. If you walk ahead of the horse, you cannot control it properly.

To make the horse halt, say "Whoa," pull back gently on the shank, and stop walking. When the horse stops, reward it by releasing the tension on the shank.

If the horse balks and refuses to move forward again, turn its head to the right until it is forced to take a step to keep its balance. Now that the horse's feet are moving, you can direct the horse forward. If it still balks, and you can see nothing ahead that may frighten it, the horse may just need some extra incentive. Let the end of the shank unravel, and with your left hand, flick the end of the shank behind you so it hits lightly against the horse's flanks. This should encourage the horse to move forward, so be prepared. Also, be careful not to swing the shank too enthusiastically, as a sensitive horse may leap forward in surprise.

Grazing

Most trainers agree that getting the racehorse out of its stall in the afternoon to graze does a great deal of good. Being a natural habit, grazing keeps the horse relaxed and in a good frame of mind. Also, fresh grass is rich in B-vitamins, and the horse's skin manufactures vitamin D when exposed to sunlight. Grazing may be more practical at certain tracks or in certain regions, depending on the facilities and the climate. However, horses should be taken out to graze whenever circumstances allow it.

Fig. 2–25. Grazing keeps a horse relaxed and in a good frame of mind.

Although it sounds simple, the practice of grazing a horse by hand is a skill which should be taken seriously. For instance, your horse should be kept at a safe distance from other horses when grazing. Your horse's personal space may become threatened by another horse and it might react by kicking out with its hind legs or attempting to bite. The groom should also avoid getting too close to another horse that is grazing to prevent personal injury as well as injury to either of the horses. When grazing the horse, carefully observe the following rules:

- The chain shank should be placed over the nose—*not* under the chin—to avoid getting a hoof caught in the shank while grazing.
- Keep the chain shank taut while the horse has its head lowered to the ground. If the shank is allowed to sag on the ground, the horse may step on or over the shank.
- Be aware of other horses grazing in the same area so as not to get too close.
- Know the sex of the horse you are grazing.
- Keep the horse away from objects it may hurt itself on such as fences, buckets, hoses, etc.
- Always be alert to anything which might "spook" the horse such as loud noises, rustling paper, or leaves blowing in the wind.

Spacing

When walking a horse, you must not allow the horse to get too close to the horse in front of you. Be aware of the sex of the horse you are walking, especially in the spring and early summer, when horses are most sexually active. A stud horse can be quite a handful if he is walking behind a mare in heat. A mare in heat urinates frequently in small amounts, holds her tail to one side, and "winks" (opens and closes the labia). A stallion that detects a mare in heat performs the flehmen response and arches his neck. He begins to prance a little and may nicker. He may try to approach the mare. Put more distance between him and the mare. If he becomes so difficult to control that he is dangerous, lead him away from the mare as quickly as possible.

You should always be alert to anything you feel will spook the horse, such as loud noises, sudden movements, loose horses, or objects blowing in the wind. You must be on your toes—or the horse may be on them.

Adjusting the Blanket

If you have a cooler or fly net on the horse, check it periodically to be sure it is not pulling on the horse's ears or slipping off. If the horse in front of you has a cooler on and it is slipping, let its handler know.

Stopping

If you have to stop with the horse to go into the stall, fix its cooler, or allow it to drink, calmly tell the handler behind you to "hold back." This indicates that you are stopping and acts as a signal to prevent a rear-end collision.

Returning to the Stall

Removing the Halter and Shank

When returning a horse from walking or exercise, always walk the horse straight into the stall to avoid hitting its hips on the door frame. Once inside the stall, turn the horse around so that the horse is facing the door (as if you were going to lead the horse out again). After removing the halter and shank (or just the shank), do not turn your back to the horse. Instead, back out of the stall so that you can keep your eyes on the horse. This procedure keeps you from getting "cornered" in the stall by the horse, and allows you enough time to get out of the stall if the horse wheels around and tries to kick after you let it loose. Also, be sure the stall door, gate, screen, or webbing is closed and latched securely before walking away.

Many trainers have their grooms remove and clean the halters and shanks after the morning grooming and training sessions are over (at about 10:30 or 11:00 A.M.) or after grazing (at about 2:00 P.M.) The grooms clean the halters and shanks, then hang them neatly outside each horse's stall door. In late afternoon, just before feeding time (around 4:00 P.M.) the halters may be put back on the horses for the rest of the night.

Fig. 2–26. Hanging the halter and shank makes for a neat appearance in the barn.

Cleaning and Storing the Halter and Shank

To clean leather items, you need a small tack sponge, warm water, and a commercial leather cleaner (such as saddle soap, Lexol®, or Murphy's Oil Soap). A metal cleaner for brass and chrome, such as Noxon® or Duraglit® may be used to clean the metal rings of the halter, the metal name plate, and the metal chain of the shank.

The process for cleaning the halter and lead shank, illustrated in Figure 2–27, is as follows:

1. Unfasten the crownpiece of the halter. Dampen the sponge with water and then apply leather cleaner to the sponge. Wipe the leather parts of the halter and shank with the sponge and leather cleaner. Rinse out the sponge, then remove all soil and excess cleaner.
2. Be sure to remove all soil and excess cleaner from the buckle holes in the halter using cotton swabs.
3. With the metal polish, polish the name plate, buckles, and rings of the halter. Polish the metal chain and snap of the shank.

To properly roll the shank, use the following procedure, as illustrated in Figure 2–28:

1. Clip the lead shank to something and stretch it out.
2. Insert the end of the leather part of the shank from underneath into the opening between the chain and leather part. Pull until a small loop is formed.
3. Begin rolling the leather portion of the shank toward the loop.
4. Tighten the loop around the rolled portion and place the shank in the crownpiece of the halter.

Hanging the clean halter and shank on the wall makes for a neat appearance in the barn.

Cleaning the Halter and Shank (Fig. 2–27.)

Step 1

Wipe the leather parts of the halter and shank with the sponge and leather cleaner.

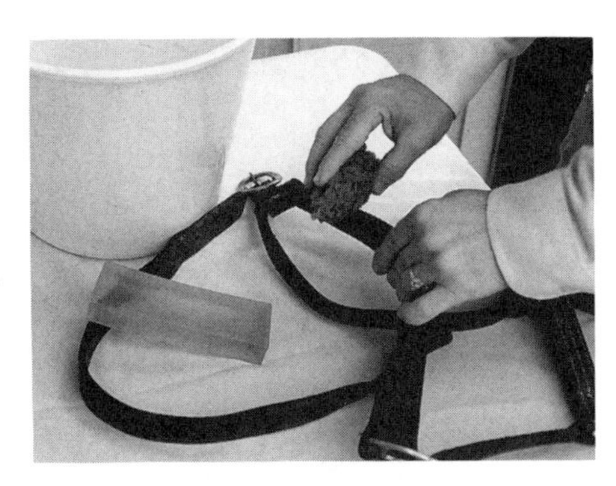

Step 2

Remove all soil and excess cleaner from the buckle holes in the halter using cotton swabs.

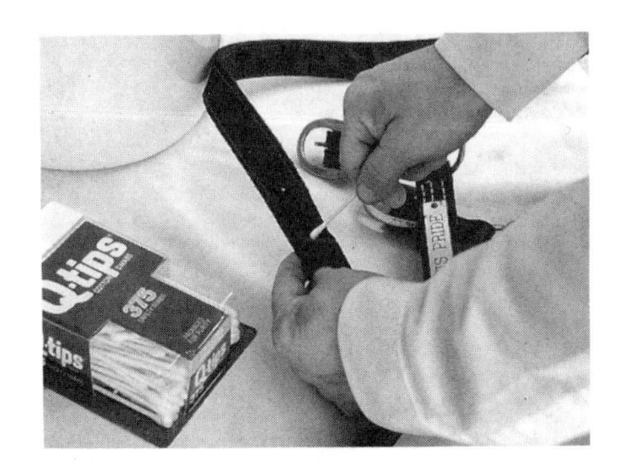

Step 3

Polish the name plate, buckles, and metal rings of the halter. Then polish the chain and metal snap of the shank.

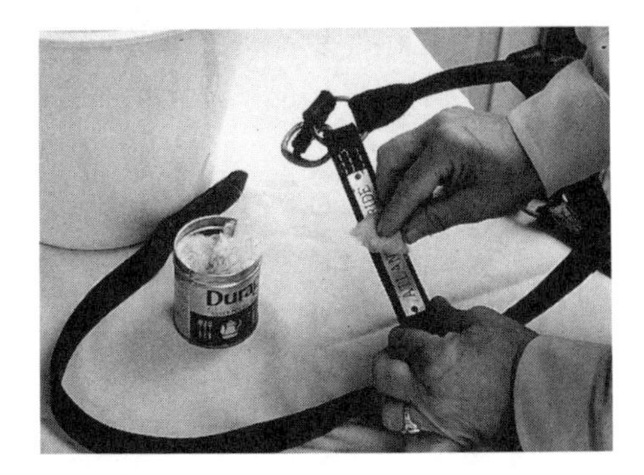

Rolling Up the Shank (Fig. 2–28.)

Step 1

Clip the lead shank to something and stretch it out.

Step 2

Insert the end of the leather part of the shank from underneath into the opening between the chain and leather part. Pull until a small loop is formed.

Step 3

Begin rolling the leather portion of the shank toward the loop.

Step 4

Tighten the loop around the rolled portion.

SUMMARY

Handling horses is not difficult once you can understand (and predict) their behavior. For instance, knowing to offer calm reassurance to a frightened horse gives you more control in this tense situation. Such control allows you to go about your daily routine in a more confident manner. Finally, as you develop more "horse sense," you will become increasingly valuable to horses, trainers, and owners.

Chapter

3

FEEDING

Proper feeding is necessary for a horse to reach its maximum potential in growth, energy, speed, endurance, body and coat condition, and reproduction. The groom plays an important role in the feeding process, making sure that the horses' food and water are fresh, that the horses are fed at the same time every day, and that each horse finishes its feed satisfactorily.

Fig. 3–1.

The horse is not physically capable of digesting a large amount of food at one time. In the wild, horses grazed almost continually throughout the day. As a result, the horse's digestive tract, particularly the stomach, can only accommodate small amounts of food eaten at frequent intervals. For this reason, the most effective feeding program for a racehorse (or any horse) would include grazing on pasture all day. Unfortunately, this is not possible at the racetrack. Trainers face the challenge of developing a feeding program that is compatible with the horse's nature yet conducive to a competitive racing stable. A few trainers have been known to feed their horses as little as once per day or as often as six times per day. Feeding at minimum three times daily (as the majority of racing stables do) is efficient, and prevents many stable vices that arise from boredom. Feeding four times daily is even better, because it is best not to have long stretches between feedings.

However often the horses are fed, it is important to stick with the schedule once it has been established. Horses are creatures of habit. Their minds and bodies gear up in anticipation of a feeding and they can become very upset, psychologically and physically, if their routines are disturbed.

One normal schedule at stables in some areas might have the morning meal at about 3:30 or 4:00 A.M., the midday meal at 10:30 or 11:00 A.M., and the evening meal at 4:00 P.M. To feed four times daily, add another meal between 8:00 P.M. and 10:00 P.M. Other stables in other areas may schedule the morning meal at 6:00 A.M., the midday meal at 11:30 or 12:00 P.M., and the evening meal at 4:00 P.M. Again, another meal can be fed between 8:00 P.M. and 10:00 P.M., which will reduce the amount of time between the afternoon and morning feedings.

The evening meal typically makes up the largest portion of the ration. If a horse has not eaten all of its feed before the groom is ready to leave in the evening, it is best to leave the feed tub in the stall. The horse should be given all night to eat and digest its feed.

(For complete, accurate information on feeding for all horses, read *Feeding to Win II*, published by Equine Research, Inc.)

CHOOSING THE PROPER FEED

While it is not the groom's responsibility to decide what each horse is fed, understanding the basis for this decision is important. The major considerations in determining the type and ration of feed are:

- the horse's size
- the horse's temperament
- the quality of feed
- the horse's age
- the type of exercise

Most horses are fed a standard ration consisting of hay and grain. In addition to this ration, some horses are fed vitamin and mineral supplements.

Hay

The natural fiber in good quality hay is an important part of the horse's diet. A horse may consume more than 20 pounds of hay daily. Hay is available in two varieties—grass hay and legume hay. Grass hay includes Bermuda, timothy, and Kentucky blue grass. Grass hay is the most common type of hay. A horse that is stabled at the racetrack should have free access to grass hay at all times, except on race days when hay intake may be limited.

Fig. 3–2. Horses should have free access to grass hay.

Legume hay, including alfalfa and clover, is "richer" (high in protein) and should be fed in limited amounts. Generally, only one flake of legume hay per day is fed to the horse to supply additional calcium and protein. Alfalfa, which is the most popular legume hay, is also available in a pelleted form.

Fig. 3–3. Legume hay should be fed in limited amounts.

Hay should always be sweet smelling and green. (The greener the hay, the more vitamin A and other nutrients are present.) The hay should be dry, but not so dry that it is dusty. Dust may cause horses to develop chronic coughing. *(See Chapter 13 for more information.)* All hay must be free of mold (moldy hay appears black with white patches) and any foreign materials such as stickers, rocks, trash, or insects.

QUALITY OF HAY	
Good quality hay is...	**Poor quality hay is...**
Sweet smelling Green Dry Free of dust Free of foreign matter	Musty smelling Yellow, brown, or black Wet or moldy Dusty Contaminated with foreign matter

Fig. 3–4. Grooms must be able to distinguish between good and poor quality hay.

The groom must be able to distinguish between good and poor quality hay. At some point the groom may be in a position to accept a delivery of hay, and he or she will have to examine the hay to make sure it is up to quality standards.

Grain

Because hay cannot supply all the energy needs of hard-working horses, grains are fed as a concentrated energy source. There are three basic grains which make up most horses' rations—oats, barley, and corn. **Note:** Grains should be fed by weight, not volume. (One coffee can filled with corn weighs more than the same can filled with oats.)

Fig. 3–5. Oats are the most popular grain fed to racehorses.

Oats—Oats are the most popular grain fed to racehorses. They are widely available, higher in protein than most grains, and are not as "rich" as other grains, which makes them safer to feed. (In this case, "rich" means high in energy.) They can be fed whole, crimped, rolled, steamed, or clipped. Whole oats are the most difficult form for the horse to digest. Therefore, most feed mills sell oats in the more digestible forms. Unfortunately, the process of making the oats more digestible and thus, more nutritious, decreases the "shelf life" of this grain.

Barley—Barley is a popular grain for racehorses in the western United States. Because it is widely grown and available in that part of the country, it may be less expensive there than oats or corn. Like oats, barley is a higher protein grain. The energy concentration of barley is greater than oats but less than corn. Barley is considered a moderately "rich" grain and is available whole, flaked, or crushed for better digestibility.

Fig. 3–6. Like oats, barley is high in protein.

Corn—Corn is the "richest," most digestible grain and has a higher energy value than oats or barley. Because one pound of corn can provide the same amount of energy as 1½ pounds of oats, corn is often used as an economical energy source. However, because corn is lower in protein than oats or barley, it is usually mixed with one or more of the higher protein grains and/or supplemented with legume hay. Corn is available whole or cracked, again for better digestibility.

Fig. 3–7. Corn has a higher energy value than oats or barley.

The type and combination of grain fed to horses depends largely on the trainer's preference. As mentioned earlier, the rations depend on factors such as the horse's age, size, and exercise program. Most racehorses are fed 10 – 15 pounds of grain (or even more) per day. Any increase or decrease in a horse's grain ration should be accomplished gradually. If the horse eats too much grain at one time, either due to human error or because it broke out of its stall and got into the feed room, the horse may founder. *(See Chapter 5.)*

Supplements

Feed supplements are often added to a horse's daily ration. Typical feed supplements include vitamins, minerals, salt, and protein supplements such as soybean meal. Feed supplements come in powder, pellet, or liquid form and are usually added to the horse's evening meal.

Typically, if a horse is fed a properly balanced ration with high quality grains and hay, there is no need to feed a vitamin/mineral supplement. In fact, over-supplementation of some vitamins and minerals can cause toxic symptoms in horses.

ENCOURAGING THE POOR DOER

Sooner or later a groom will be faced with a horse that is considered a "poor doer." This is a horse that regularly fails to eat its full ration and therefore loses important nutrients. Feed consumption is very important to the racehorse. If the horse does not eat properly, it cannot be expected to perform to its maximum abilities. Any of the following factors might contribute to a poor doer:

- lack of proper exercise
- boredom
- stress
- nervous temperament
- unpalatable feed
- vitamin or mineral deficiency
- tooth or mouth problems
- illness
- internal parasites

Trainers have devised many techniques to stimulate the appetite of a poor doer. The groom should be familiar with each of these methods.

Feeding a bran mash in the evening—Feeding a bran mash makes the ration more palatable, provides a gentle laxative, and increases the horse's water intake. To prepare a bran mash, mix the normal evening ration in a feed tub, then add two quarts of wheat bran to the ration. (Wheat bran is made from the outer portion of the wheat kernel, which is very fibrous.) Saturate the entire mixture with hot water. Cover the feed tub with a clean towel and allow the mash to steam and cool for approximately two hours before feeding. Before the mash is placed in the stall, it is important to mix the entire ration with your hands. This distributes the bran evenly and ensures the temperature is not too hot for the horse's mouth. If it is too hot to touch, it is too hot for the horse to eat.

Fig. 3–8. Wheat bran is used to make a hot bran mash.

Fig. 3–9. Adding molasses to grain improves its taste.

Some trainers feed bran mashes year round. Others feed them only in the winter. A few trainers do not feed them at all unless directed to do so by a veterinarian. The groom should not feed a bran mash unless instructed to do so by the trainer.

Feeding a pre-mixed sweet feed—Sweet feed is a mixture of grains with molasses added to reduce the dust and improve the taste. In most sweet feeds, the primary grains are oats and corn. Some sweet feeds also contain vitamin or mineral supplements.

Adding a B-complex vitamin (such as thiamine) to the diet—B vitamins are known to stimulate the appetite. They may be added to the feed in the form of brewer's yeast or administered by a veterinarian through injections.

Adding treats—A horse can often be enticed to eat simply by adding treats to the feed, such as carrots, honey, molasses, or sugar. Apple cider vinegar is often added to improve taste and digestibility. A few tablespoons of salt daily (perhaps one spoonful at each feeding) improves appetite and water intake.

WATERING

Many people fail to understand the importance of clean, fresh water as part of the horse's daily requirements. Horses can live much longer without food than they can without water. Water should be as fresh as possible—changed frequently throughout the day—and should be free of hay and other foreign material. The water should be cool (not cold) in warm weather. In cold weather, a little warm water can be added to the horse's bucket periodically to take the chill out of the water. Water makes up ⅔ of the horse's body mass and fulfills the following bodily functions:

- quenches thirst
- maintains body temperature within a normal range
- aids in the processes of digestion and excretion
- ensures adequate blood flow to working muscles

A healthy horse should be allowed free access to clean, fresh water under normal circumstances. An average-sized racehorse in training may drink 8 – 15 gallons or more of water per day. Of course, this amount varies depending on the climate, the horse's diet, and the level of exercise. When conditions are not normal, it is the groom's duty to water the horse responsibly. For instance, horses should not consume any water when it is nearly time for racing. (The trainer will give instructions on when to draw the water.) A stomach filled with water presses on the diaphragm, restricting expansion of the lungs and resulting in impaired breathing.

Horses deprived of water for any length of time, such as while traveling or after a strenuous race or workout, should not be allowed to fill themselves with water. The groom must make sure the horse is cooled out and "watered off" over a period of 45 minutes to one hour. *(See Chapter 9 for more information on cooling the horse out.)*

EQUIPMENT

Feeding Tubs

The best kind of feeding tub is the removable kind, as the groom is able to remove the tub daily to clean it. However, if the feed tub is permanently affixed to the stall, the groom should still clean it out after each feeding (as feed left in the tub can spoil or mold).

Standard Plastic Feed Tub

This type of feed tub is very popular as it is made of a tough, durable plastic. It is usually 9 inches deep and 18 inches in diameter. It is made with three holes so it may be secured to the stall wall with screw-eyes and double-end snaps.

Fig. 3–10. A standard plastic feed tub can be attached to the stall wall.

Slow Feeder Metal Tub

This type of metal tub is ideal for a horse that bolts (gulps) its feed. An iron ring fits into the tub and is secured to the tub by screws. The horse must work its mouth around the iron ring, thus slowing its feed intake and preventing digestive disorders.

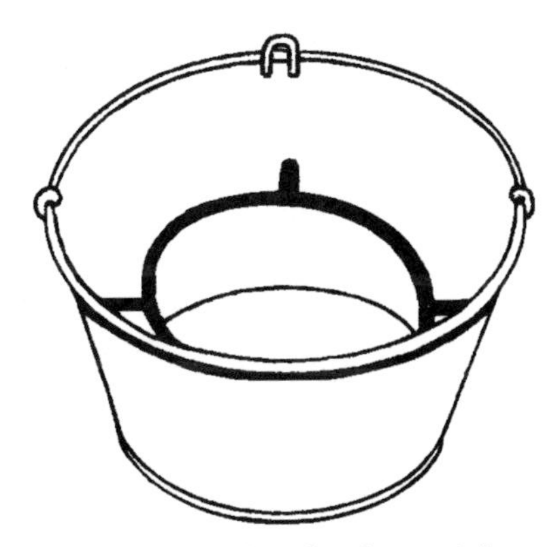

Fig. 3–11. A slow feeder metal tub slows a horse's feed intake.

Reinforced Metal Feed Tub

The reinforced metal feed tub is usually made of galvanized sheet steel with a roll-top rim. The roll-top rim prevents the sheets from coming apart and creating a hazard which might injure the horse. Due to the galvanized sheet steel construction, the tub will not rust or corrode if you use salt in the horse's feed.

The normal capacity of this type of tub is about six gallons. There are three holes in the tub so that it can be secured in the corner of the stall with double-end snaps and screw-eyes.

Collared Feed Tub

Fig. 3–12. A collared feed tub prevents a horse from spilling its feed.

A collared feed tub looks similar to the standard plastic feed tub except for a four-inch collar or lip extending down into the tub at a slight angle. The collar design prevents a horse from throwing its feed out of the tub and thus avoids waste. If the horse attempts to throw its feed about, the feed strikes under the collar and bounces back into the tub.

Feeding Accessories

Feed Scoop

This scoop is designed for measuring oats, sweet feed, corn, and other foodstuffs. It is usually made of galvanized metal or plastic. The feed scoop is available in an oval or tubular design, in one, two, or three quarts.

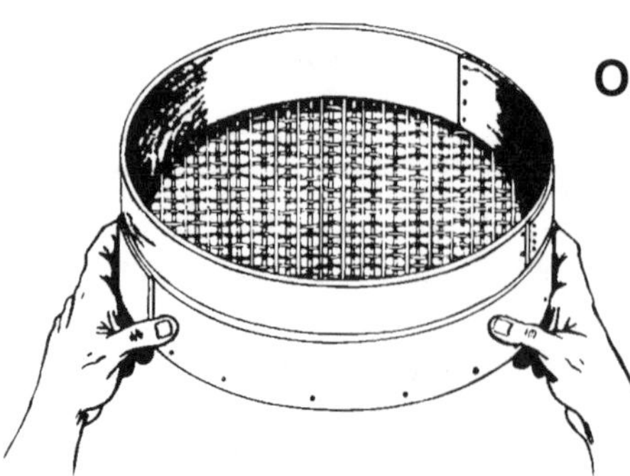

Fig. 3–13. An oat sieve is used to sift chaff and foreign materials from the horse's oats.

Oat Sieve

The oat sieve is used to sift oats, removing chaff and foreign material. It is made of wood with a fine metal mesh screening set in a circular pattern, available in a 14-inch or 18-inch diameter. To use the sieve, take (for example) two dry quarts of oats and place them in the sieve. With a circular rotating movement, shake the oats in the sieve until the foreign materials fall through the bottom of the sieve and the clean oats remain. Continue this process until you have the desired amount of oats. Then place the freshly sieved oats in a clean container until it is feeding time.

Water Buckets

Water should be made available to the horse in metal or rubber water buckets. (The capacity of most water buckets is 3 – 4 gallons.) Most trainers use one bucket per horse, but it is not uncommon to find two water buckets in a stall, or one bucket inside and another immediately outside the stall where a horse may drink while its head is sticking outside the stall.

Fig. 3–14. A water bucket may be flat on one side so that it can be hung against the stall wall.

Most water buckets are designed with a flat back to allow them to hang flat against the wall and prevent rolling and spilling. They are usually placed in one corner of the stall and secured to the wall by the use of a double-end snap and screw-eye. All water buckets should be removed from the stall daily and cleaned with a stiff brush to remove any film, hay, or grain which might build up in the bucket. Three important points to remember when working with water buckets are:

- Never use a drinking water bucket to wash horses or laundry. Soap residue may accumulate on the bucket and make the water distasteful to the horse.
- Never let a horse drink out of another horse's water bucket. Sharing buckets encourages the spread of viruses, bacteria, etc.
- Always hang a water bucket high enough on the wall so that the horse is unable to defecate (pass manure) in the water bucket. The actual height, of course, depends on the size of the horse.

Electric Water Heater

This is an excellent tool for heating water to use in the stable (perhaps to cook oats or bran when making hot mash feed). It is designed to stand upright in a metal bucket or large metal trash can. *Do not use the electric water heater in a plastic bucket.* The heating element is usually encased in a copper tube and works on either AC or DC current.

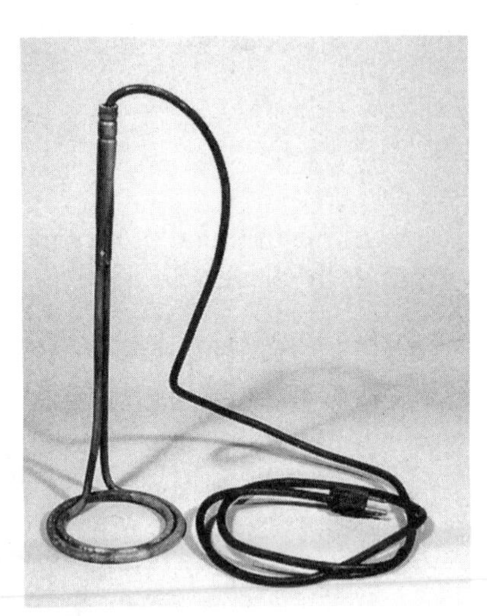

Fig. 3–15. An electric water heater can be used to make a hot mash feed.

The groom fills the container with cold water and submerges the coil of the heater in the cold water—with the cord and plug above the water level. After putting the coil in the water, the unit should be plugged into an outlet. *Make sure the plug is dry when inserting it in the socket.* The water in the container gradually becomes hot. When the water is at the desired temperature, unplug the electric water heater, wait a minute or two, then remove it. Then mix the hot water with grain to create a hot mash feed. *(Be sure to unplug the heater before removing it from the water, or the heater will short-circuit and break.)*

Feeding Hay

There are several ways to feed a horse hay. Grooms should use whatever feeding method their trainers prefer. Some trainers prefer to feed hay from the floor of the stall. The advantage of feeding hay on the floor, some trainers argue, is that the horse is in the most natural position for eating—with its head lowered to the ground as nature intended. The horse gets less dust in its airways when eating hay from the floor. Also, many trainers feel that this method minimizes digestive problems of the horse (such as choking).

However, there are some disadvantages to feeding a horse hay from the floor of the stall. The hay is exposed to dirt, as well as bacteria and parasites if it becomes contaminated with urine or manure. Also, it is sometimes thrown about the stall by the horse and is wasted as it becomes mixed with the bedding.

Many trainers prefer to feed hay in hay nets and racks. The hay is kept free of contamination from dirt, manure, urine, etc. It is not wasted, as any uneaten hay is left in the net or rack rather than mixed with the bedding. Finally, hay nets and racks allow the horse to munch on hay while it is tied to the wall for grooming, tacking, etc.

Hay nets and racks have some disadvantages, too. Many horsemen feel that the stretching of the neck and head upward to eat hay from the net or rack is an unnatural eating position for a horse and may cause digestive disorders. Also, when a net or rack is incorrectly placed inside the stall, the horse may get a foot or leg caught in it and be severely injured.

Hay Net

A hay net is designed to hold approximately 25 pounds of hay. It is usually made of nylon or cotton cord knotted together with a purse string method of opening and closing. The groom should load the net with "flakes" of hay until the net is full. (There are usually 10 – 12 flakes in one bale of hay.)

Many trainers use the hay net *outside* the stall door to prevent the horse from getting hung up (catching a foot or leg in the net). When a hay net is outside the stall, it should be taken down when a horse enters or exits the stall. This practice allows a clear stall entrance and thus avoids injury to the horse's hips.

When placing the hay net inside or outside the stall, it should be tied high enough so that the horse will not get hung up on it. At the same time, the screw-eye from which the hay net is hung should be at a height where the groom can reach it with little difficulty.

Always use a slip knot when securing the hay net to the wall or in a van or trailer. Pulling the single end of the slip knot opens it immediately, and the hay net drops from the wall. This safety measure is important in an emergency situation, such as when a horse has caught its foot in the net. Practice tying a hay net up and releasing it until this task becomes routine. This procedure is illustrated in Figure 3–16 on the next page.

Hanging a Hay Net (Fig. 3–16.)

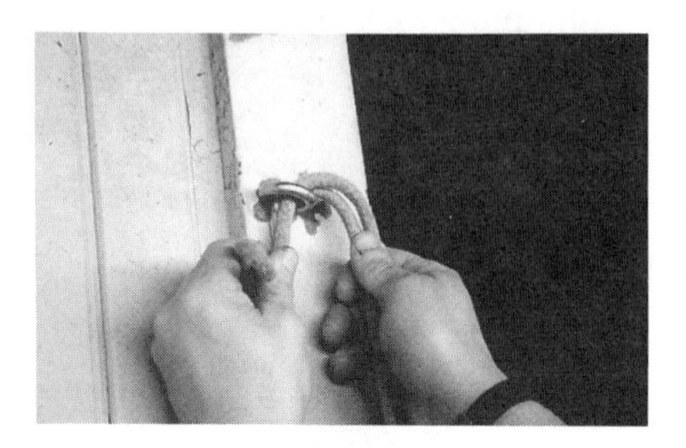

Step 1

Thread the long, knotted end of the net through the screw-eye.

Step 2

Pull the end downward (which pulls the hay net up) as far as it will go.

Step 3

With the hay net up, loop the long end through the mesh in the front of the net and pull up toward the screw-eye.

Step 4

With the excess end, make a loop over the line extending from the screw-eye.

Step 5

Pull the end of the loop under the line, forming a slip knot.

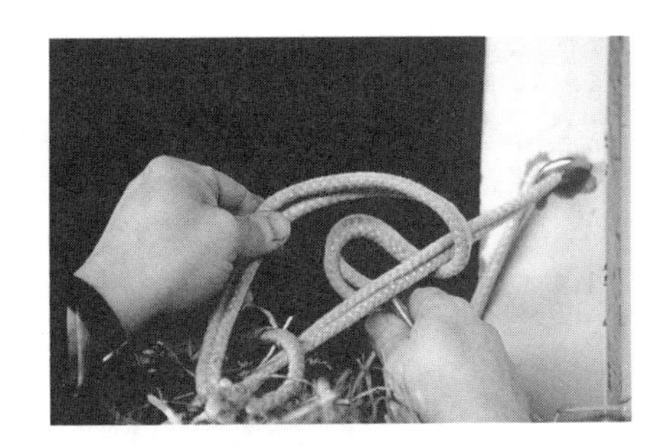

Step 6

Pulling the net downward tightens the knot, resulting in a knot with a single end.

Step 7

Pull the excess rope up so it does not hang where the horse can pull it out with its teeth.

Step 8

Turn the entire hay net so that the slip knot is against the wall. This measure prevents the horse from chewing on the knot.

Metal Hay Rack

The metal hay rack is a stationary fixture which is permanently held in the corner of the stall by screws. It should be positioned high enough in the stall corner to prevent a horse from getting a foot or leg caught in it. The metal hay rack is usually 36 inches high and 30 inches wide from wall to wall. The groom simply places hay in the rack when it is empty.

Fig. 3–17. On the left, a metal hay rack. On the right, a chain hay rack.

Chain Hay Rack

This type of hay rack is placed in the corner of the stall similar to the metal rack. It is secured to the corner of the stall by four small snaps on each side fastened to two small screw-eyes on each wall. It is usually made of a lightweight galvanized linked chain. Unlike the metal rack, the chain rack may be removed from the stall with little difficulty.

Fig. 3–18. A salt block holder allows the horse free choice salt.

Salt Block Holder

The salt block holder holds a mineral or salt block. The metal holder is secured to the stall wall and the four-pound mineral or salt block is inserted in the holder. Then the horse has free access to the mineral or salt block.

FEED STORAGE

The feed room (or extra stall where the feed is stored) should be kept tidy by the grooms. Hay and straw should be stored on wooden pallets. The wooden pallets allow air to circulate under the stored bales to prevent moisture absorption and formation of mold.

Grain should be stored in metal bins or large plastic trash cans with secure lids to keep the grain fresh, dry, and safe from rodents. The grain containers should also be labeled according to their contents. Vitamin and mineral supplements, feed additives, etc. should be stored on shelves in sealed containers. Feeding accessories such as feed scoops, feed tubs, and buckets should be cleaned on a daily basis and stored neatly in the feed room when not in use.

FEEDING PROCEDURE

As stated earlier in this chapter, racehorses are usually fed three or four times per day. With some trainers, the first feeding is done by a designated feed person (in some stables, the night watch person), as the grooms have not yet arrived at the stable. The grooms are responsible for the second and third feedings. With other trainers, the grooms are responsible for all three feedings. Typically, the grooms work as a team to feed and water all the horses as quickly as possible. This is important in large stables when some horses may become upset when they see other horses being fed and their feed has not arrived yet.

Most trainers follow the same basic feeding rules:

- Follow a feeding program based on experience and sound scientific principles.
- Feed only the best quality hay and grain available.
- Change feed gradually (over a week), if it is necessary to change feed at all.
- Evaluate and feed each horse according to its particular needs and nutritional requirements.
- Feed horses at the same times every day.
- Allow at least two hours after feeding before working a horse for the feed to be fully digested.
- Allow the horse free access to salt, grass hay, and clean, fresh water at all times under normal circumstances.
- Maximize the benefits of good feeding by deworming on a regular basis.

- Clean the feed tubs and water buckets on a daily basis.
- Do not feed grain or allow free access to water unless the horse is completely cooled out.
- Have the horse's teeth examined regularly for sharp edges, which can cause pain in the mouth and inhibit digestion.

EATING HABITS

Abnormal eating behavior is one of the first indicators of disease or other disorders. Therefore, as a groom, it is very important for you to become familiar with the horse's eating habits. The more familiar you are with the horse's normal behavior, the more quickly you can recognize and report anything abnormal to the trainer. Use the following guidelines to evaluate the eating habits of each horse:

- Is the horse a slow or fast eater?
- Does the horse become excited at feeding time?
- Does the horse bolt its feed and gobble it down with very little chewing?
- Does the horse knock its feed tub about in an attempt to force the feed to fall to the ground?
- Does the horse prefer to drink water before or after it eats?
- Does the horse normally eat up every bit of feed or does it leave some uneaten?
- Does the horse normally eat all of its hay?
- Does the horse prefer to eat its food from the stall floor?

When a horse that typically finishes its feed shows no interest in food or completely refuses to eat, you can be sure that something serious is wrong. Other problems that interfere with proper nutrition may be easier to remedy. For instance, a horse that attempts to eat, but spills most of its food while chewing probably has problems with its teeth. Or, a horse that does not drink its water may have a contaminated bucket.

Grooms should watch their horses carefully, reporting any abnormal behavior to the trainer. Something as simple as a change in feeding methods can cause this behavior. Remember, grooms should not make any changes in feed or routine unless instructed to do so by the trainer.

Treats

Some caretakers like to give their horses treats occasionally. The most common treat given to horses is carrots. Horses also like apples, but be careful not to give too much. The proper way to feed a horse a treat is to cut or break it up into pieces no bigger than your finger. Put the treat in the horse's feed tub or offer it to the horse on the flat of your palm. If you are feeding the horse from your palm, be sure to keep your fingers out of the way. Never tease a horse with treats or feed.

Fig. 3–19. Feed the horse a treat on the flat of your palm, with your fingers well out of the way of the horse's teeth.

SUMMARY

Because feeding is such a routine task, it is easy to become casual about it. Unfortunately, careless feeding practices are the reason many horses get sick. To avoid this mistake, make a habit of examining all feed for freshness and check to make sure each horse is getting the correct type and amount of grain *at every feeding.* Also, watch the horse carefully during its meal and after it finishes eating to ensure that the animal is behaving in a healthy manner.

(This chapter is a general guide to feeding racehorses and cannot provide complete feeding information for all horses. *Feeding to Win II,* published by Equine Research, Inc. is the most comprehensive book available on feeding and nutrition for horses.)

Chapter
4

Grooming, Clipping, & Washing

Fig. 4–1.

A sleek horse is one sign of a good groom. But grooming does more than just make the horse look good. Brushing stimulates circulation in the skin and muscles, which makes the horse feel good. Washing cleans the hair coat and helps prevent skin diseases and parasites. Keeping the horse clipped in cold weather helps it to cool down more quickly after a race or workout. Good grooming also helps prevent saddle sores—in fact, there is positively no excuse for the person who allows a horse to get a saddle sore because of a dirty hair coat, saddle cloth, or girth.

The horse should be groomed daily, before and after a workout or race. The lighting inside the barns is usually sufficient to groom, but it is even better to groom or wash a horse outside the stall under ideal weather conditions.

HALTERING & TYING THE HORSE

Before doing any grooming, you must first halter and tie the horse. When entering the stall to catch the horse, make sure the horse is facing you. If the horse has its rump toward you, get the horse to turn around and face you by calling its name. *Never enter the stall when the horse is lying down.* The horse might be sleeping and if you startle it, it could hurt itself or you. Get the horse to stand up by making a "chirping" sound. Give it plenty of time to get up; do not rush immediately into the stall. Once the horse is up on its feet and turned to face you, enter the stall, talking softly as you approach.

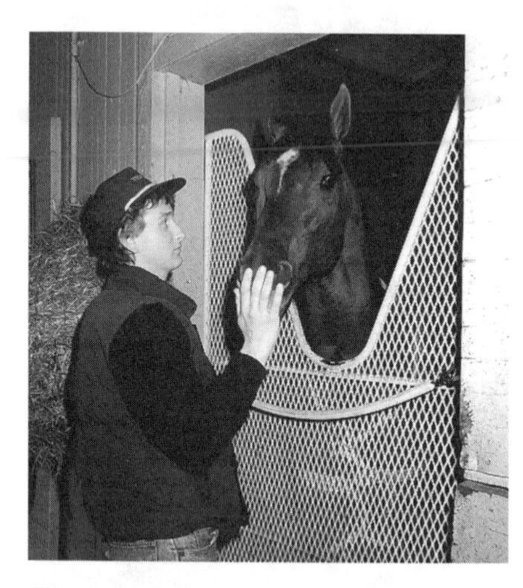

Fig. 4–2. Get the horse to turn and face you before you enter its stall.

Many stables leave the halters on their horses at night. If that is the case, then the halter should already be on the horse. If not, it should be hanging on the wall outside the stall door. *(See Chapter 2 for instructions on haltering the horse.)* Once haltered, the horse may be tied in one of the following three ways.

Tie Chain

A tie chain consists of a strip of rubber about two to three feet long and about one inch wide, with a metal snap at either end. One end of the rubber tie chain is snapped onto the center ring of the horse's halter and the other end is snapped onto a screw-eye high on the wall. (The screw-eye should be high enough so that if the horse becomes excited and rears up, it cannot get its front legs over the rubber tie chain.)

The rubber tie chain has enough flexibility to give slightly (but not break) if the horse becomes frightened and pulls back. This is ideal for young horses, particularly Thoroughbreds, that may be nervous about being tied up.

Cross-Ties

Cross-ties consist of two chains (or ropes)—one attached to a screw-eye high on the wall on either side of the stall or aisleway. The chain on the left should be attached to the lower near ring on the horse's halter, and the chain on the right should be attached to the lower off ring. If the cross-ties are in the horse's stall, remove the chains from the stall after grooming so the horse does not injure itself.

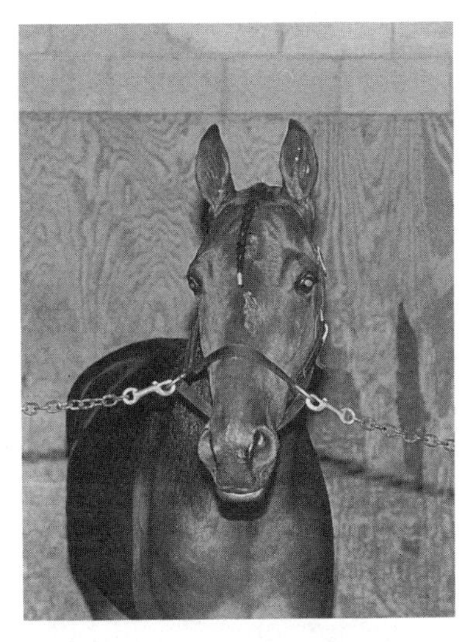

Fig. 4–3. Cross-tying the horse works well for grooming.

Cross-tying is one popular way to tie a horse for grooming. (Cross-ties are also used in many horse vans and trailers to stabilize the horses during shipping.) On cross-ties, the horse cannot move more than a few steps forward, backward, or to either side, and the groom can easily move around the horse to work.

Lead Shank

Some trainers have their grooms tie horses to the stall wall (or to another sound structure) with a lead shank. If the horse is trained to tie this way and stands quietly, this method is acceptable. When tying a horse in this manner, it is preferable to use a rope shank. Leather and nylon chain shanks are more difficult to tie and break more easily if the horse pulls back suddenly. *Never tie a horse up while the chain shank is attached over the nose, under the jaw, through the mouth, or over the upper gum.* Always make sure that the rope is not long enough for the horse's head to reach the ground. If it is, the horse can get a leg caught in it, which can lead to serious injury.

Follow the trainer's instructions on how to tie each horse. If the trainer has no preference, it is best to tie the horse using whatever method with which the animal is most familiar and comfortable.

GROOMING TOOLS

Early on, the groom should become familiar with all the equipment needed to properly groom a horse. Each grooming tool has a specific purpose.

Rub Rag—This is a linen cloth or towel used to wipe the dust from the coat. The rub rag is also used to polish the coat.

Curry Comb—A curry comb is used in a circular motion to remove dried mud, loose hair, and dirt from the coat. Its use is confined to the neck and torso, but should probably not be used on the flank unless the hair is caked with mud (a rare occurrence with racehorses, as they are usually kept in a stall). The curry comb should not be used on the head or legs. Rubber, plastic, and metal curry combs are available. The plastic curry comb is considered by many horsemen as being too harsh on the horse's skin, but is sometimes used to comb the mane and tail. The metal curry comb is too harsh for either purpose. Most professionals use only the rubber curry comb on the horse's skin.

Stiff Brush—(Dandy brush) This brush is used in a sweeping motion to remove dirt from the body. It may also be used to brush the mane and tail. The stiff brush, like the rubber curry comb, should not be used on the head, but may be used on the flanks and the outside of the legs.

Soft Brush—(Body brush) This brush is used in a sweeping motion over the entire body to remove dirt. The soft brush may be used on the horse's face and legs.

Large Natural Sponge—The sponge is dampened and used to clean the corners of the eyes, inside the nostrils, under the tail, and between the hind legs.

Mane Comb—The mane comb is used to comb out knots in the mane or tail. It is also used when braiding or pulling the mane or tail.

Hoof Pick—This grooming tool is used to remove dirt and manure from the bottom of the foot.

Foot Brush—A foot brush is used to clean the bottom of the foot and the heels.

There are various methods of grooming horses, but whatever system you use, it should be efficient. The grooming method described in the next few sections is used successfully by many professional grooms at the racetrack.

Fig. 4–4. The rub rag may be used anywhere on the horse's body.

Fig. 4–5. Do not use the curry comb on the shaded areas: the head, legs, or flanks.

Fig. 4–6. Do not use the stiff brush on the shaded areas: the head, or the inside of the legs.

Fig. 4–7. The soft brush may be used anywhere on the horse's body.

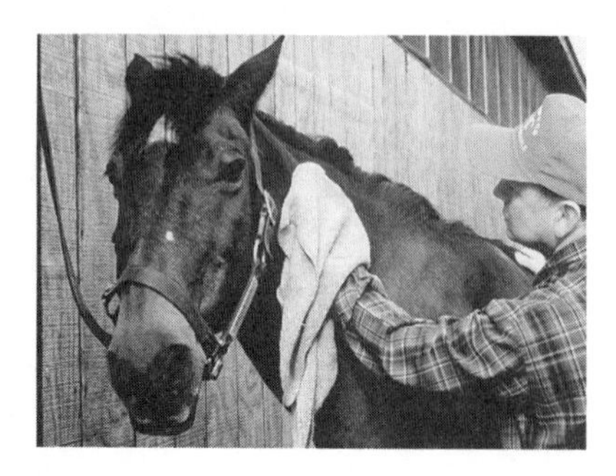

Fig. 4–8. The rub rag removes the visible dirt from the surface of the horse's coat.

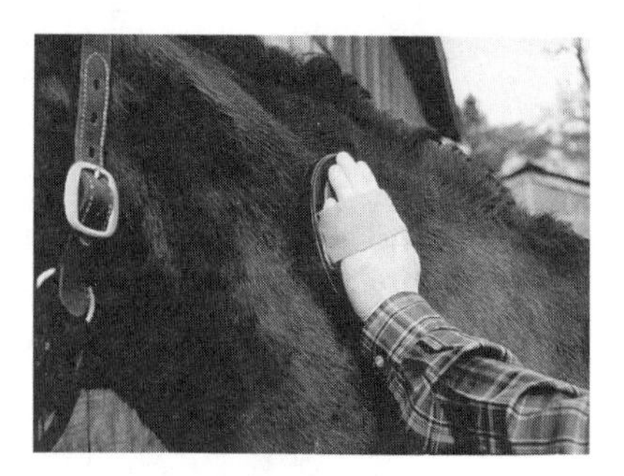

Fig. 4–9. The rubber curry comb brings ground-in dirt, loose hair, and dandruff to the surface.

BODY & HEAD

Using the Rub Rag

First, remove any visible dirt from the horse's face. Then wipe the near side of the horse's body, including the legs. Repeat the process on the off side. Periodically shake the rub rag to remove any dirt. *Be careful, as "popping" the rag may frighten the horse.* When moving from one side of the horse to another, it is probably better to cross in front of the horse. When you do, watch that the horse does not try to bite you or strike out with its front feet. If you do cross in back, stay close to the horse and hold its tail, as this usually discourages kicking.

Using the Curry Comb

Now take the rubber curry comb in your left hand and begin at the top of the neck on the near side. Rub the curry comb in a circular motion. This brings the ground-in dirt to the surface, along with any loose hair or dandruff. Continue this procedure down the neck, chest, shoulder, forearm, withers, back, side, belly, rump, and outside of the gaskin. Do not use the curry comb on bony areas or on the flank—horses are sensitive there and may try to kick you. Proceed to the off side and repeat the procedure.

Using the Stiff and Soft Brushes

Hold the halter with your left hand, and in your right hand hold the soft brush. Begin brushing the face, following the direction of the hair. Be sure to brush the forehead and pay special attention to those areas under the halter. Brush the forelock as well, working the bristles into the roots with a gentle side-to-side motion to dislodge any dirt or dandruff.

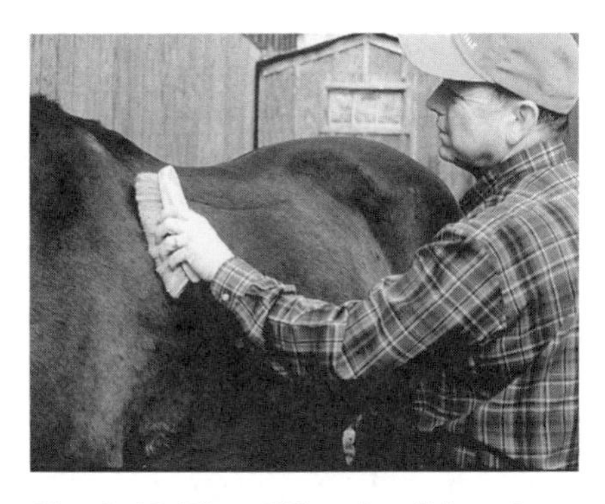

Fig. 4–10. The stiff and soft brushes whisk away the dirt that the curry comb brought to the surface.

Place the stiff brush in your right hand and the soft brush in your left hand. Brush away the dirt that has surfaced from the use of the rubber curry comb, beginning at the top of the neck on the near side. With short, sweeping strokes, remove the dirt from the body with the stiff brush. Follow this action with the soft brush by brushing over the same area. After a while, you will develop a definite rhythm with these brushes.

Rub the bristles of the two brushes together during the grooming procedure to remove any dirt that has accumulated on them. Continue brushing down the neck, chest, shoulder, withers, back, side, belly, flank, and rump. When brushing the flank, be sure to *carefully* brush in the same direction as the hair. Continue this procedure on the off side of the horse. When on the off side, switch brushes so that the stiff brush is now in your left hand and the soft brush is in your right hand. When grooming the off side, be sure to brush the mane with the stiff brush and do not forget the area under the mane.

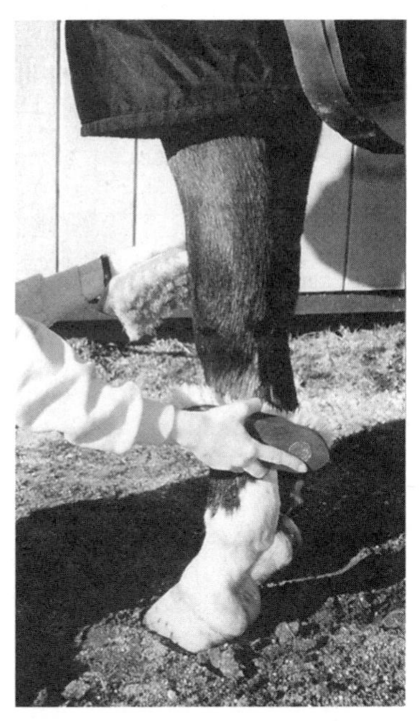

Fig. 4–11. Use the stiff brush on the outside of the leg and the soft brush on the inside.

When grooming the legs, face the rear of the horse and brush the legs thoroughly with the stiff brush on the outside of the leg and the soft brush on the inside. Remove any dirt from the heels with the soft brush. Do not worry about brushing over the chestnuts (the small horny growths on the inside of the horse's legs). Also, it will not hurt the horse if you peel a layer of the chestnut off, nor will it hurt if you leave it alone.

Using the Sponge

Saturate a sponge with clean, warm water and squeeze the excess water from it. Hold the halter with your left hand and the sponge with your right hand. Begin by wiping the face and eyes with the sponge. Wipe the sponge down over the upper lid of the eye to close the eye while removing any crust or dirt from the corners. Rinse the sponge, then wipe out each nostril.

Fig. 4–12. Left: Use the sponge to clean out each nostril. Right: To clean the anal area, the underside of the tail, and between the thighs, stand on the near side of the horse and lift the tail with your left hand.

Move to the off side of the horse. Wipe the mane in a downward motion, rinsing the sponge out periodically. Then move around to the near side of the horse, facing the hindquarters (trail a hand along the horse's body as you go). Standing close to the near hind leg, lift the tail with your left hand. With the sponge in your right hand, clean the anal area, the underside of the tail, and between the thighs. This does not have to be a dirty job. If the horse is washed on a daily basis, these areas stay relatively clean.

Polishing the Coat

The final step in grooming is to use a clean rub rag all over the horse's body. Begin at the head, and rub the coat on the near side of the body briskly. Continue to the off side. Remember to rub in the direction of the hair growth, especially on the flanks. The rub rag provides that final sheen to the horse's coat.

For an extra sparkle on race day, use some baby oil or coat conditioner on the rub rag. (Do not saturate the rub rag, just moisten it.) Rub the oil or conditioner in until the horse's coat is smooth and shiny.

MANE & TAIL

Brushing the Mane and Tail

A healthy mane and a full, long tail are desirable on a horse, which is why you should brush the mane and tail carefully with the stiff brush. The mane comb may be used first, to remove any knots in the mane and tail before brushing them with a stiff brush. When brushing the mane, be sure to brush underneath the mane as well as on top, dislodging dirt and dandruff from the roots with a stiff brush.

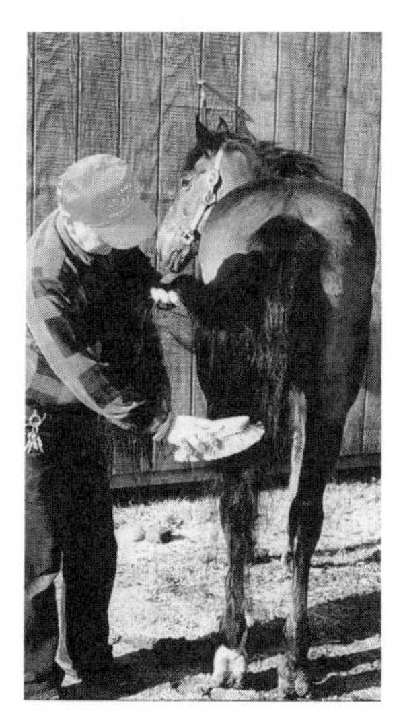

Fig. 4–13. Stand to the near side of the horse to brush the tail.

Fig. 4–14. To remove dirt and bedding from the center of the tail, flip the tail so that the hairs are draped over your left hand.

When brushing the tail, your left hand should be holding the tail and the stiff brush should be in your right hand. Do not stand directly behind the horse; there, your chances of getting kicked are increased. Brush small segments of the tail as opposed to brushing the entire tail at once. To remove dirt and bedding material from the center of the tail, merely flip the tail so that the tail hairs are draped over your left hand, and brush downward with the stiff brush or with your fingers. When using the stiff brush, be careful not to brush so hard that you rip hair out of the horse's tail.

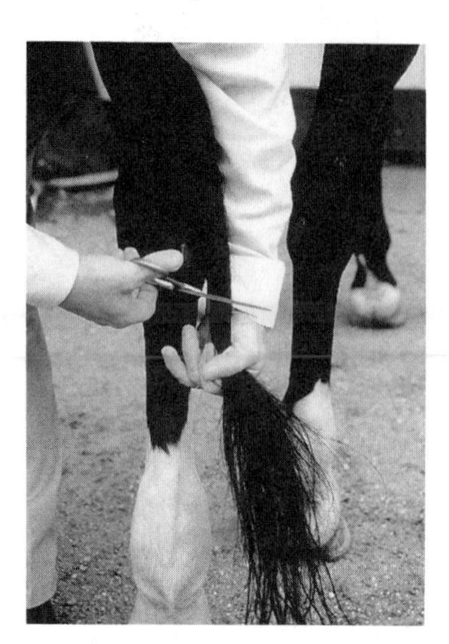

Fig. 4–15. Cut the horse's tail midway between the hocks and fetlocks. The end of the banged tail is parallel to the ground.

Banging the Tail

The groom should be familiar with the English practice of "banging" the tail. Banging consists of cutting the end of the tail off midway between the hocks and fetlocks. To cut, hold the tail down firmly and trim the ends of the tail hairs evenly across. Ideally, the end of the banged tail should be parallel to the ground.

Banging the tail is a common practice in England, Europe, and South America, but it is not as common in the United States. With that in mind, ask your trainer before cutting a tail this way.

Pulling the Mane

Pulling is the most common method of thinning and shortening a mane. When the mane is pulled, it appears more natural than if it is cut with scissors. Pulling also encourages the mane to lie flat on the neck. The ideal length of the mane ranges from four to six inches. Use common sense as to the amount of hairs to be pulled. In the case of a thin mane, it may be necessary to cut the mane first with scissors, then pull only a small amount to give it a natural look. Pulling small amounts of mane is not painful. *However, do not try to pull a whole fistful of hair out of any mane at once—this will be painful for the horse.*

The materials used to properly pull the mane are:

- plastic tooth curry comb
- stiff brush
- mane comb
- scissors (optional)

The following is the procedure for pulling the horse's mane *(see Figure 4–16):*

1. Comb and brush the mane to make sure it is free of soil and knots.
2. Start with the mane closest to the withers. Grasp the long hairs with your right hand.
3. Comb the shorter hairs upward, or "back comb" with the mane comb until only a few long strands remain in your hand.
4. Wrap the remaining long strands around the mane comb and pull downward sharply.
5. Comb the area just pulled until the mane lies flat. Continue this procedure all the way up the mane toward the poll.

Note: Some experienced grooms choose to pull the mane with their hands instead of using a mane comb. To do this, grasp a section of mane, "back comb" using the fingers, wrap the remaining strands of hair around two or three fingers, and pull. The hair should come out fairly easily. If the horse has a thick, course mane, pulling may be too painful to use the hands, in which case you would be wise to opt for the mane comb. In any case, using the mane comb makes the job go faster and the mane more even when completed.

Pulling the Mane (Fig. 4–16.)

Step 1

Comb and brush the mane to make sure it is free of soil and knots.

Step 2

Start with the mane closest to the withers. Grasp the long hairs with your right hand.

Step 3

Comb the shorter hairs upward, or "back comb" with the mane comb until only a few long strands remain in your hand.

Step 4

Wrap the remaining long strands around the mane comb and pull downward sharply.

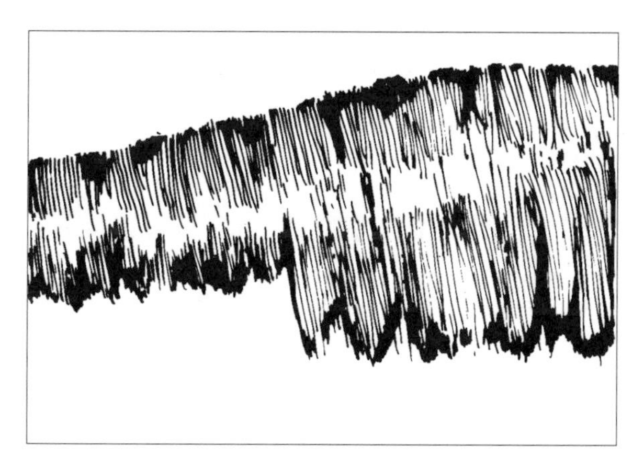

Step 5

Comb the area just pulled until the mane lies flat. Continue this procedure all the way up the mane toward the poll.

Braiding the Mane

Braiding is a skill which takes awhile to learn, but it can be mastered with a reasonable amount of practice. There are several reasons for braiding the mane. One reason is to train an unruly mane to lie flat against the neck on the off side. Another reason to braid the mane is for appearance. Or, a Standardbred groom may braid the horse's forelock in a single braid (left unfolded) to keep the hair out of the horse's eyes for a race or training session.

The materials needed to perform the basic braid are as follows:

- mane comb
- braiding rubberbands
- small sponge
- scissors

Before braiding the entire mane, comb it thoroughly. Dampen the mane with water, using the small sponge. (You may have to dampen the mane periodically while braiding.) Starting with the area closest to the ears, separate the mane into sections approximately 1½ inches wide. Use the following steps to braid the mane properly *(see Figure 4–17 on the following pages):*

1. Divide the section of mane into three equal strands.
2. Begin braiding by crossing the left strand over the middle strand. (The left strand now becomes the middle strand.) Then cross the right strand over the middle strand.
3. Continue this process with a steady downward pull until you have about ½ inch of hair left.
4. Secure the end of the braid with a braiding rubberband.
5. Fold the braid under and in half so that the end with the rubberband rests at the base of the mane.
6. Place a second rubberband around the folded braid, close to the base of the mane.
7. The finished braids should be of equal size and evenly spaced, lying flat against the horse's neck.

If the braid appears too long, you can fold it under a second time and secure it with a third rubberband. This type of braid has a "button" appearance on top of the neck.

Braids should not be left in for more than one day as they will cause the hair to break. Also, do not rip the rubberbands out of the mane when undoing the braids. Use scissors and carefully cut them out of the hair.

Braiding the Mane (Fig. 4–17.)

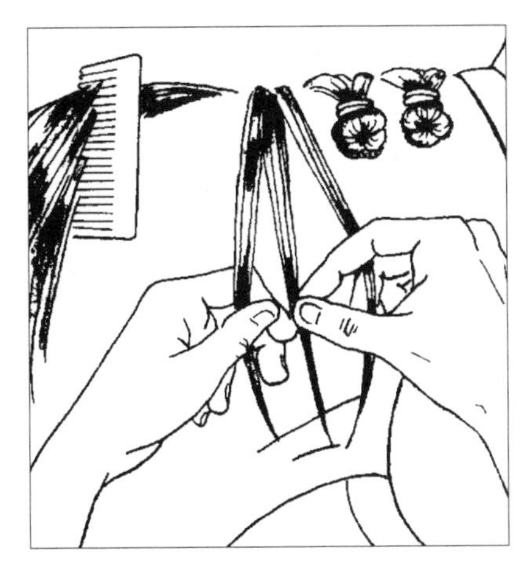

Step 1

Divide the section of hair into three equal strands. You can use a hair clip or the mane comb to keep excess mane out of the way.

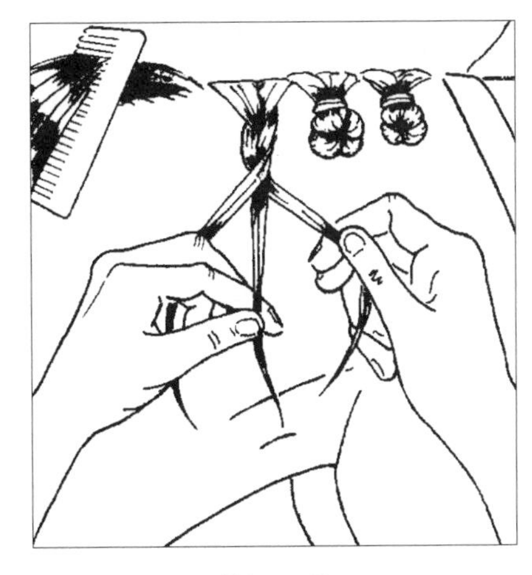

Step 2

Begin braiding by crossing the left strand over the middle strand. Now cross the right strand over the middle strand.

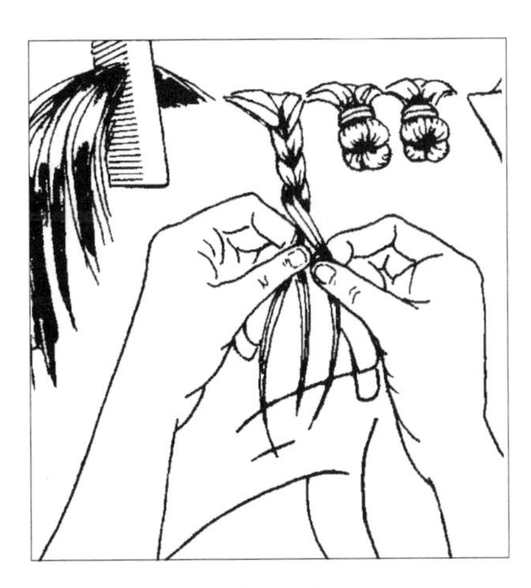

Step 3

Keeping a steady tension on the braid, continue crossing strands over until you run out of hair.

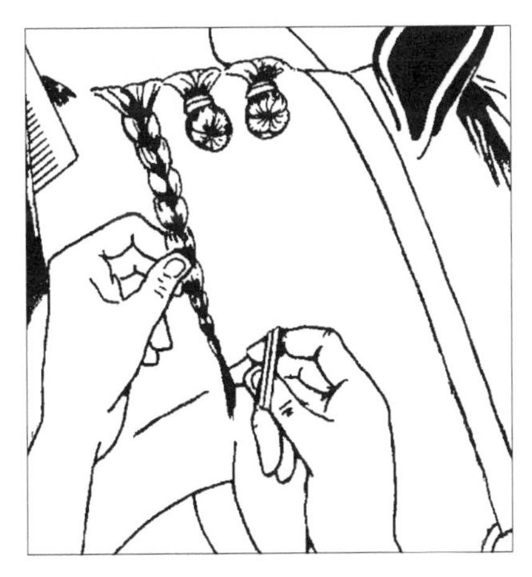

Step 4

Secure the end of the braid with a braiding rubberband.

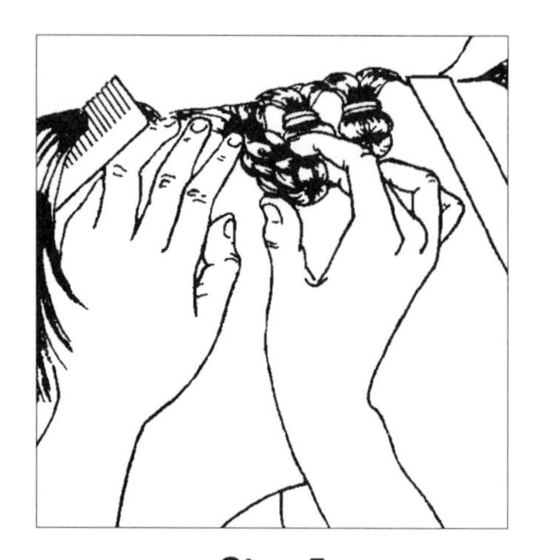

Step 5

Fold the entire braid under one time.

Step 6

Place another rubberband across the folded braid for security.

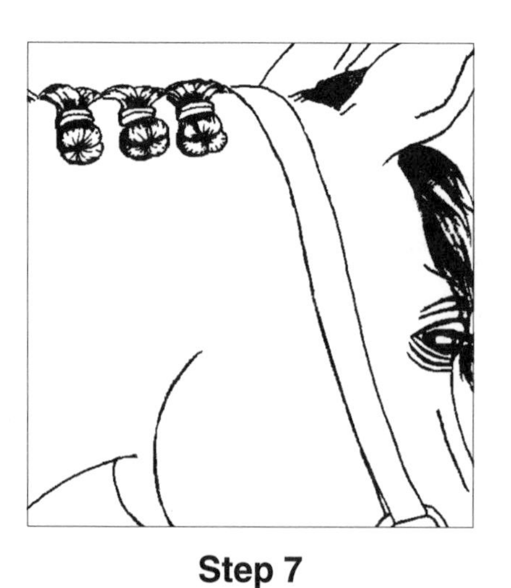

Step 7

The finished braids should be of equal size and evenly spaced, lying flat against the neck.

Braiding the Tail

Tail Rope

After a trotter or pacer is harnessed and hitched to the jog cart or sulky, the driver may decide to braid the horse's tail to keep it out of the way. (Some horses have a tendency to swish their tails when the driver uses a whip.) To braid the tail in this manner, attach a thin rope to the tail by tying a knot directly beneath the end of the tail-bone. Begin braiding the tail at this point, weaving the rope into the braid. When you reach the end of the tail, secure the end of the braid by tying a single or double knot with the rope. (There should still be rope left at the end of the braid.)

Some drivers choose to tie the rope to the cross bar of the sulky to keep the tail out of the way when racing. Most drivers prefer to sit on the end of the rope so that the horse will not be tied to the sulky by its tail in the event of a wreck.

Mud Knot

A mud knot prevents the tail of a running horse from becoming laden with mud when the horse runs on a wet and sloppy racetrack. Putting the tail up also makes it easier to clean after a race or work-out in such conditions.

The materials needed to make a mud knot are:

- stiff grooming brush
- braiding rubberband
- electrical tape

The following steps explain how to braid the tail in a mud knot *(see Figure 4–18):*

1. Brush the tail free of knots, matted areas, and foreign materials such as wood shavings, straw, or manure.
2. Divide the tail into two sections.
3. Tie the two sections together twice to form a square knot.
4. Fold these two sections underneath the tail and begin to crisscross them toward the top of the tail.
5. When you have about 12 inches of tail remaining, divide the hair into three sections and make one small braid. Secure the end of this braid with a rubberband.
6. Wrap electrical tape around the entire tail to keep the mud knot from coming loose. Be careful not to wrap it too tightly, as this can cut off circulation to the dock (base of the tail).

Braiding the Tail in a Mud Knot (Fig. 4–18.)

Step 1

Brush the horse's tail free of knots and debris.

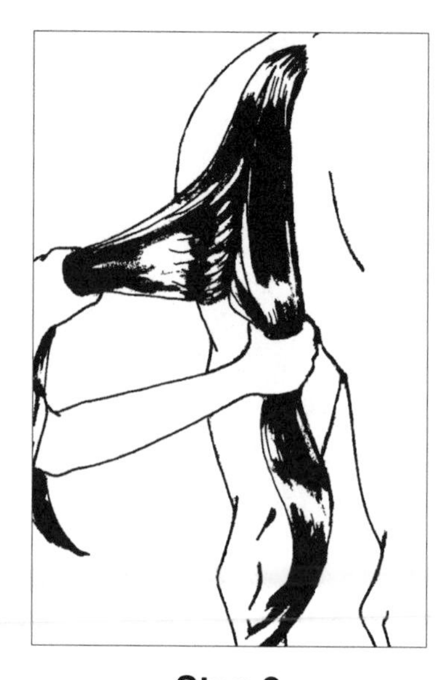

Step 2

Divide the horse's tail into two distinct sections.

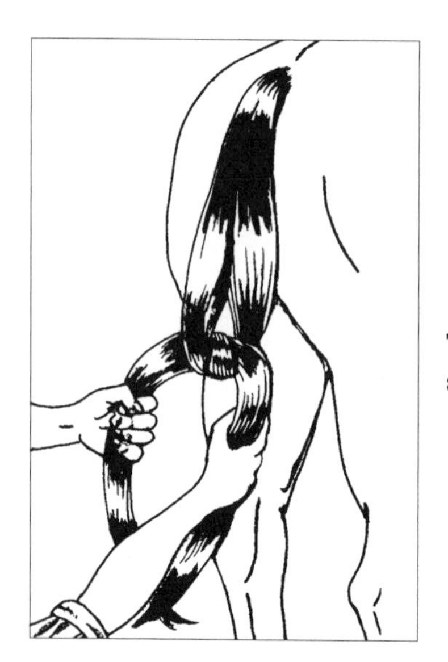

Step 3

Tie the sections together twice to form a square knot.

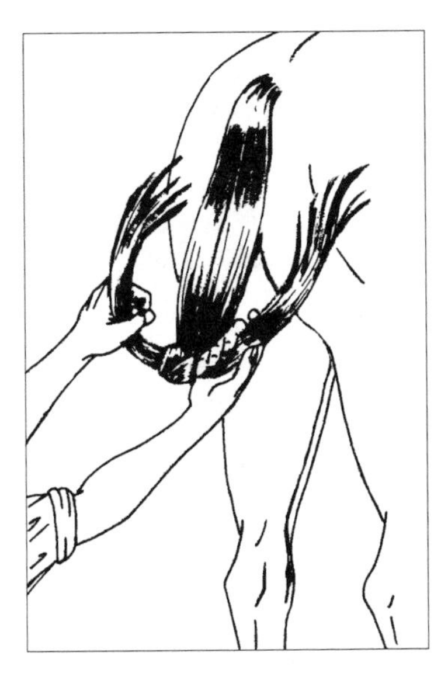

Step 4

Fold the two sections under the tail and crisscross the sections back toward the top of the tail.

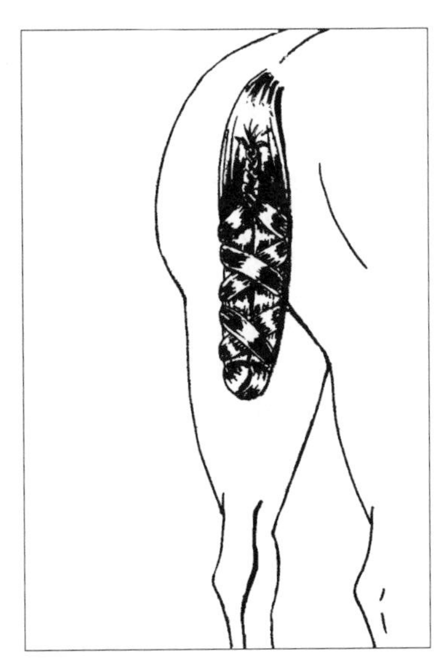

Step 5

Divide the remaining hair into three sections and make a small braid. Secure the end with a rubberband.

Step 6

Wrap electrical tape around the entire tail several times to prevent it from loosening.

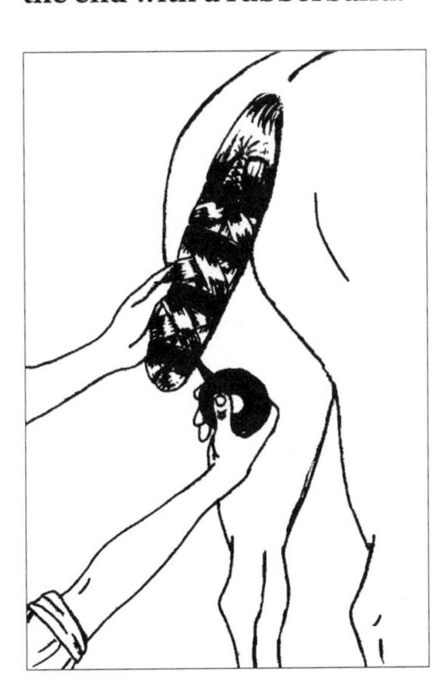

CLEANING THE FEET

Cleaning the feet is a very important step in grooming a horse. Most trainers require that the horses' feet be cleaned both before and after a race or workout. The materials for this job are a hoof pick and foot brush.

Many racehorses are used to having their feet cleaned from the near side. Tie the horse to the wall or have someone hold it. Stand on the near side facing the rear of the horse. Begin with the near front foot, then the off front foot, the near hind foot, and finally, the off hind foot. There are several ways to clean a horse's feet; the following is one method that has proven successful *(see Figure 4–19):*

1. Run your left hand down the near front leg until it reaches the back of the ankle. With your thumb and forefinger, give a steady pull on the fetlock hairs. This pull causes most horses to pick up the foot. (If this does not work, squeezing the nerve at the back of the horse's knee will cause the leg to bend so that you can lift the foot.)
2. When the horse lifts its foot, support it with your left hand.
3. Use your right hand to clean the dirt and manure from the bottom of the foot, picking from heel to toe.
4. Brush the foot and heels clean with the foot brush. Set the foot down gently—do not drop it. Now reach over to the off front foot. Pull on the fetlock hairs with your left hand until the horse lifts this leg. Again, support the foot with your left hand while cleaning with your right. Move to the rear of the horse on the near side. Remember to pat the horse on the side and hindquarters to let it know what you plan to do.
5. Place your left hand on the inside of the near hind leg, slide it to the back of the ankle, and pull on the fetlock hairs. When the horse lifts its foot, support it with your left hand. It is important that you stretch the hind leg back so the horse is comfortable. Repeat the cleaning process.
6. To clean the off hind foot, reach across and pull on the fetlock hairs with your left hand. When the horse lifts its foot, swing your body so that you are in a sitting position (but not bearing down heavily) on the hock of the opposite hind leg and the right hind foot is resting comfortably on your left thigh. This position discourages the horse from kicking and is comfortable for you. Repeat the cleaning process.

Cleaning the Feet (Fig. 4–19.)

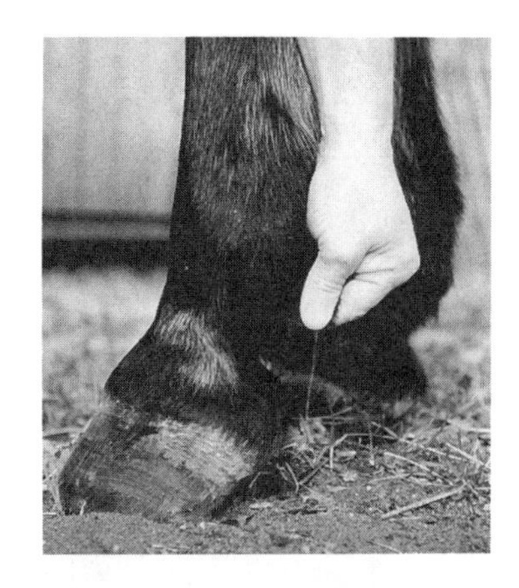

Step 1

Run your left hand down the back of the front leg on the near side. With your thumb and forefinger, give a steady pull on the fetlock hairs.

Step 2

When the horse lifts its foot, support it with your left hand.

Step 3

Begin cleaning the dirt and manure from the bottom of the foot, picking from heel to toe.

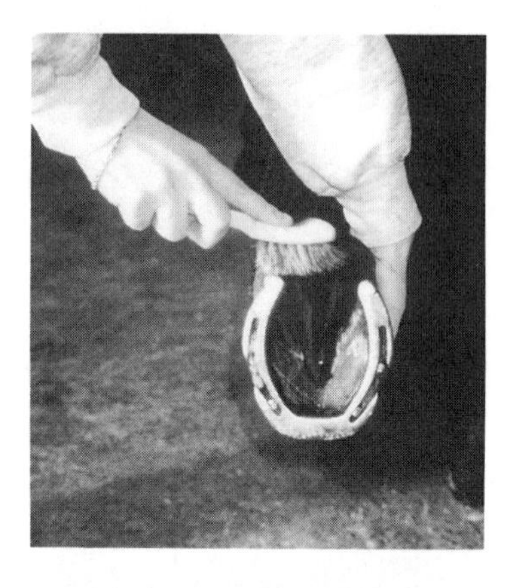

Step 4

Take the foot brush in your right hand and brush the bottom of the foot and the heels clean. Repeat the cleaning process with the front foot on the off side.

Step 5

Move to the rear of the horse and pick up the near hind foot. Hold the hind leg out behind the horse and repeat the cleaning process.

Step 6

To clean the off hind foot, swing your body so that you are in a sitting position on the near hock with the off hind foot resting on your left thigh. Repeat the cleaning process.

Some trainers prefer to have the grooms clean the off forefoot and hind foot from the off side. To clean the feet in this manner, begin on the near side, following the above procedure for cleaning the near forefoot. Next clean the near hind foot. Then move around the horse to the off side. Beginning with the off forefoot, follow the same procedure you did on the near side, except lift (and support) the foot with your right hand while using your left hand to pick the dirt out. Then clean the off hind foot in the same manner.

(See Chapter 5 for more information on care of the horse's feet.)

CLIPPING

There are several reasons why a horse might be clipped. If a horse is racing during the winter months, its body may be clipped to allow the horse to cool out and dry faster. Clipping also prevents the coat from becoming matted with sweat, mud, or oil, making it easier to keep the horse clean. Another reason the winter coats of racehorses are often clipped is if the animals are being shipped to a warmer climate for winter racing.

Clipping makes a horse more attractive for events such as horse sales or photo sessions. Preparation for these events includes trimming the ears, muzzle, bridle path (area of the mane just behind the ears), jaws, back of the legs, fetlocks, and coronet (any hair that hangs down over the hoof).

Types of Clips

There are several different types of clips that may be used on racehorses. The type of clip depends on the temperature and the activity the horse will be performing. The groom should be familiar with each clipping pattern (see Figures 4–20 through 4–23).

Body Clip—This type of clip removes the long hair from the entire body except the forelock, mane, and tail.

Hunter Clip—This type of cut removes the hair from the entire body except for the legs, saddle area, forelock, mane, and tail.

Blanket Clip—This type of clip removes the hair from the head, neck, sides, and belly. (Hair is left on the back, loins, legs, rump, forelock, mane, and tail.)

Fig. 4–20. A body clip—hair is removed from the entire body.

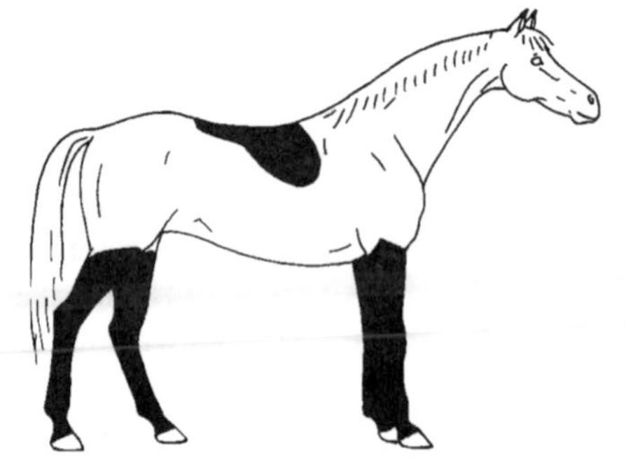

Fig. 4–21. A hunter clip—hair is left on the shaded areas: the legs and saddle area.

Fig. 4–22. A blanket clip—hair is left on the shaded areas: the legs, back, and sides.

Fig. 4–23. A trace clip—hair is left on the shaded areas: the legs, back, sides, neck, and head.

Trace Clip—This type of cut removes the hair only on those areas of the body that sweat the most. These areas include the front of the neck, the chest, the belly, and the sheath area.

Clipping Safety

- If the horse has never been clipped before, allow it to look at and smell the clippers before you begin.
- Turn the clippers on to allow the horse to get used to the sound. Hold the clippers over the body without actually clipping the coat. This accustoms the horse to the sound and vibration of the clippers so that when clipping actually begins, the horse will not jump suddenly and cause you to injure the horse or clip the wrong area.
- Use only a heavy duty extension cord with a three-prong ground plug to prevent electric shock to both the horse and its handler.
- Wear shoes with rubber soles to prevent the conduction of electricity and electric shock.
- Always clip the horse in a dry, well-lit area, preferably on dirt or with a non-slip floor covering.
- Keep tools and clipping materials near the front of the horse, but off to the side.
- Never sit or kneel under, behind, or in front of the horse during clipping (or at any other time).
- Clean and lubricate the clipper blades frequently while clipping. Otherwise, the blades will stick and pull at the horse's hair, causing pain and an uneven clip. Blades can be lubricated with a commercial clipper oil, clipper spray, or kerosene.

If at all possible, clipping should be done with two people—one person to perform the actual clipping and the other person to assist. The following duties should be performed by the assistant:

- holding the horse
- restraining the horse with hands and body restraining methods, or with a twitch if necessary (always use the least amount of restraint necessary)
- keeping the horse as quiet as possible
- holding and pulling the front legs forward to allow the elbow areas to be clipped (be careful when clipping around the loose folds of skin in this area)
- keeping the electrical cord away from the horse during the clipping process

Clippers

The groom should know enough about clippers to properly use and maintain them. For instance, there are a variety of blade sizes for clippers. The larger the size (the number on the clipper blade), the closer the clip will be. In other words, a small numbered blade would be better for a body clip, whereas a larger numbered blade would work better for trimming around the ears. Also, the clipper blades must be removed and sharpened frequently to avoid pulling on the horse's hair.

Large Animal Body Clippers

Electric body clippers are essential in every stable. They are used to clip the winter coat of the horse, to clip the hair around wounds, and to clip hair prior to surgery. Electric clippers are also used to remove the hair coat of horses that are infested with external parasites such as lice or mites.

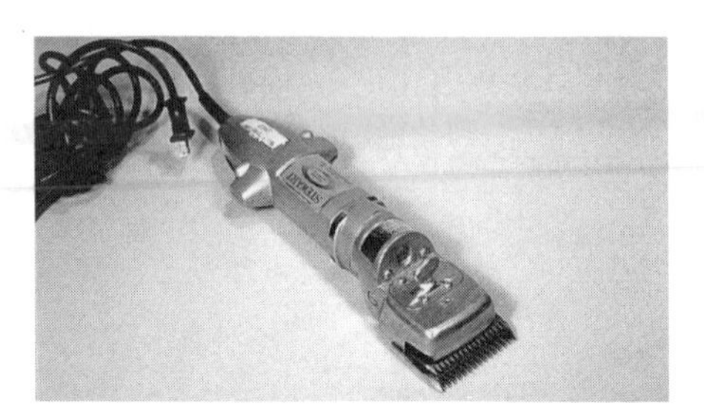

Fig. 4–24. Large animal body clippers.

Clippers are driven by an electric motor which operates on 110 – 120 volt AC or DC current. Some models are available with variable speeds. The electric cord is 18 feet long to allow free movement about the horse.

Electric Ear and Fetlock Clippers

These clippers are smaller and quieter than large animal electric clippers. They are used primarily on the horse's ears, face, jaws, and fetlocks. Electric ear and fetlock clippers are ideal for clipping around wounds. These clippers are powered by an electric motor which operates at 110 – 120 volt 60 cycle AC current. They are available in a cordless model or with a seven- to eight-foot electric cord.

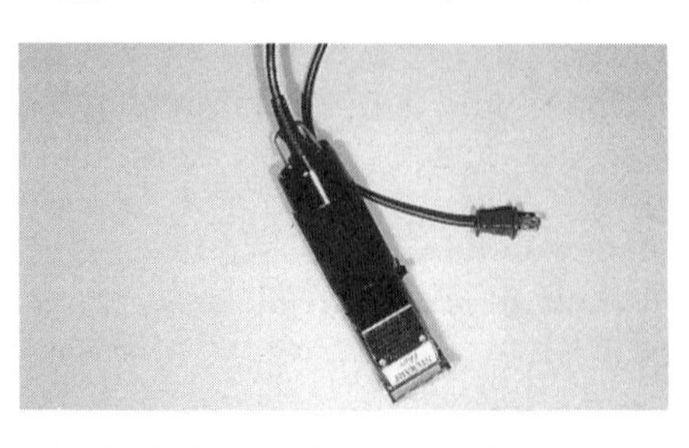

Fig. 4–25. Ear and fetlock clippers.

Fetlock Trimmers

This grooming tool can be used to trim the hair on the fetlocks and other parts of the body when electric clippers are impractical or impossible. Fetlock trimmers are usually made of steel with a double spring action. When you squeeze the handles, the cutting blades move across each other and cut the hair. One disadvantage of fetlock trimmers is that they are difficult to keep sharp.

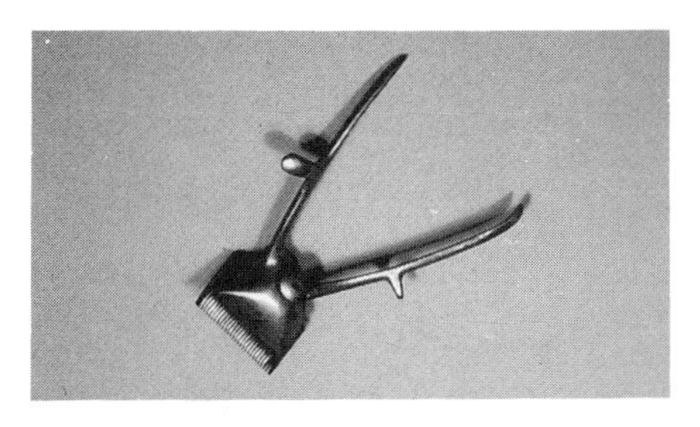

Fig. 4–26. Fetlock trimmers are used when clippers are impractical.

Curved Scissors

These stainless steel scissors are usually about eight inches long and are curved at the end for safety. (The pointed tip should always curve away from the horse's body.) They can be used to trim the hair coat on the legs, head, and other parts of the horse's body. They are also used to remove hair from around a wound. The curved scissors are ideal for trimming the bridle path and muzzle area when clippers cannot be used.

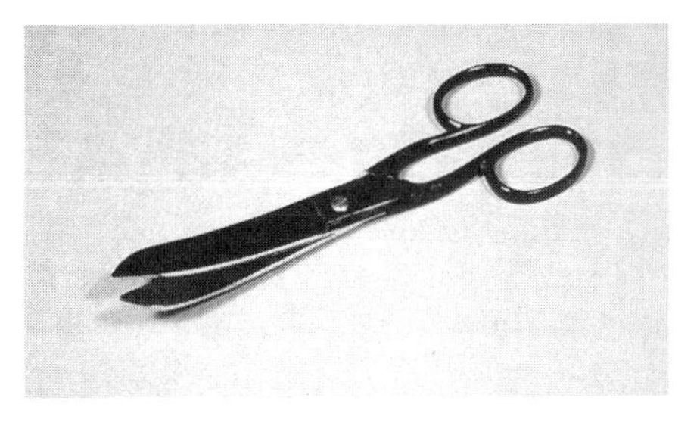

Fig. 4–27. Curved scissors can also be used instead of electric clippers.

Other Clipping Materials

For most thorough body clipping jobs, the following materials are necessary:

- grooming tools
- saddle pad and chalk
- large animal body clippers
- heavy duty three-prong grounded electrical extension cord
- coolant spray for clipper blades
- screwdriver for changing dull blades
- twitch
- small ear clippers

- blanket (for cold weather)
- stiff brush for cleaning blades
- container of blade wash
- lubricating oil for clippers

Preparation for Clipping

Groom the horse's body with a rubber curry comb and brushes or with an electric vacuum grooming machine to remove dirt from the coat. Nothing dulls clipper blades faster than a dirty coat.

Horses that are too frightened of the clipping process must be sedated before clipping. An interesting phenomenon that occurs with some sedatives is that the horse's hair stands up a little, as if the horse is cold. It is necessary to wait 15 – 20 minutes for this reaction to subside before beginning clipping.

Clipping Procedure

Before you begin clipping the horse, draw an outline of the area to be clipped (except in the case of a full body clip). For instance, if you are doing a hunter clip, trace the outline of a saddle pad on the horse's back with a piece of white chalk. On the front legs, follow the natural muscle line. On the hind legs, establish a guideline from the point of the stifle to below the curve of the gaskin.

Keeping the large animal body clippers flat against the body, move the clippers in long, even strokes against the lay of the hair. Have your assistant lift and pull the front legs forward so you can clip the elbow area. Use your free hand to smooth the skin in other wrinkled areas to obtain a neat, clean clip and avoid cutting the skin.

Stop clipping occasionally to spray coolant on the blades and to remove hair from the cooling vent of the clippers.

Most horses allow you to clip their bodies without much of a fuss, but some object to having their faces and especially their ears clipped. If the trainer wants you to clip the horse's ears, you may have to apply a twitch *(see Chapter 15 for more information)* and use the small ear clippers. If the horse will tolerate it, plugging the horse's ears up before clipping around them is a good idea. Some grooms put a wad of cotton inside a pair of knee-high pantyhose and gently insert this into each ear as a plug. This plug not only lessens the irritating noise of the clippers, but it also prevents hair from falling into the ear canal. The pantyhose simply makes for easy removal of the plug after clipping.

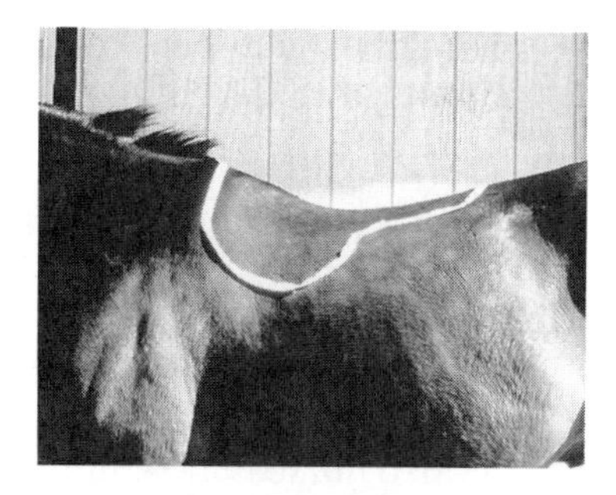

Fig. 4–28. Draw an outline of the area to be clipped using a piece of white chalk and a saddle pad.

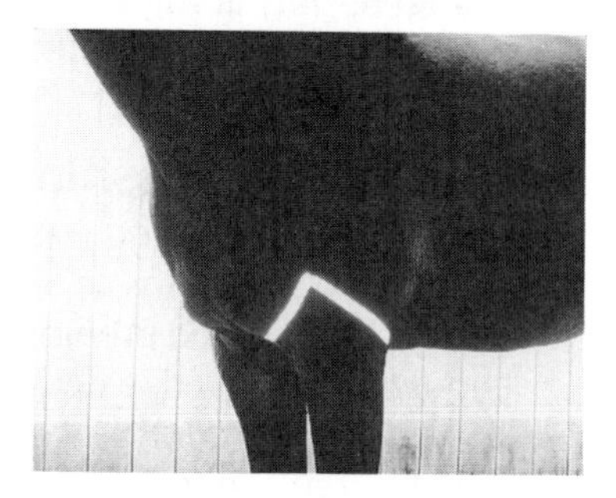

Fig. 4–29. On the front legs, follow the natural muscle line.

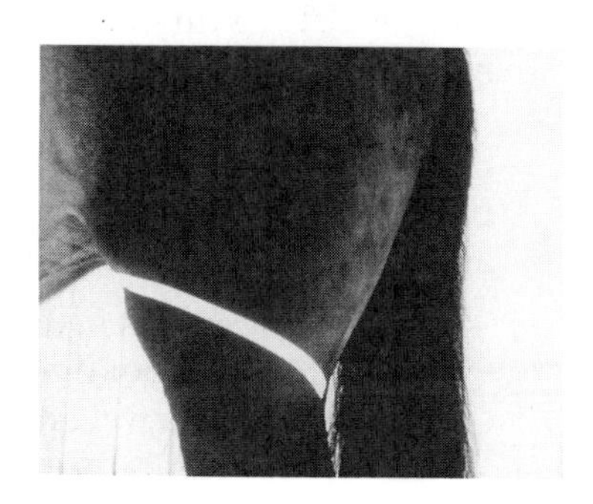

Fig. 4–30. On the hind legs, establish a guideline from the point of the stifle to below the curve of the gaskin.

Post-Clipping Procedure

Groom the entire body of the horse. With the clippers, go over any areas that need touching up. Always blanket a horse after clipping in cold weather to prevent a chill. After clipping the muzzle and pasterns, some grooms like to apply a very thin layer of baby oil or Vaseline® to soothe the sensitive skin there.

Clean the clippers thoroughly with a stiff brush to remove hair, dirt, etc. Submerge the blades in a blade wash compound. Then wipe them dry and apply a thin film of lubricating oil to the blades and motor. Store the clippers in a dry place until the next use.

Special Needs of the Clipped Horse

Blanketing

A horse that has had its coat clipped is very sensitive to cool weather and drafts. Depending on the season and the climate, the clipped horse may require a blanket all the time (except when exercising) or only at night. It is also very important that you cool down and dry the horse completely before putting on a blanket.

Grooming and Washing

While grooming and washing can be easier on a clipped horse, such a horse requires as much, if not more, attention as one that is not clipped. All the grooming implements, including the curry comb, should still be used, whether the horse is clipped or not. *Never leave a clipped horse damp.* Dry it off with towels if necessary, paying particular attention to the lower legs. If the horse is not properly cleaned and dried, the skin may become infected. This condition is commonly found in the saddle area.

A similar skin infection called "grease heel," (also called scratches, cracked heels, or white pastern disease) occurs at the back of the pasterns, especially on horses with white leg markings. Grease heel is aggravated by the following conditions:

- the horse stands in damp bedding for 6 – 8 hours
- the stall has poor drainage
- the weather is wet and humid
- the horse works on an abrasive track surface, such as sand

The bacteria which cause these infections are found naturally on the skin of normal horses. However, a clipper burn combined with inadequate grooming and washing and/or one of the above conditions will likely result in a painful infection.

The horse's skin in the affected area becomes hot, inflamed, and has many tiny, pus-filled bumps, like pimples. If this condition is found on a horse, bathe the horse daily using an iodine shampoo until the condition clears up. Infected pasterns should also be soaked in warm, soapy water until the scabs are soft and can be removed. Be careful, as removing the scabs, while necessary to allow the infected area to dry thoroughly, is often painful to the horse. If the skin is too painful to touch, a veterinarian may give the racehorse antibiotics.

WASHING THE RACEHORSE

During its racing career, a racehorse must be washed almost daily, and the groom is responsible for this task. Usually, the groom has an assistant who holds the horse during the washing process.

There are several reasons for washing a horse:

- to maintain the health of the skin
- to remove dirt and sweat
- to reduce body temperature to normal after a race or workout

It should be understood that while bathing makes the horse clean, nothing replaces good grooming to get a horse's coat healthy and shiny. Bathing should always complement grooming as part of maintaining the overall health and appearance of the horse.

Materials

The most important product necessary for washing the horse is a good equine shampoo. The basic function of any shampoo is to clean and condition the coat and skin. Some shampoos contain lanolin and protein for the coat and skin while others contain iodine to help prevent or cure bacterial and fungal skin infections. Most horse shampoos are highly concentrated and create a rich lather.

Other materials and tools required to wash the horse in cold and warm weather are as follows:

- two or three large water buckets (about 3 – 4½ gallon capacity) filled with warm water
- large body sponge
- blanket (woolen cooler or sheet)
- clothespins
- sweat scraper
- bath towels

Some trainers believe using a pressurized water hose to wash the horse is quicker and more effective than using buckets of water and a sponge. Also, many horses seem to enjoy the feel of the water massaging their skin. However, most horses dislike water spraying in their faces, and when using the hose there is the added danger of getting water in the ears. There is also the possibility that a horse will become tangled up in the hose and get hurt trying to get loose. Using buckets, this problem can be avoided.

While both methods are common, washing the horse with buckets and a sponge is probably the safer route. Therefore, this method is the one that will be described here.

In Warm Weather

Most stables have a special wash area where their horses can be washed in pleasant weather. When washing outside in warm weather, follow this process, as illustrated in Figure 4–31:

1. Use a large body sponge to wash the horse. Beginning at the head, saturate the body sponge with clean, warm water and squeeze the water over the horse's head. Wipe the eyes free of water and be careful not to get water in the ears. (It is essential to use clean water in this first step to avoid irritating the eyes with shampoo.) Sponge around the horse's mouth to remove any sweat or saliva.
2. Clean each nostril out thoroughly using a damp sponge.
3. Pour shampoo into one of the buckets. Stand on the near side of the horse, facing the rear. Hold the bucket with shampoo in your right hand and the sponge in your left hand. Beginning at the top of the neck, create a lather over the body in the following order: neck, chest, front leg, back, side, belly, rump, hind leg, between the hind legs, the scrotum (for males), the tail, and the anal area underneath the tail.
4. Now move around to the off side. Facing the horse's rear, hold the bucket with shampoo in your left hand and the sponge in your right hand. Wash the off side in the same manner as the near side, including the mane. It is not necessary to wash the tail and in between the hind legs again, as these areas were washed and rinsed from the near side. Returning to the near side, saturate the sponge with clean water. Rinse the shampoo off thoroughly by squeezing the soaked sponge over the horse's topline. Do not forget to rinse under the tail and in between the hind legs. Move to the off side of the horse and continue rinsing until all soap residue is gone. Every month, check the sheath or udder and if necessary, give

it a good cleaning. (See the section entitled, "Cleaning the Sheath or Udder.")

5. Gently run a sweat scraper along the horse's body, removing the remaining water. Do not use this tool on the legs or head as it will irritate the horse. It is not necessary to press very hard with the scraper.
6. After scraping, squeeze the sponge as dry as possible and wipe the entire body to free it of any excess moisture. Be sure to wipe the fetlocks and head with the sponge, as dripping in these areas may annoy the horse and cause it to become skittish. It is also good to apply a small amount of talcum powder to the heels after wiping them dry, as excess moisture can cause grease heel.
7. It may be necessary to put an anti-sweat sheet on the horse after it has been washed and sponged off. *(See Chapter 16 for more information on sheets and coolers.)* Some trainers elect not to put any blanket on their horses in hot, humid weather.
8. If a blanket is placed on a horse after washing in warm weather, it usually remains on the horse for about 5 – 10 minutes while walking. The blanket absorbs some of the moisture from the body and aids in the drying process. The horse is then walked without the blanket until dry.

In Cold Weather

Now that racing is a year-round activity, sometimes the racehorse must be washed in cold temperatures. During the winter, or whenever the weather is unusually cold and windy, the horse should be washed indoors if possible. Some barns are equipped with special indoor wash stalls with hot and cold running water. If such facilities are not available, an empty stall may be converted to a wash stall. A heat lamp installed overhead in the stall keeps the horse from getting a chill and helps the stall floor dry more quickly.

If there is no washrack and no extra stall, and you must wash the horse in its own stall, remove most or all of the bedding before beginning. Be sure to use a minimal amount of water so the floor does not become flooded. When you are finished washing the horse, remove any wet bedding and lay down dry bedding to ensure a fresh, comfortable bed for the horse.

Washing the Racehorse (Fig. 4–31.)

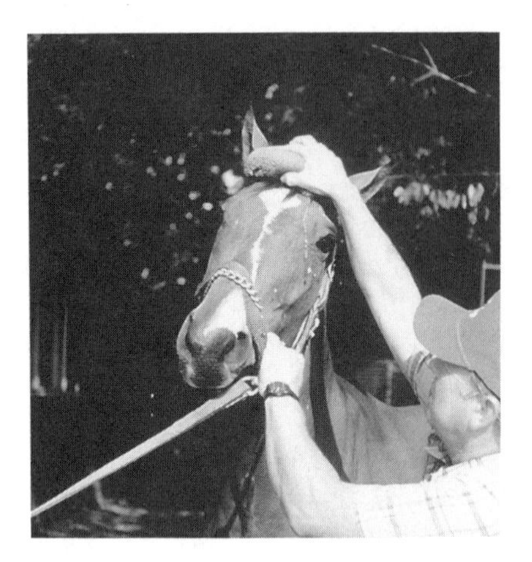

Step 1

Squeeze the wet sponge over the top of the head so the water runs down the face.

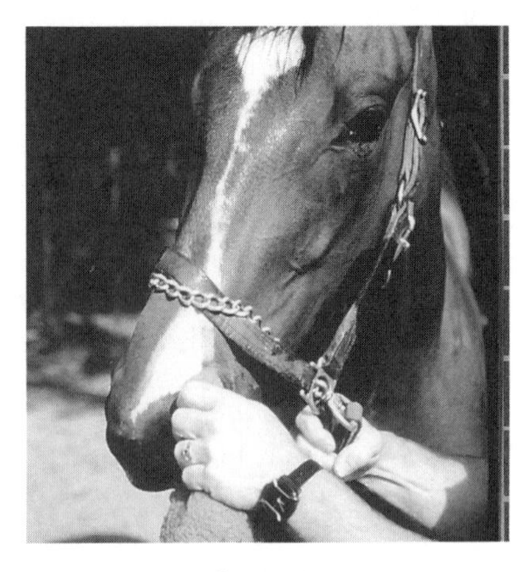

Step 2

Wipe each nostril out thoroughly with a clean sponge.

Step 3

Beginning on the near side, create a lather over the entire body, beginning at the neck.

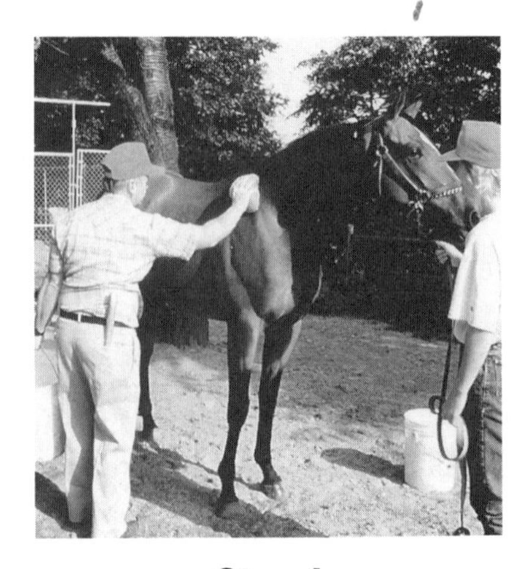

Step 4

Move to the off side and continue to lather. Rinse the horse with the sponge and clear water.

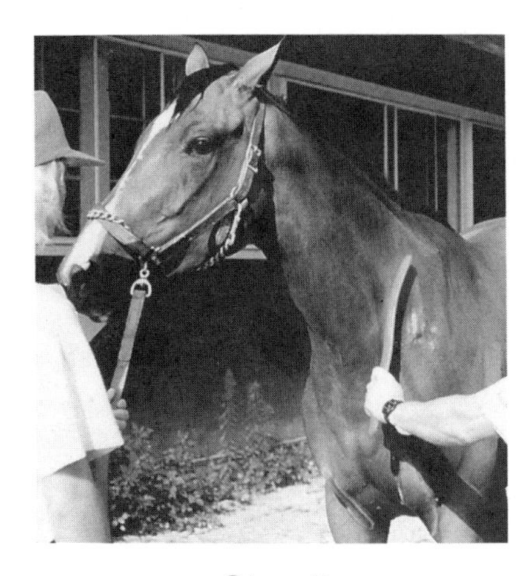

Step 5

Use the sweat scraper to scrape excess water from the horse's body.

Step 6

Use a wrung-out sponge to soak up the excess water from the horse's head and heels.

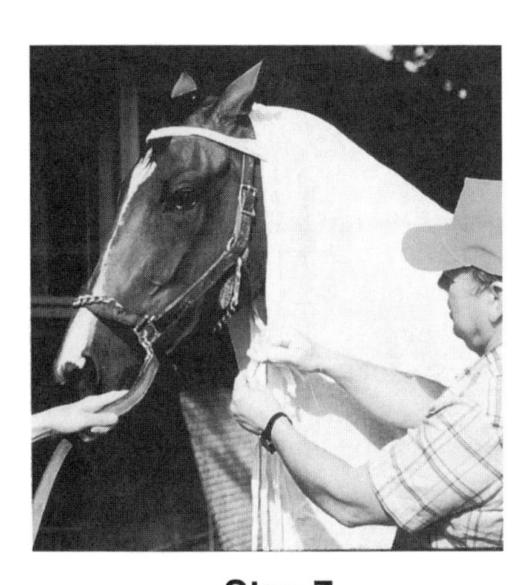

Step 7

Put a light sheet or fly net over the horse. Tie the strings together under the jaw to close the front.

Step 8

The horse should be walked until it is dry. (The sheet may be removed after 5 or 10 minutes.)

Use the following procedure to wash a horse in cold weather:

1. Wet the animal completely using a large body sponge and warm water.
2. If necessary, apply shampoo to the body and work up a good lather.
3. Rinse the entire body with warm water.
4. Quickly scrape the body to remove the water from the coat. Do not use the scraper on the legs or head.
5. Rub the horse's entire body in a circular motion with clean, dry towels.
6. Place two heavy woolen coolers over the horse, covering its body from behind the ears to the top of the tail.
7. Close the front section of the coolers and clamp them together with a clothespin. Clamp the coolers closed under the belly in the same manner. Allow the horse to stand in the stall tied to the wall with access to hay.
8. Every 15 minutes, unclamp the coolers on the bottom and rub a dry towel over the body under the coolers.

The entire washing time should be about 10 minutes and drying time should be about one hour. Obviously, in cold weather it would be better to wash and dry the horse as quickly as possible to avoid any drafts or chill which may cause the horse to become ill.

General Washing Tips

- Wash the horse as quickly as possible.
- When a horse is covered with mud from a sloppy track, be sure to have extra buckets of water to remove all the mud.
- Spraying a horse with a garden hose can sometimes be dangerous as the horse may become tangled in the hose or may be frightened by the water pressure.
- When washing between the hind legs, set the bucket down out of the way to keep the horse from stepping on it.
- Be sure to wash the anal area and the underside of the tail.
- Be sure the mane is laying smoothly on the off side before putting the cooler or sheet on the horse.
- A thorough job in scraping and sponging off the excess water after washing aids the drying process a great deal.
- Avoid using freezing cold or scalding hot water. Freezing water may cause the horse to become chilled or result in muscle

cramping. Extremely hot water can scald the skin and cause a great deal of discomfort to the horse. Ideally, the water should be lukewarm.

- Be sure to rinse the horse thoroughly, particularly in areas where there is little hair (soap residue irritates the skin).

CLEANING THE SHEATH OR UDDER

The sheath of the male horse and the udder of the female horse are two areas that are subject to the accumulation of dead skin, soil, and natural body secretions commonly called smegma. A dirty sheath or udder becomes evident in one or more of the following ways:

- physical examinations of sheath, penis, or udder
- constant tail rubbing due to itching
- spraying of urine (male horse) instead of a steady stream
- attempts to bite at the sheath or udder

Cleaning the sheath of the male or the udder of the female is a task which should be performed, on average, every two to three months, although the sheath or udder should be checked at least once a month for cleanliness. Cleaning this area can be included as part of the grooming process or can be done while washing the horse. The materials needed to wash the sheath or udder are:

- rubber or plastic gloves
- small sponge
- two buckets of warm water
- Castile soap or another mild soap

Castile soap is a mild everyday soap made of pure animal fat and coconut oil. Castile soap is gentle—it will not irritate the horse's skin because it does not contain any harsh chemical ingredients. Thus, it is excellent for cleaning sensitive areas.

Most horses do not like to be touched around their sheath or udder. However, washing these areas is necessary to keep the horse clean and comfortable. The groom can make the experience tolerable for the horse by using warm water and a clean, soft sponge. It is also recommended that the groom wear latex or rubber gloves when cleaning the sheath or udder, for hygiene purposes.

The following is the procedure for washing the sheath or udder:

1. Place a halter and chain shank on the horse. Have someone hold the horse for you. Also, the horse should be standing

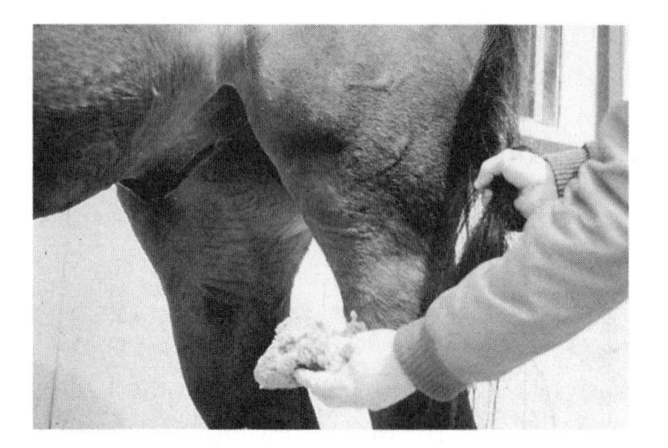

Fig. 4–32. Pull the horse's tail downward with your right hand. Hold the soapy sponge in your left.

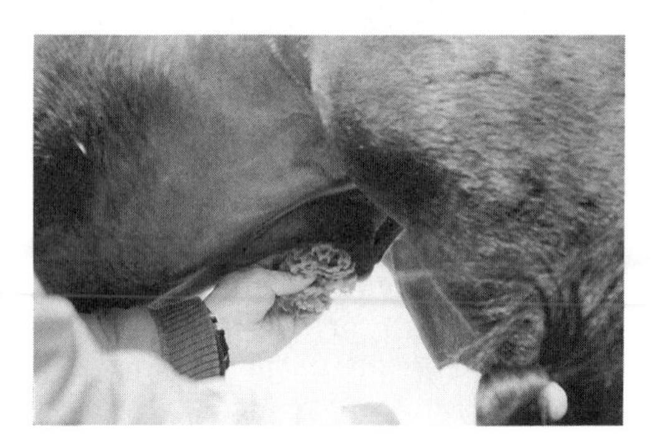

Fig. 4–33. Stroke the belly first, then clean the inside of the sheath (or the outside of the udder). Rinse well with clean water.

against a solid wall to restrict movement. The person holding the horse should stand on the horse's near side, prepared to pull the horse's head sharply towards the near side if it tries to kick. This automatically swings the horse's hind end away from the groom and handler. *(See Chapter 15 for more information on restraining the horse.)*

2. Stand on the near side of the horse. Wearing gloves, hold the tail in your right hand and the sponge in your left hand.
3. Saturate the sponge with soap and water. With your right hand, pull the horse's tail downward. This method forces the horse to put its weight on the hind legs and helps to prevent the horse from kicking. With the sponge in your left hand, gently stroke the belly to allow the horse to become accustomed to the wet sponge. Then move the sponge upward and wash the sheath or udder thoroughly with soap so that all of the grime and soil is removed.
4. Rinse the area thoroughly with clean water. Be sure to remove all of the soap residue.

The sheath is easier to clean when the penis is in the dropped position. Many horses get used to having their sheaths cleaned and will drop the penis. (Intact horses are more likely to drop down than geldings.) However, if the horse does not drop its penis, it is still possible to clean the shaft while it is retracted within the sheath; you simply have to gently push the sponge in as best you can.

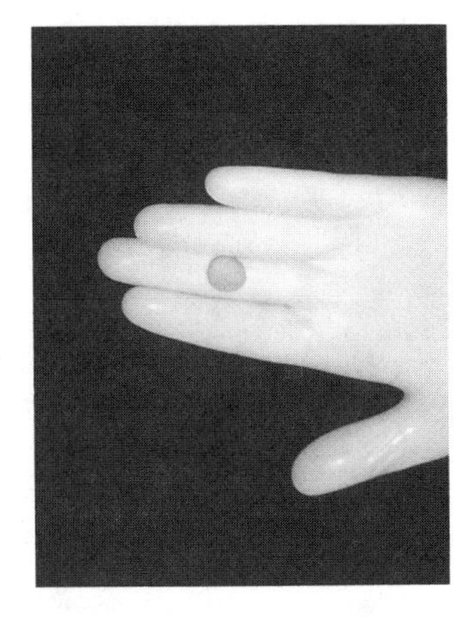

Fig. 4–34. A smegma "bean" that was removed from a male horse's sheath.

If a male horse sprays when he urinates, he most likely has a smegma "bean." This bean is a mass of glandular secretions, dirt, and urine that has formed into a hard pebble. It collects just inside the opening of the penis in a small pouch between two folds of skin. While young horses are unlikely to have this type of build-up, if a bean is found, it can be gently removed with a gloved finger.

APPLYING FLY REPELLENTS

During warm weather it is often necessary to apply fly repellents after grooming to keep the horse comfortable and free from the irritation of flies. Fly repellents are available in various forms.

Wipe On—a liquid designed to be applied to a cloth mitt or towel and then wiped onto the body of the horse.

Water Base Liquid—a liquid fly repellent that will not leave an oily residue on the coat to attract dust. It is usually applied with a pump spray bottle.

Aerosol Spray—sprays directly onto the horse's body. Be sure the horse is used to the "hissing" sound of the spray before using the spray. *Do not spray the face or head.*

Stick—a solid fly repellent in a stick applicator. This type of fly repellent is usually applied to the face and around the edge of wounds to repel flies.

Roll-On—a liquid fly repellent in a roll-on applicator. It is used about the face and around wounds.

Ointment—an ointment used directly on open wounds, cuts, and sores to aid healing by repelling flies.

CLEANING GROOMING EQUIPMENT

Even if you have groomed only one horse, it is certain that your grooming tools have accumulated dirt. It is impossible to clean a horse properly with dirty grooming tools. Therefore, most grooming kits need daily cleaning. (A groom should have an extra set of brushes and rub rags to use while the first set is drying.)

Fig. 4–35. Set brushes in the sun with the bristles facing up to dry.

Wash the brushes by soaking the bristles in warm water with a mild detergent. Then soak them in disinfectant. While still wet, rub the bristles of the stiff and soft brushes together, then rinse the bristles. Set both brushes in the sun with the bristles facing up to dry. Hoof picks should be soaked in disinfectant to minimize the spread of thrush. Rub rags and sponges should also be soaked in disinfectant.

Use a chlorine bleach solution (five parts water to one part bleach) for disinfecting equipment, as this product kills bacteria and most viruses. Follow the directions on the product label to determine how long to soak the brushes for the disinfectant to work. After soaking, be sure to rinse the grooming equipment thoroughly with clear water (especially sponges) before using them, as chemical residue can irritate the horse's eyes and skin. *(See Chapter 11 for more information on washing and disinfecting tools and equipment.)*

SUMMARY

As this chapter indicates, grooming involves more than just brushing the dust off of the horse's back. In fact, much of the groom's time is delegated to brushing, braiding, clipping, and washing the horse. One of the most important parts of grooming is caring for the horse's feet. While this chapter discussed cleaning the feet before and after exercise, this element of grooming is so important that the following chapter is devoted entirely to care of the feet; it provides more detailed explanations of proper foot care for the horse.

Chapter
5

Care of the Feet

Probably the most common phrase used by horse people is "No Foot—No Horse." This statement seems more true for racehorses than for any other type of horse.

The feet are important because they are responsible for both support and locomotion. As these two factors are the foundation of a horse, it is essential to clearly understand the anatomy of the horse's feet, how to properly care for them, and how to detect common ailments of the foot. The groom should also be familiar with common types of horseshoes and shoeing emergencies.

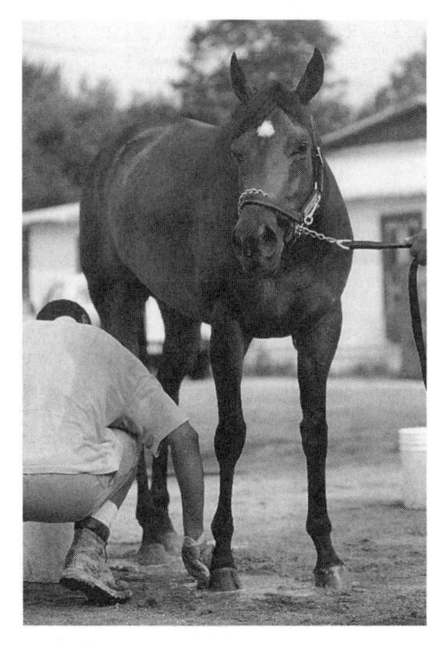

Fig. 5–1.

(Complete anatomy and bone diagrams are located in the Appendix, starting on page 431.)

PARTS OF THE FOOT

Hoof Wall

The hoof wall is the outer covering of the hoof. It is approximately ⅛ – ¼ inch thick at the toe, although hoof wall thickness varies with different breeds of horses. For instance, Thoroughbreds are known to have thin hoof walls as compared to Standardbreds. Normally, the hoof wall is thickest at the toe and gradually becomes thinner at the heels, growing downward from the coronary band at a rate of approximately ¼ inch per month.

The function of the hoof wall is to bear the horse's weight and protect the more sensitive inner parts. While the hoof wall may appear hard and dry, approximately 25% of the overall moisture of the foot is contained in the hoof wall. This moisture must be maintained to keep the hoof from cracking when it hits the ground hard. For instance, the heels must be elastic enough to spread slightly (absorbing concussion) upon contact with the ground.

Coronary Band

The coronary band is a ring of specialized tissue (not visible in the illustration) from which the hoof wall grows, similar to a person's cuticle. The coronet is that part of the horse's foot (between the pastern and the top of the hoof wall) where the coronary band is located.

Any physical damage to the coronary band usually results in a deformed hoof wall. For this reason, precautions should be taken to protect the coronary band from injuries when trailering or in the case of a horse over-reaching and stepping on its foot in the coronet area.

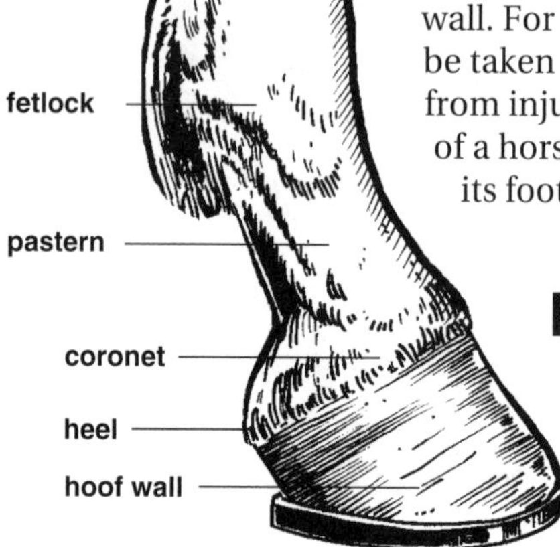

Fig. 5–2. The parts of the lower leg and foot as viewed from the side.

Perioplic Ring

The perioplic ring cannot be shown in Figure 5–2 because it is actually a ring of cells immediately above the coronary band (toward the pastern). These cells secrete a waxy, whitish substance

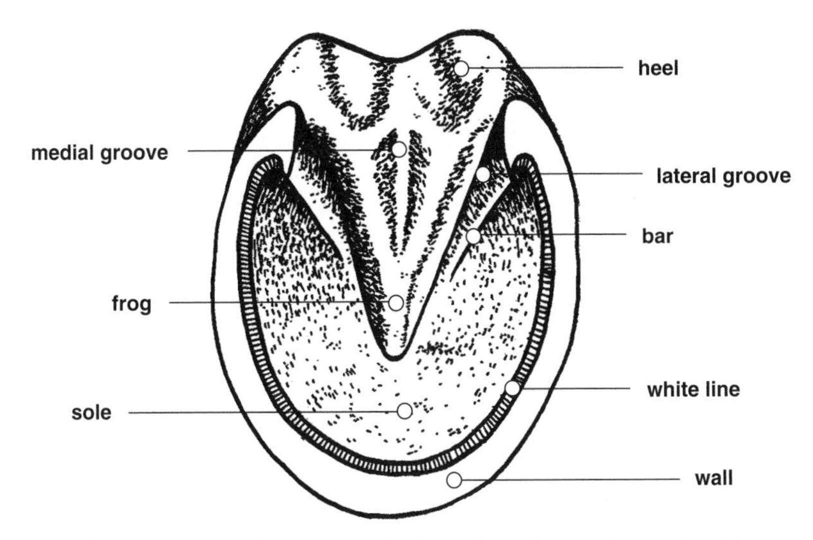

Fig. 5–3. The parts of the horse's foot as viewed from the underside.

called periople over the top of the coronary band. You can see this periople if you lift the hairs away from the hoof wall and look underneath. This substance extends only about ½ inch below the coronary band, and its function is to maintain the moisture content of the band. The perioplic ring is also responsible for secreting a clear, varnish-like substance down over the entire hoof wall to prevent excessive moisture loss.

A loss in moisture content in the foot can result in serious abnormalities, including cracks in the hoof wall. For this reason, the farrier must *not* rasp away this protective varnish from the outside of the hoof wall after shoeing a horse. A competent farrier will never rasp the outer hoof wall above the nail line.

Frog

The frog, which is a soft, triangular shaped horny growth on the bottom of the foot, is important for two reasons. First, it acts as a cushion to absorb concussion. The frog contains approximately 50% of all the moisture of the foot. When you apply pressure to the frog there should be a little give to it. Horses with large, healthy frogs have better "shock absorbers" than horses with small, shrunken frogs. By keeping the frogs clean and free from thrush infections (discussed later), you are preserving the horse's shock absorbers.

The second function of the frog is equally important. As the frog comes in contact with the ground, it aids in pumping blood throughout the foot and back up the leg. In this sense, the frog is sometimes referred to as the "second heart" of the horse.

On each side of the frog are the lateral grooves (where most of the dirt collects). In the center of the frog is the medial groove.

Bars

The bars, which are the raised areas on the outside of each lateral groove, are hard like the sole. They provide strength to the wall as the foot contracts and expands when it comes into contact with the ground. They also provide extra support to keep the sole from coming into contact with the ground.

Sole

The sole is the hard area on the bottom of the foot between the wall and the lateral grooves of the frog. It should be slightly concave and should never come into contact with the ground. The sole contains approximately 25% of all the moisture in the foot and protects the internal structures of the foot from the ground.

White Line

As viewed from the bottom of the foot, the white line is the junction between the wall and the sole. Its name comes from a visible "white line" following the circumference of the hoof wall. The farrier uses this white line as a guide for driving the horseshoe nails into the horny wall and not into the sensitive area of the wall. Because it is a point of division, the white line is more sensitive and vulnerable to intrusions (such as a nail puncture) than the sole.

In addition to the parts discussed above, the foot can also be divided into three sections: toe, quarter, and heel. Breaking the hoof down into these sections makes it easier to refer to the correct area.

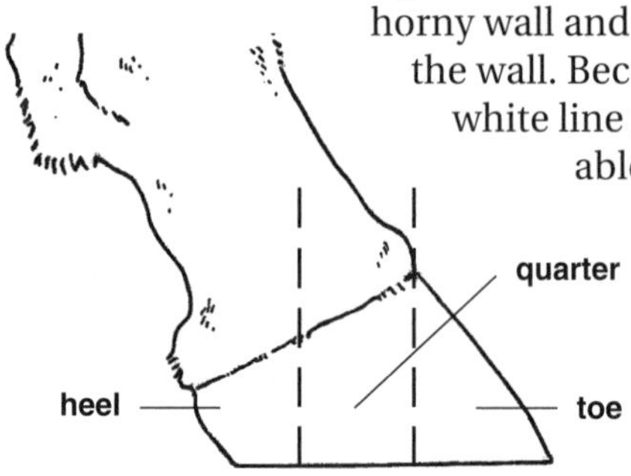

Fig. 5–4. The outside of the hoof can be divided into three sections.

CLEANING THE FEET

It is the groom's duty to properly care for the horse's feet. As discussed in Chapter 4, each foot should be cleaned with a hoof pick and foot brush before the horse goes out on the racetrack for a morning workout. After the horse comes back from its workout, the groom cleans the feet again, but this time he or she also washes the feet with warm water and a sponge, and dries them with a towel. The horse's feet should be cleaned before and after a race as well. This daily attention to the feet helps prevent bruises, foot abscesses, and thrush.

The following materials are needed to wash the horse's feet:

- hoof pick
- foot brush
- bucket of warm water
- foot sponge
- cloth towel

The steps for cleaning and washing the feet are illustrated in Figure 5–5. Chapter 4 contains more general information on how to pick the horse's feet.

1. Using the hoof pick, clean the dirt and manure from the bottom of the foot as described in Chapter 4.
2. Clean the outside of the foot with the foot brush.
3. Brush the bottom of the foot clean with the foot brush. It is important to gently brush the heels free of dirt also.
4. After you have picked and brushed the bottom of the foot, place the bucket of warm water under the foot to prevent water from spilling onto the stall floor. Soak the foot with warm water by saturating the foot sponge and squeezing it over the foot. Be sure to rinse the grooves of the frog thoroughly.
5. Rinse the foot sponge and wash the outside of the hoof.
6. After washing it, dry the outside of the foot, particularly the heels, with the cloth towel.

When washing the feet inside the stall, it is best to place the bucket of water under the raised foot to catch the water from the sponge. If the horse is nervous, it is best to work outside the stall and not place the bucket under the horse's foot. If you must work with a nervous horse inside its stall, clear away the bedding from the stall floor where you are working, and after you are finished, replace it. This way the horse is not forced to lie in wet bedding.

Cleaning and Washing the Feet (Fig. 5–5.)

Step 1

Using the hoof pick, clean the foot as described in Chapter 4.

Step 2

Scrape the dirt off the outside of the hoof wall using the foot brush.

Step 3

Use the foot brush to brush away the dirt from the sole and heels.

Step 4

Squeeze warm water over the bottom of the foot, using the foot sponge.

Step 5

Wipe the hoof wall clean with the foot sponge.

Step 6

Use a towel to soak up excess water from the outside of the foot, especially the heels, after washing.

PACKING THE FEET WITH MUD

Once the feet have been washed, begin packing the feet with mud. Packing the feet with mud draws out any heat in the foot and makes the frog softer and more pliable. Mud is usually applied to the bottom of the foot after a race, workout, shoeing, and in cases of founder. In each of these situations, there may be heat present in the horse's foot.

Materials

The following materials are needed to pack the feet with mud:

- empty bucket
- Bowie® clay
- 4-gallon bucket with water
- trowel
- Epsom salts
- distilled white vinegar or mineral oil
- burlap feed bag
- paper sheets (8 inches x 6 inches, cut to the approximate shape of the bottom of a horse's foot)

Mixing Mud

Gather the above materials from the tack room (where they most likely will be stored). Use the following procedure to mix the mud correctly:

1. Fill the empty bucket about half way with dry Bowie clay.
2. Slowly add water to the clay and stir the mixture with the trowel until the clay reaches a "pasty" texture and sticks to the trowel. If the clay is too loose, it will not stick to the bottom of the foot.
3. Add ½ cup of Epsom salts. Epsom salts enhance the drawing power of the clay and help to maintain its moisture.
4. Add 2 cups of distilled white vinegar or mineral oil to the mixed clay to keep it from drying out and putting undue pressure on the horse's sole.
5. Place a wet burlap feed bag over the top of the bucket. This also keeps the clay from drying out.

Applying Mud to the Feet

1. Wash the feet thoroughly as described earlier in this chapter.
2. Start with the left front foot. Pick the foot up and hold it with one hand, and with the other hand place the mud bucket next to your feet. Take a small amount of mud with the trowel, and press it onto the bottom of the foot.
3. Place a paper sheet over the bottom of the foot and place the foot down on the ground. The paper keeps bedding from entering the clay and causing the mud to fall out of the foot.
4. Repeat steps 2 and 3 for the other three feet.

Fig. 5–6. Press mud into the bottom of the foot with a trowel.

Fig. 5–7. Place paper over the mud to hold it against the foot.

After the feet are packed with mud, the horse should be confined to the stall. If the horse is walked or grazed while the feet are packed with mud, the mud will tend to fall from the bottom of the feet. Therefore, it is best to wait until the horse has finished all its exercise and grazing for the day before packing the feet. The next morning, before the horse leaves the stall for exercise, remove the clay from all four feet with a hoof pick and foot brush.

CONFORMATION FAULTS

Conformation is defined as the structural foundation of the horse—its "build." It is the shape, size, and angle of the bones which make up the horse's skeleton. Faulty conformation of the foot eventually leads to faulty conformation of the limbs. To recognize faulty conformation, however, it is first necessary to know what is correct or ideal.

Nature has provided the horse with a wide, round front foot and a more narrow, upright hind foot. The hoof angle of the front foot should be about 45° – 47°, while the hoof angle of the hind foot should measure 50° – 52°. The sole of the front foot is flatter (not completely flat, though) than the sole of the hind foot, which is more concave. Too long and sloping a pastern causes weakness because it puts undue strain on the tendons, sesamoid bones, and suspensory ligament. On the other hand, a short, upright pastern increases concussion and trauma to the fetlock and hoof. *(See Chapter 12 for information on, and illustrations of, lameness caused by poor conformation of the feet and legs.)*

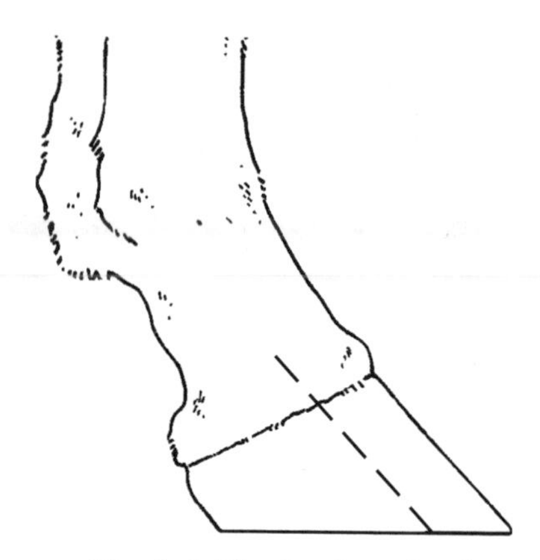

Fig. 5–8. The hoof angle of the front foot should be 45° – 47°.

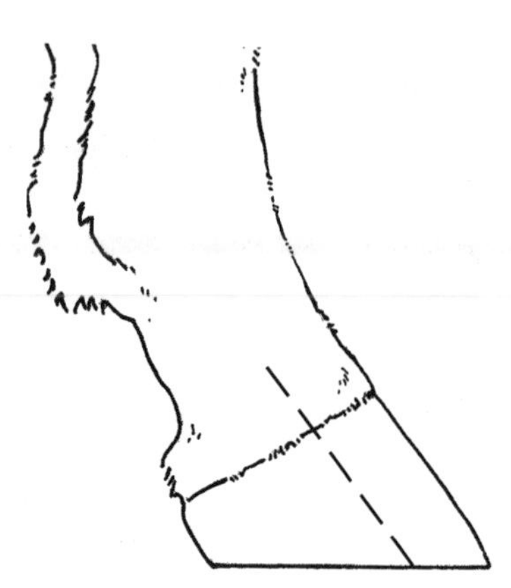

Fig. 5–9. The hoof angle of the hind foot should be 50° – 52°.

Viewed from the front, the horse should stand "square." (You should be able to draw an imaginary line from the front of the shoulder down the center of each leg and hoof.) Horses with deviating conformation, namely splay footed and pigeon toed horses, are more prone to lameness because of the uneven distribution of weight on their legs.

Splay Footed

This conformation fault is where the toes of the feet point outward, away from each other. Another term for splay footed is "toes out."

Horses that are narrow in the chest have a tendency to be splay footed. Depending on the severity of the toeing out, this fault may be improved through proper trimming or corrective shoeing.

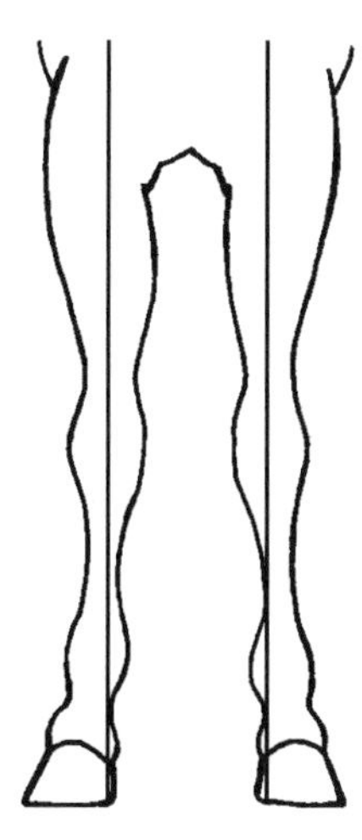

Fig. 5–10. A horse that is splay footed, or toed out.

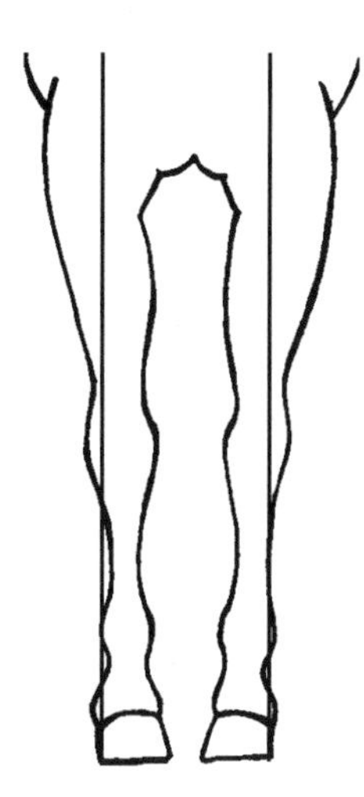

Fig. 5–11. A horse that is pigeon toed, or toed in.

Pigeon Toed

This is the opposite of splay footed, because the toes point inward, toward each other. Another term for pigeon toed is "toes in." In this case, horses that are excessively wide in the chest may be prone to toeing in. This fault may also be improved by proper trimming and corrective shoeing.

Flat Footed

This is a conformation fault of the sole. The sole should normally be concave. Horses that have flat feet are subject to undue pressure on the sole when the foot comes into contact with the ground. The sole then develops bruises which can cause lameness. This condition is more common in the front feet than in the hind feet. Sometimes, horses with flat feet must wear shoes with pads over the sole to prevent bruising and to decrease concussion.

Club Foot

Fig. 5–12. A club foot has an extreme hoof angle of 60° or more.

This condition is where the hoof wall is almost perpendicular to the ground. (A club foot has an extreme hoof angle of 60° or more.) The heels are high, the ankle pitches forward, and most of the horse's weight is on the toe. While this is a severe conformation fault, an experienced farrier can improve a club foot with corrective shoeing.

A club foot might be considered by many people to be uncommon with racehorses, but nonetheless, it does occur. For the most part, racehorses with this condition are unsuccessful due to their uneven gaits and shortened strides. In extreme cases, stumbling may be apparent due to the contraction of the superficial flexor tendon and the deep flexor tendon behind the cannon and pastern, which often accompanies a club foot condition. *(See Figure 5–18 for an illustration of these tendons.)*

AILMENTS AFFECTING THE FOOT

A horse's caretaker should be familiar with various foot ailments, particularly the signs and symptoms associated with them. It is the groom's duty to detect ailments and report them to the trainer as soon as possible.

Thrush

Thrush is a bacterial infection affecting the frog (particularly the medial and lateral grooves). It may occur in all four feet.

Causes of Thrush

Thrush is usually caused by general neglect of the feet—allowing the horse to stand in unsanitary conditions such as manure- and urine-soaked stalls and paddocks. Also, excessive frog growth due to infrequent trimming may contribute to thrush.

Symptoms of Thrush

Symptoms of thrush include a foul odor, extensive moisture of the frog, signs of deterioration, and heat. In advanced cases, a thick, black, liquid discharge from the grooves of the frog may be present, and lameness may also be evident.

Treatment of Thrush

Thrush is easily curable if treated as soon as it is diagnosed.

1. Have an experienced farrier remove all dead frog tissue.
2. Wash the bottom of the foot thoroughly with warm water and a mild disinfectant such as Betadine®.
3. Saturate the bottom of the foot with iodine, formaldehyde, or any other commercial liquid thrush remedy.

For a severe case of thrush:

1. Have an experienced farrier remove all dead frog tissue.
2. Wash the bottom of the foot thoroughly with warm water and a mild disinfectant such as Betadine. Rinse well.
3. Pack the infected grooves of the frog with sterile cotton. Saturate the cotton with iodine, formaldehyde, or any other commercial liquid thrush remedy. (See Figures 5–13 and 5–14.)
4. Apply a foot bandage or Easyboot® to keep the bottom of the foot clean.
5. The next morning, remove the saturated cotton from the foot before the horse leaves the barn for its morning training session.

Fig. 5–13. Pack the grooves of the frog with cotton.

Fig. 5–14. Saturate the cotton with any thrush remedy.

Prevention is the best defense against thrush. The groom must clean out the horse's feet thoroughly (several times a day), and the stall should be kept clean and dry. *No competent horse person allows a horse in his or her care to develop a severe case of thrush.*

Founder

This condition (also called laminitis) is a degeneration of the laminae. The laminae are part of the sensitive inner portion of the foot which covers the coffin bone. The laminae connect the coffin bone to the inner structure of the hoof wall.

As a result of founder, the laminae weaken and the coffin bone may break away from the hoof wall and rotate downward. Eventually, it may even protrude through the sole. Founder can occur in one foot or all four feet.

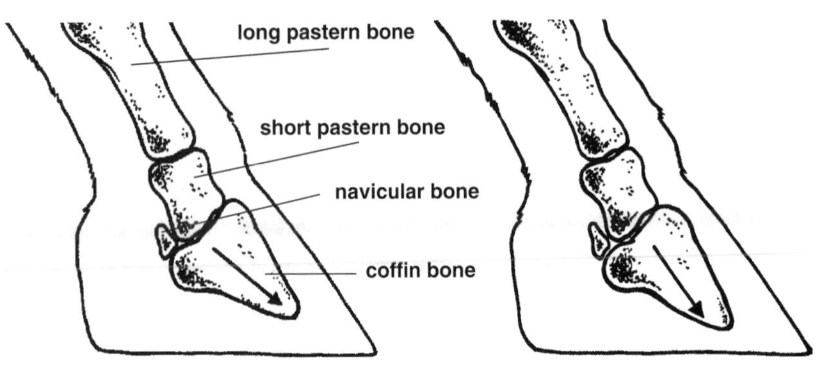

Fig. 5–15. A normal coffin bone before founder.

Fig. 5–16. A coffin bone that has rotated downward due to founder.

Causes of Founder

There are many causes of founder of which the caretaker must be aware. Excessive concussion to the feet from work on hard surfaces can cause a horse to founder. Founder may also be caused by overeating after working, or when a horse breaks out of its stall, gets into the feed room, and gorges itself on grain. Excessive weight on one limb can cause founder. For example, if a horse is recovering from a limb ailment, it shifts its weight away from the hurt leg, causing excessive weight on the other three legs.

Symptoms of Founder

It is important to recognize symptoms of founder as quickly as possible. The foot is hot to the touch, and the sole is sensitive. The horse takes its weight off the foundered foot and appears distressed. Often, both front feet are affected and the horse shifts its weight from one foot to the other. In many cases, the horse stands stretched out, is unwilling to walk and, in severe cases, the horse wants to lay down constantly.

Treatment of Founder

1. A veterinarian should be summoned immediately.
2. Have the horse stand with its feet in cool water to reduce inflammation and pain.
3. Have the horse bedded with deep sand or shavings (at least eight inches) so that it may "dig" its toes in and support more weight on its heels. This also reduces pressure on the sole.

Some trainers pack the foot of a foundered horse with mud or a poultice to help reduce inflammation within the foot. This treatment can be effective if the mud remains soft, but if it hardens, the mud increases pressure against the horse's sole and causes more pain.

It is very important that a foundered horse not be made to move anywhere. Movement can cause premature tearing of the laminae and accelerate the downward rotation of the coffin bone. In severe cases of founder, rings form on the hoof wall as it grows downward. Founder is a very serious ailment; moreover, once a horse founders, it is prone to do so again.

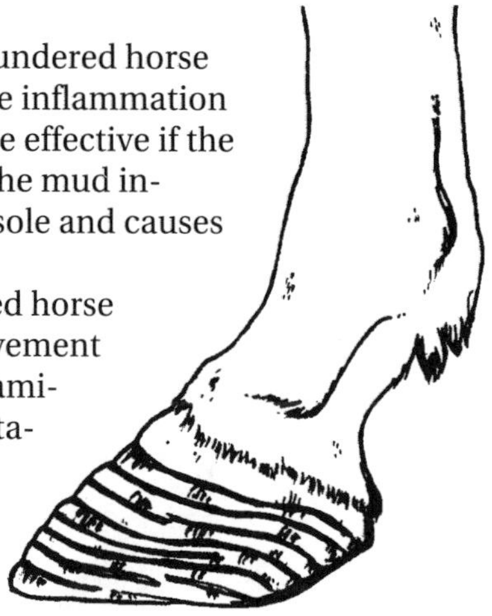

Fig. 5–17. Rings appear on the hoof following severe cases of founder.

(More detailed information and illustrations on this and any other form of lameness can be found in *Equine Lameness,* published by Equine Research, Inc.)

Navicular Disease

This condition is serious, and is, unfortunately, fairly common with racehorses. It usually occurs in the front feet only, primarily because they carry 60% of the horse's weight. The disease takes its name from the navicular bone (a small bone in the back of the foot), although it could involve any of the structures in that area.

Causes of Navicular Disease

The exact cause of navicular disease is unknown, but several contributing factors seem to be common to most navicular horses. Horses with straight, upright pasterns and feet seem prone to navicular. Horses with very low heels and long, sloping shoulders also seem especially vulnerable to navicular. Heredity and inadequate blood supply to the navicular bone may be contributing factors.

Symptoms of Navicular Disease

A horse with navicular disease attempts to walk on its toes. Lameness is evident in severe cases. While a horse is standing in the stall, it tends to "point" the affected forefoot. If both forefeet are affected, the horse alternates pointing both feet.

Treatment of Navicular Disease

There is not much that can be done to alleviate the pain associated with navicular disease except to make the horse comfortable and summon a veterinarian. Although navicular is not an emergency, it is important that the horse be made comfortable as soon as possible. It is up to the veterinarian to determine what treatment to pursue.

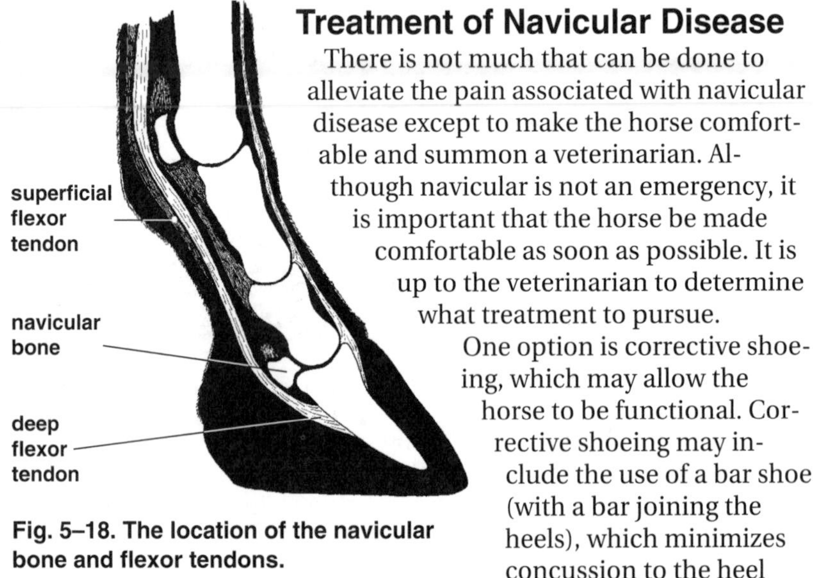

Fig. 5–18. The location of the navicular bone and flexor tendons.

One option is corrective shoeing, which may allow the horse to be functional. Corrective shoeing may include the use of a bar shoe (with a bar joining the heels), which minimizes concussion to the heel area. Or, a wedge pad (thicker at the heel than at the toe) between the shoe and the hoof wall reduces the tension on the deep flexor tendon and reduces concussion as well. Often, a horse with very low heels has been improperly shod and the heels have become hot and sore. This minor inflammation is sometimes mistaken for more serious navicular disease. Heel growth should be encouraged by the farrier over the next month or so to alleviate this problem.

Another option to treat serious cases is a procedure called a neurectomy, which alleviates the pain. To do this, the veterinarian severs the nerve of that area of the foot. This surgery eliminates all feeling to the back third of the foot. It is unlikely that a horse that has had navicular disease will ever race again.

Foot Abscesses

A foot abscess is an infection in the soft tissues of the foot, most commonly under the sole or hoof wall. Because the soft tissues of the foot are encased in a rigid box (the hoof wall and sole), the pus caused by an infection has no easy avenue of escape. Pressure builds up, which causes pain and further disruption of the soft tissues. Because the pus tends to move along the path of least resistance, most abscesses break out at the coronet.

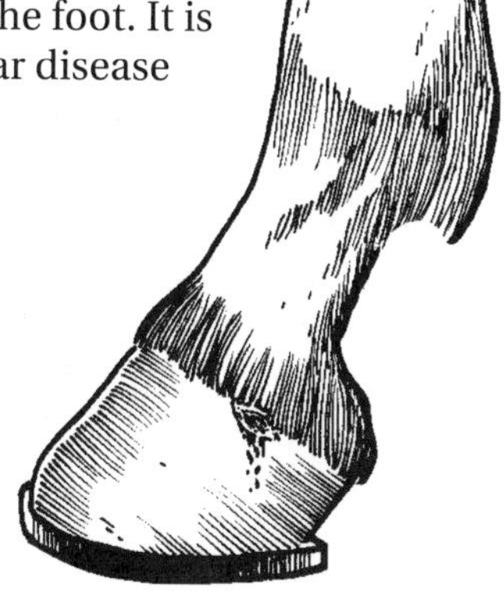

Fig. 5–19. A foot abscess may erupt at the coronet.

Causes of Foot Abscesses

Gravel

Most abscesses start with a hoof crack or separation (defect) of the wall and sole at the white line. Dirt, mud, manure, tiny bits of gravel, and water can enter the defect. This material is forced into the defect when the horse puts weight on the foot. As more material packs in, the defect becomes deeper, and causes more and more disruption of the laminae of the hoof wall, or of the white line. This debris also introduces bacteria that multiply in the sensitive tissues of the wall or sole, and infection results. Gravel may occur in all four feet but is found primarily in the front feet.

Sole Bruises and Puncture Wounds

Bruising of the sole can also lead to an abscess, particularly if the horse is forced to stand in wet, unsanitary conditions. The hoof wall and sole are somewhat porous, and will absorb water. If they become waterlogged, surface bacteria can invade the bruised area and create an abscess. Puncture wounds can also cause abscesses by introducing bacteria directly into the foot. If a nail pierces the sensitive tissues of the hoof wall, bacteria has entered, and infection may be the result.

Quittor

Quittor is an old term that is used for chronic infection of the lateral cartilage. This infection may be caused by a cut that involves or exposes the lateral cartilage, a puncture wound in the area of the cartilage, or (uncommonly) another kind of foot abscess in the back part of the foot. Typically, quittor causes a persistent or recurrent drainage of pus at the coronet near the affected cartilage. Other symptoms include swelling and pain over the affected cartilage, and lameness. Successful treatment requires surgical removal of the infected or dead cartilage.

Symptoms of Foot Abscesses

Onset of lameness in the affected foot may be gradual, or sudden and severe enough for you to suspect a fracture. The horse may stand with only the tip of its toe on the ground if the abscess is toward the back of the foot. For abscesses at the front of the foot, the horse is more likely to hold the affected foot completely off the ground. The hoof wall is often noticeably hot to the touch over the abscessed area, and the pulse in the arteries at the back of the fetlock are usually pounding. (This also occurs with other serious foot conditions, such as founder or a fracture.) There is almost always pain over the affected area when the veterinarian or farrier presses on it.

Treatment of Foot Abscesses

The most important part of treating an abscess is to establish drainage. The veterinarian or farrier may have to open up the sole or hoof wall to allow drainage. Once the abscess has broken out (or has been cut open), the groom can help by soaking the foot in warm water and Epsom salts, and by poulticing the foot, including the coronet. *(See Chapter 12 for instructions on how to apply a poultice.)*

Continue soaking and poulticing the foot as often as the veterinarian instructs you to do so. This treatment ensures complete removal of all infected material. If the hoof wall or sole has been opened to drain the abscess, the area should be kept covered with a waterproof dressing or protective boot until it has filled in with horn. Abscesses can recur if they do not heal properly the first time.

Corns

This condition is a bruise of the sensitive tissue under the sole on the bottom of the horse's foot. It is often characterized by a dark red or purple discoloration in the area between the bars and the wall. (This discoloration may be more obvious on light-colored hooves.)

Causes of Corns

Corns are generally the result of improperly fitted shoes (a shoe should never put pressure on any part of the sole) or shoes that have loosened. Or, they may be caused by waiting too long before reshoeing. Excessive concussion, especially on rough ground, can also contribute to corns. A horse with flat soles and weak bars is especially susceptible to corns.

Symptoms of Corns

A horse that has corns may not display obvious lameness. Instead, the horse may merely race slower or not be able to handle distances that it once could. Although corns are difficult to see until the shoe is removed, the groom should examine the bottom of the foot carefully every day for any discoloration or bruises.

Treatment of Corns

Corns may be treated by either the farrier or the veterinarian. If incorrect shoeing is the problem, the horse is either left barefoot or properly fitted with a new shoe. The sole should not be pared, as thinning of the sole may predispose the horse to re-injury.

The veterinarian may recommend anti-inflammatory drugs, such as "bute" (phenylbutazone) to relieve pain and reduce inflammation and fever in the foot. Soaking, poulticing, and protecting the bottom of the foot may also be recommended, although drainage is usually unnecessary. A deep, serious corn can take up to six months to heal. However, most corns respond to treatment within a few weeks. It is important for the groom to check the horse's shoes daily for missing nails or looseness. This subject is discussed later in this chapter, under the section entitled, "Shoeing."

Hoof Cracks

This term refers to any crack occurring in the hoof wall. Hoof cracks include toe cracks, quarter cracks, and heel cracks, and can occur in all four feet. Most hoof cracks are vertical, and originate on the ground surface; they are commonly referred to as sand cracks. In rare cases, a vertical crack will occur at the coronary band, as shown below. However, most cracks originating at the coronary band are horizontal.

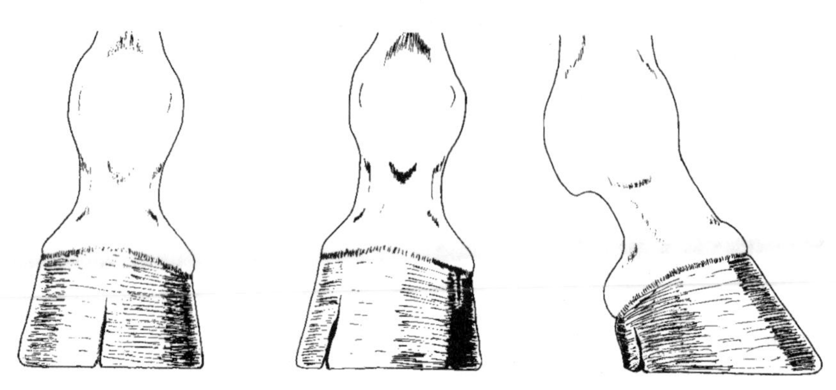

Fig. 5–20. From left to right: a toe crack, two quarter cracks, and a heel crack.

Toe Crack—This is a crack in the toe of the hoof wall. It usually starts at ground level and works upward toward the coronet.

Quarter Crack—This crack is found in the quarter of the hoof wall. When a quarter crack is left untreated, it spreads upward toward the coronet more quickly than a toe crack or a heel crack. This is due to the significant expansion of the quarter upon contact with the ground. Vertical quarter cracks may also originate at the coronet and spread downward.

Heel Crack—This crack is found in the heel of the hoof wall. Like quarter cracks, heel cracks that start at the ground surface typically spread upward more quickly than toe cracks because the heel expands more than the toe upon contact with the ground.

Causes of Hoof Cracks

The primary cause for any crack in the hoof wall is dry, brittle hooves. Pressure and concussion to the feet also contribute to hoof cracks. A normal, moist foot will spread at the heels and quarters to absorb concussion. Under the same concussion, a dry, brittle hoof is likely to crack. Hoof cracks can become a serious matter if they are allowed to become deep and expose the sensitive inner tissues of the foot to infection.

Symptoms of Hoof Cracks

Although hoof cracks may extend downward from the coronet to the toe, quarters, or heel, most cracks originate at the ground surface and spread upward. Lameness results in cases of deep cracks.

Treatment of Hoof Cracks

The treatment of cracks varies with the seriousness of the condition. Deep quarter cracks are probably the most difficult to treat because they spread more quickly than the other two. A quarter crack takes several months to grow out and heal completely.

A veterinarian and farrier work together to treat hoof cracks. The veterinarian cleans and disinfects the crack thoroughly to prevent infection, and the farrier may fit the foot with a corrective shoe, which prevents the crack from spreading. Or, the farrier may patch the crack with plastic, epoxy glue, or fiberglass, and the horse can usually return to work immediately. It is the groom's job to follow the instructions of the veterinarian and to keep the foot clean while it is healing.

Prevention of Hoof Cracks

To prevent cracks, the hoof wall should be moist and pliable. Many horsemen apply a commercial hoof dressing to the hoof wall regularly to keep it flexible. Shoeing will not prevent cracks from occurring. But applying a commercial, lanolin-based hoof dressing once a week (or more often if recommended by the farrier) will definitely aid in preventing the foot from drying out and developing cracks. *(See the section entitled "Foot Care Products" for more information on hoof dressings.)*

Contracted Heels

This condition is characterized by a narrowing of the hoof at the heels. Contracted heels may occur in all four feet but are most commonly found in the front feet.

Fig. 5–21. A normal heel.

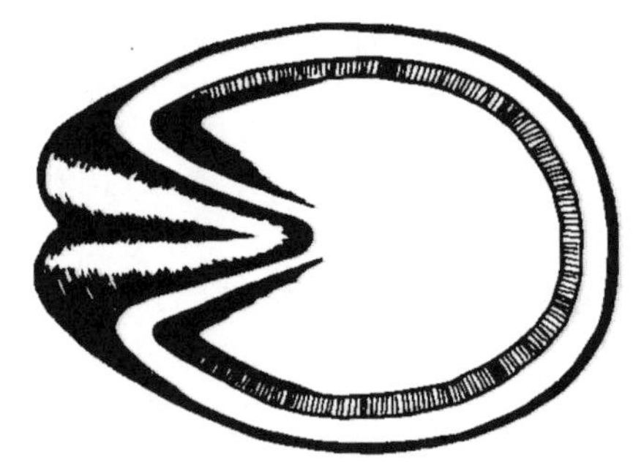

Fig. 5–22. A contracted heel.

Causes of Contracted Heels

The basic causes of contracted heels are faulty conformation or improper shoeing that reduces the spreading of the heels and frog contact with the ground.

Symptoms of Contracted Heels

The frog seems to shrivel up, appearing smaller and "pinched" between the heels. The hoof wall also loses its "spring" and becomes narrower at the heels. The entire hoof wall is dry and brittle. The horse stumbles frequently and tries to walk on its toes when it first comes out of the stall in the morning.

Treatment of Contracted Heels

Corrective shoeing by a competent farrier is necessary for a horse with contracted heels. The farrier can apply special shoes to help spread the heels. Also, applying a commercial hoof dressing to the hoof wall keeps the heels and frog softer and more pliable. **Note:** A horse with contracted heels is also more prone to thrush because the lateral and medial grooves are deeper and can trap more debris.

INTERFERENCE

Interference occurs when a horse hits one leg with another while in motion. It can occur in both the front and hind legs. The horse usually hits its leg in the areas of the coronet, fetlock, or cannon. Interference is usually caused by faulty conformation. A mild case of interference is generally referred to as "brushing." A severe case is very painful and can lead to reduced racing performance and lameness.

Running horses are most likely to interfere in the turns at high speed. But while it is not uncommon in any breed of racing horses, interference occurs most often in Standardbreds because of the nature of their gaits. The Standardbred groom must carefully inspect the legs and feet of the horses each day and note any swellings, cuts, scrapes, or abrasions. If any of these conditions are present, it is certain that the horse is experiencing a defect in its way of going or a "faulty gait." The groom may also notice that the horse is not jogging correctly when being exercised each day.

The Standardbred racehorse may suffer from interference arising from many different types of gait faults. You should become familiar with these various problems so that you are able to detect them while grooming or exercising the horse and notify the trainer immediately, before they become serious. The following types of interference are the most common gait faults affecting the horse:

Ankle-Hitting—When a front foot hits the ankle of the opposite front leg, it is referred to as "ankle-hitting." This problem is often found in harness horses that are splay footed.

Knee-Hitting—When the foot hits the knee of the opposite front leg, it is called "knee-hitting." This problem is also common in harness horses that are splay footed.

Elbow-Hitting—When the foot hits the elbow of the same leg, it is referred to as "elbow-hitting." This problem is occasionally found in trotters.

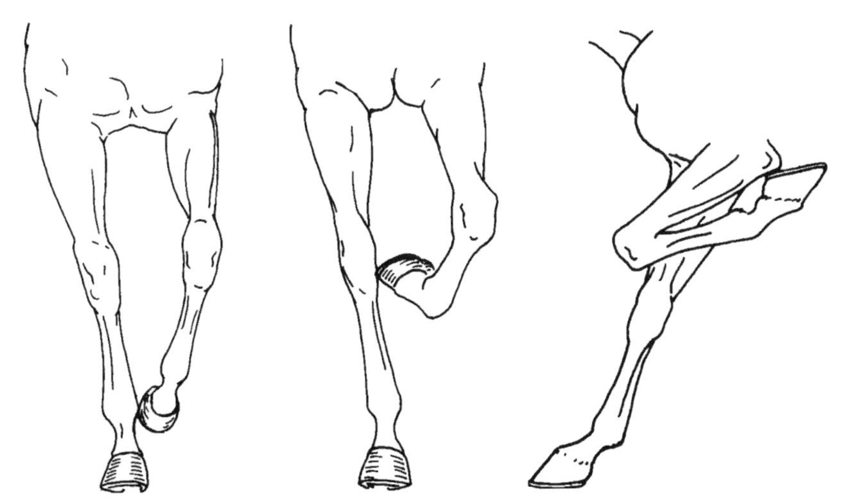

Fig. 5–23. Left: ankle-hitting. Middle: knee-hitting. Right: elbow-hitting.

Forging—This gait fault usually occurs when the toe of a hind foot strikes the bottom or side of the front foot on the same side.

Over-Reaching—This is a gait fault whereby the toe of a hind foot catches the heel of the front foot on the same side. The toe of the hind foot can actually grab the shoe of the front foot and cause it to become loose. In severe cases, it is possible for the shoe to be pulled completely from the front foot.

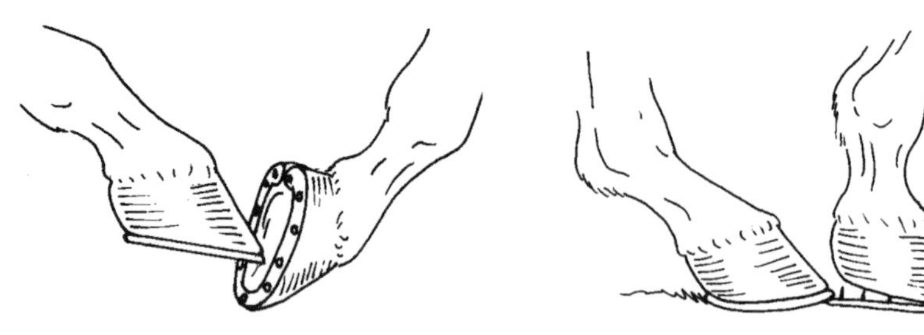

Fig. 5–24. Forging.

Fig. 5–25. Over-reaching.

Scalping—This gait fault usually occurs in trotters. The toe of a front foot strikes the coronet of the hind foot on the same side.

Speedy Cutting—This gait fault is similar to scalping. The toe of a front foot strikes the fetlock of the hind foot on the same side.

Cross-Firing—This gait fault occurs in the pacer. It involves striking diagonal feet. For example: the inside of the toe of a hind foot strikes the inside of the front foot on the opposite side.

In addition to placing various types of exercise boots on the horse *(see Chapter 10)* to protect the feet and legs, the Standardbred trainer can have the farrier correct some of these faulty gaits to an extent through corrective shoeing.

The farrier is a vital asset to the Standardbred racing stable. However, it is the groom who should be able to detect any gait abnormalities while exercising the horse. Also, when you are actually grooming the horse in the stall, you should be able to detect any minute swellings, abrasions, cuts, or loose shoes.

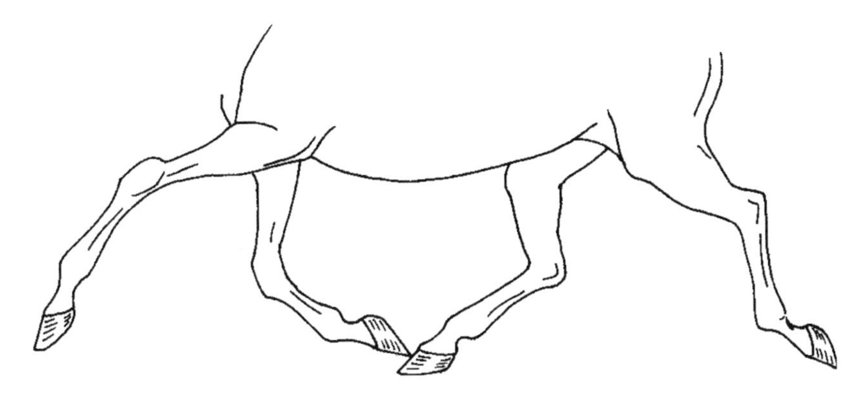

Fig. 5–26. Scalping.

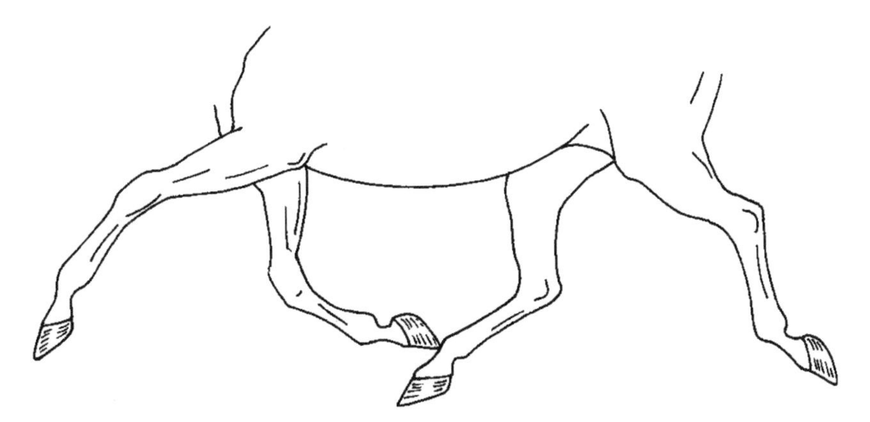

Fig. 5–27. Speedy Cutting.

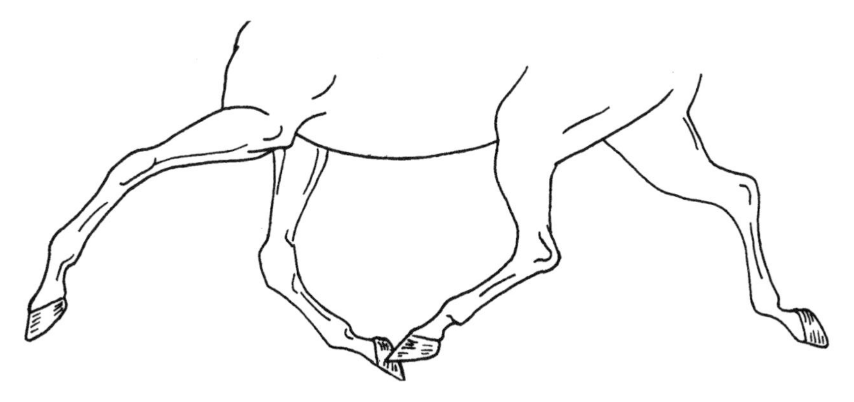

Fig. 5–28. Cross-Firing.

SHOEING

The primary reason for shoeing a horse is to prevent the hoof from breaking and wearing away faster than it grows. In addition, the racehorse wears shoes for traction on the various types of racing surfaces. Horseshoes are also used to correct conformation and movement problems. When dealing with the trotter or pacer, the type of shoe worn has a definite bearing on the horse's action as well as its overall performance. In fact, the trainer regards the hoof angle and toe length as two of the most important aspects of training the Standardbred racehorse. The slightest deviation in these factors may result in a significant change in the horse's performance. Trainers of trotters and pacers depend on very specialized shoeing to develop and maintain their horse's gaits properly, and to avoid painful interference. Caring for the feet of the trotter or pacer is therefore of utmost importance.

Fig. 5–29. Trainers of trotters and pacers depend on very specialized shoeing to develop and maintain their horses' gaits properly.

It is for this reason that Standardbred trainers are very particular as to the ability of the farrier engaged to shoe the horses in their stables. Only a farrier with a great deal of knowledge and experience in balance and hoof angles is qualified to properly shoe the Standardbred racehorse: there is little margin for error.

Horseshoes

Horseshoes are made of aluminum, steel, or plastic. Most running horses wear very light, aluminum racing plates. The pacer may only wear aluminum shoes on the front feet. The trotter usually wears heavier steel shoes all around (on all four feet). Harness horses are shod about every two to three weeks, and running horses every three to four weeks, depending on the type of surface the foot is exposed to, as well as the type of shoe worn.

You should be familiar with some of the common racing and training shoes used on racehorses. If you want to become a trainer, it is especially important to pay attention when one of the horses in your care is being shod.

For Runners

These horseshoes are illustrated in Figure 5–30.

Regular Toe Front—The most widely-used plate.
Outer Rim Front—Provides a level grip and is used on turf courses.
Jar Calk Front—Used primarily when a horse runs on a muddy track to give the horse better traction.
Sticker Hinds—Available with either a left or right sticker. It is used on a horse that tends to lose its footing on muddy tracks and turns.
Inner Rim Block Heels—Corrects running down at the heel.
Block Heel Hind Sticker—Provides better traction on wet and dry surfaces as well as corrects running down.

For Trotters and Pacers

These horseshoes are illustrated in Figure 5–31.

Plain—Flat all around. It is generally used on a horse that does not interfere or have a faulty gait. The weight of the shoe varies from 5 to 12 ounces depending on the size, width, thickness, and material used to make the shoe.
Full Rim—The raised metal rim on the outside edge of the shoe promotes better traction.
Cross-Firing—Used on the hind feet of the pacer to correct cross-firing. The shoe usually has a trailer and a heel calk on the outside of the shoe and a flat diagonal toe on the inside of the toe where the horse breaks over (rolls the hoof forward and lifts the foot from the ground heel first).

Common Horseshoes for Runners (Fig. 5–30.)

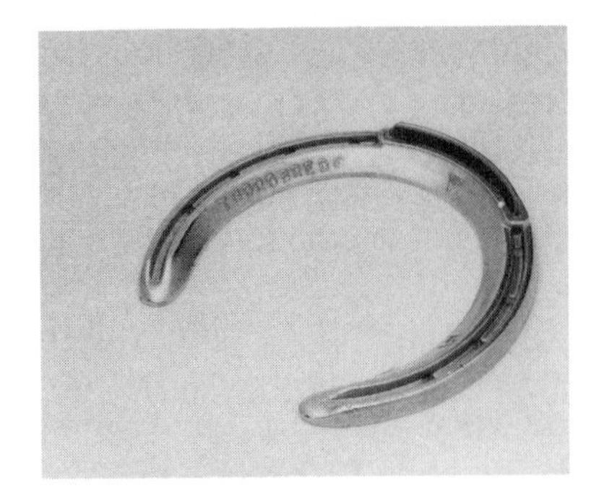

Regular Toe Front.

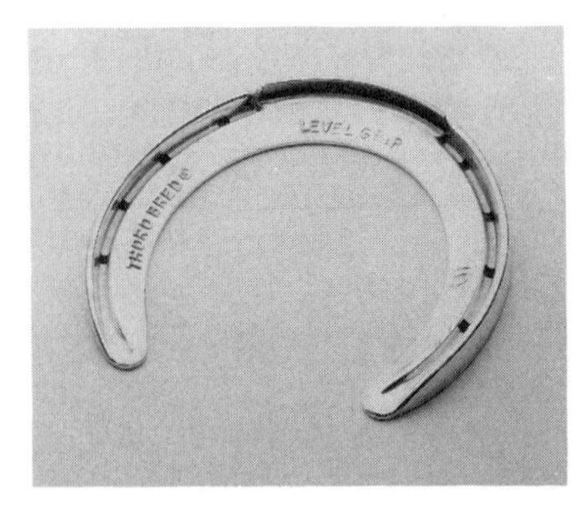

Outer Rim Front.

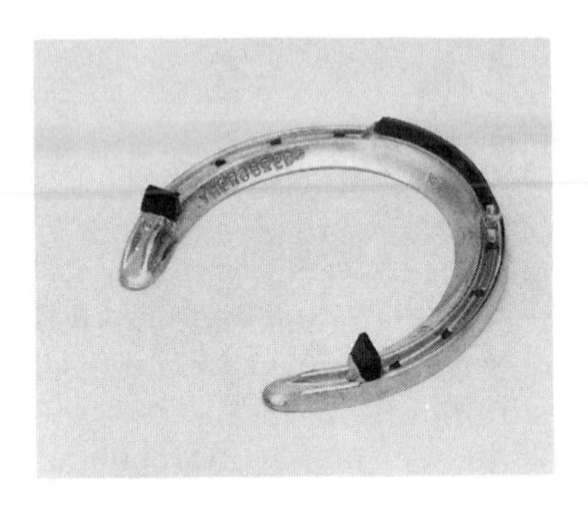

Jar Calk Front.

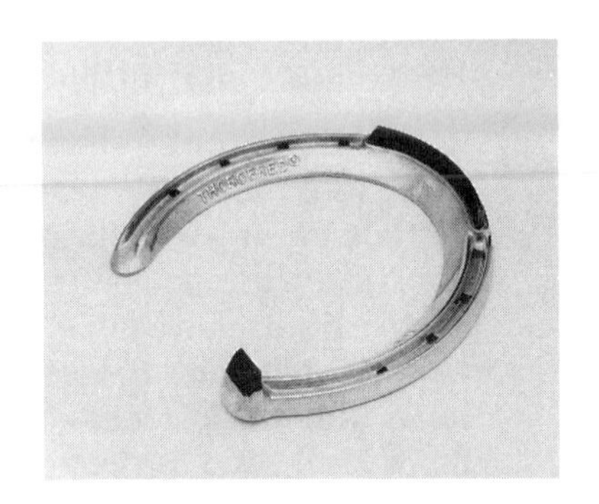

Sticker Hinds.

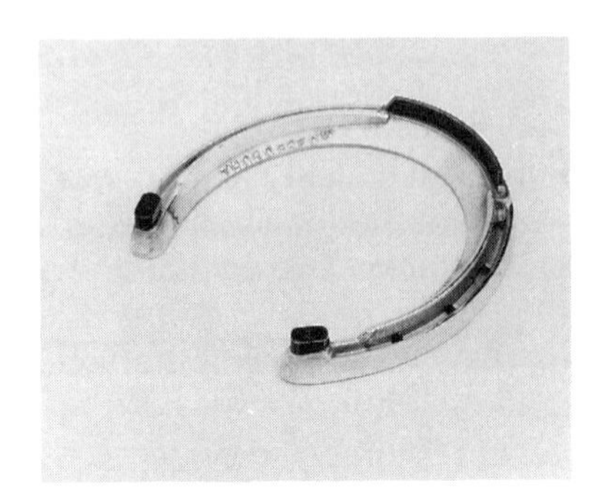

Inner Rim Block Heels.

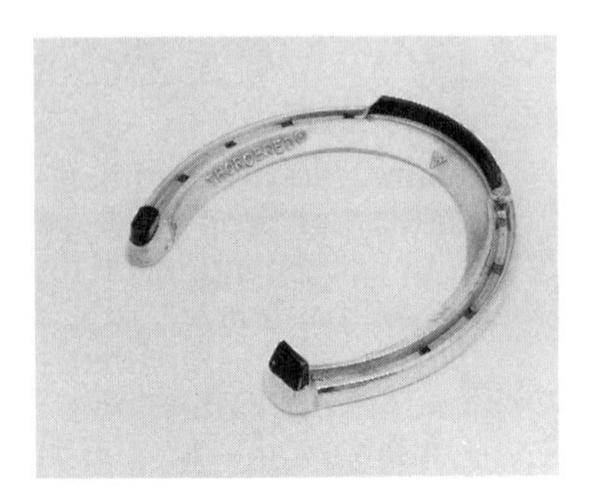

Block Heel Hind Sticker.

Common Horseshoes for Trotters and Pacers (Fig. 5–31.)

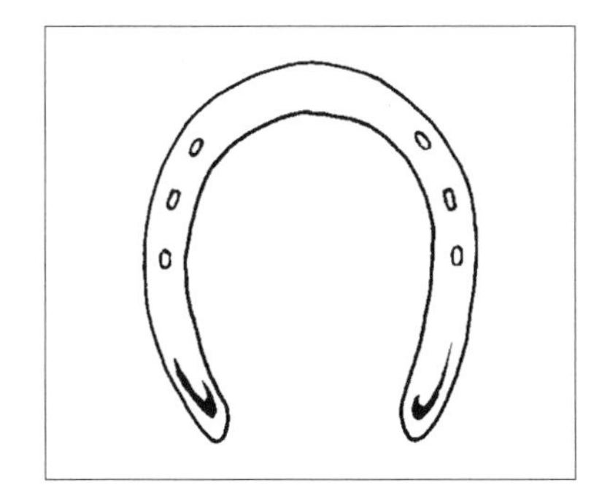

Plain.

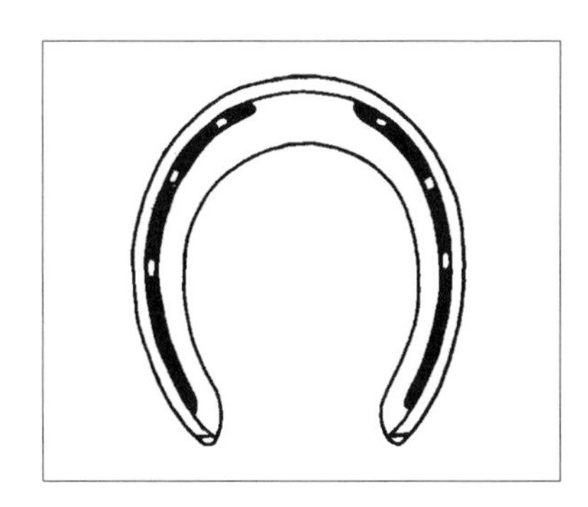

Full Rim.

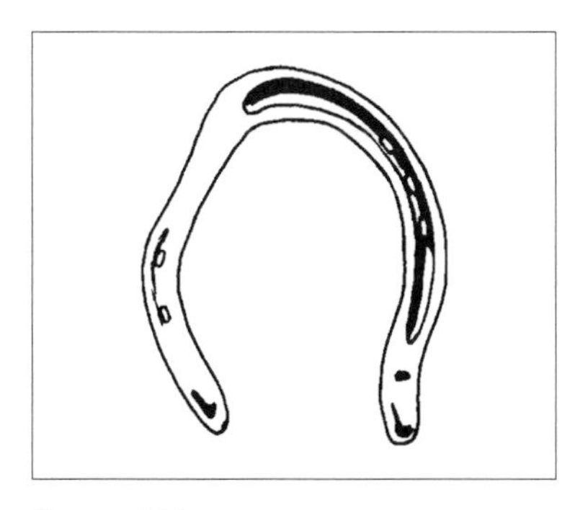

Cross-Firing.

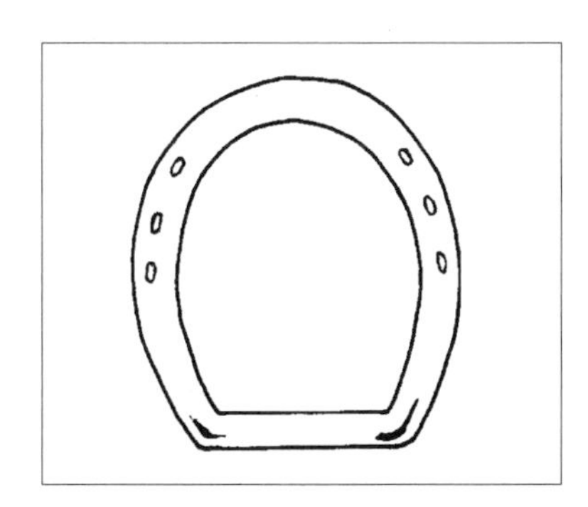

Bar.

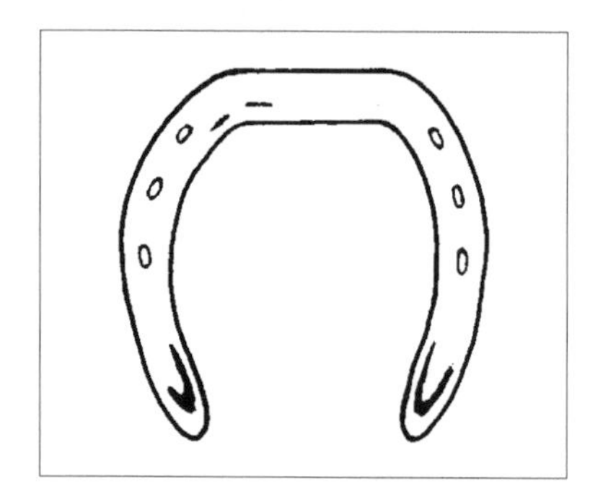

Square Toe.

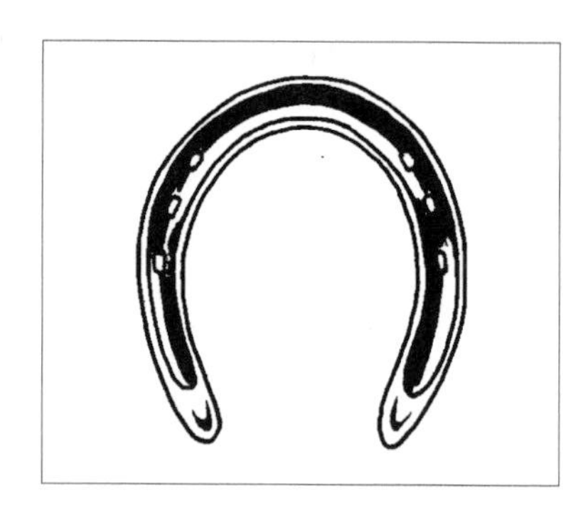

Full Swedge.

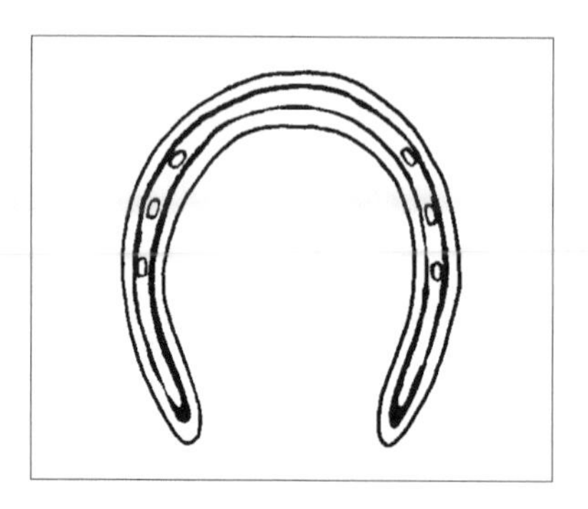

Half Round.

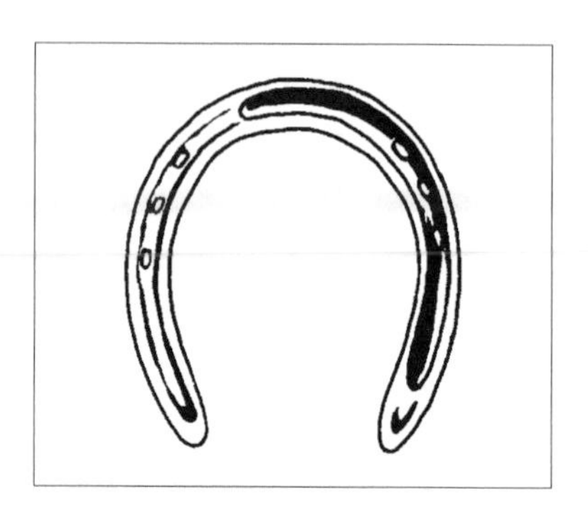

Half Round and Half Swedge.

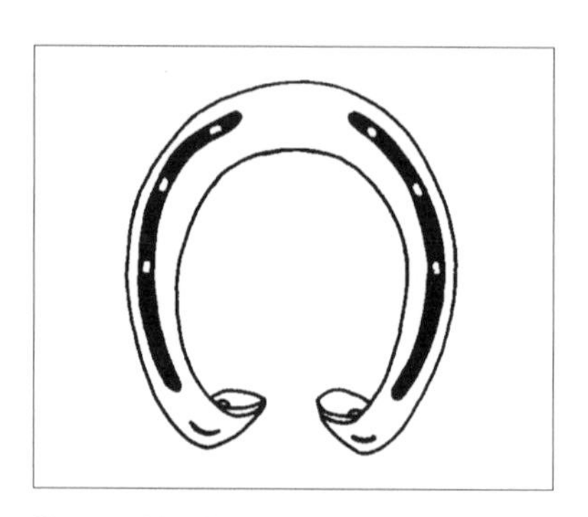

Spoon Heel.

Bar—Characterized by a metal bar connecting both heels of the shoe. It may be used with a pad on the bottom of the foot. In this case, the bar shoe prevents rocks and dirt from becoming lodged under the pad.

Square Toe—Quickens breakover to correct various interferences and gait faults. For example, when applied to the hind feet it corrects forging and over-reaching by allowing them to break over quickly.

Full Swedge—Characterized by a deep groove in the center of the track surface of the shoe. The groove extends from heel to heel. The full swedge shoe slows the action of the foot.

Half Round—The track surface of this shoe is rounded and the surface which lies against the hoof wall is flat. The rounded edge forces the foot to break over quicker when it strikes the ground, resulting in a quicker gait.

Half Round and Half Swedge—Characterized by a swedge starting at the heel and continuing half way around to the toe of the shoe. The remaining half of the shoe is round. This shoe is effective in balancing a horse and correcting interference.

Spoon Heel—The heels of the shoe are turned upward, to protect the horse's heels. It is usually worn on the front feet, and in cases of over-reaching, it prevents the hind foot from pulling the shoe off.

Shoe Weights

Shoe weights may be added to the hind shoe of the pacer on the outside heel. This enables the horse to develop a rolling action which helps it to pace. Trotters may be aided by the addition of toe weights on the front shoes. Shoe weights encourage trotters to extend their strides. They may weigh anywhere from two to four ounces.

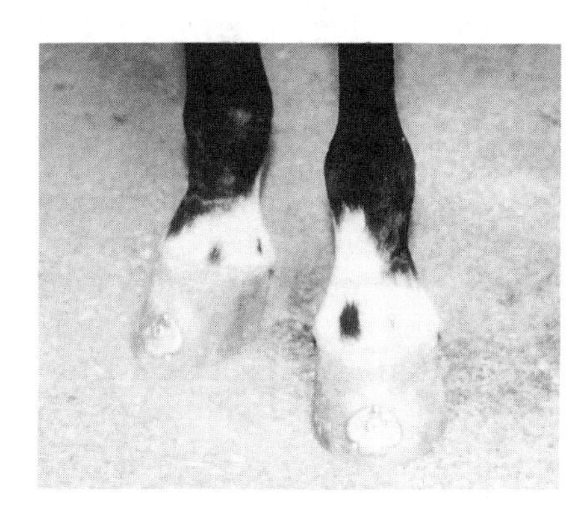

Fig. 5–32. Toe weights on the front of a trotter's shoe.

The Groom's Role

Typically, the trainer, assistant trainer, or stable foreman keeps each horse on a regular shoeing schedule; the groom is not responsible for making an appointment with the farrier for his or her horses. When the farrier comes to shoe one of the stable's horses, a good groom aids the farrier by holding the horse and keeping it quiet while it is being shod.

It is important to be able to evaluate a proper shoeing job. The major points of consideration are as follows:

- The feet should be trimmed evenly and level.
- The angle or axis of the hoof and pastern should be the same.
- The nail clinches should be filed even and smooth. None should be sticking out.
- The shoe should be properly fitted to the foot, not the foot to the shoe.
- There should be no air spaces between the shoe and the nails.
- The sole should be pared enough to prevent sole pressure.
- Nails should be no more than ¾ inches above the ground.
- There should be no rasping above the nails.
- The shoes should be solidly secured to the foot and not loose.
- A properly fitted shoe should extend at least to the back of the wall, or up to ⅛ inches past that point.
- The feet should all be in balance with each other.

Shoeing Emergency

A shoe that has shifted or become loose is considered a shoeing emergency. If the shoe falls off the foot, this is also considered an emergency. *No horse should be permitted to exercise or race with a loose or missing shoe.* If a loose shoe is overlooked, a horse could develop a career-halting injury during a workout or race. That is why it is very important to check the feet carefully first thing in the morning, and again after exercise. The value of racehorses is such that these animals must receive the best of care and judgement from those entrusted with their well-being.

It is easy to determine if a shoe is loose by attempting to move the shoe with your hands. If the shoe shifts in any way, it is considered loose and should be mended immediately. While cleaning the feet one day, you may notice that a shoe has shifted slightly or that a nail is missing from the shoe.

Pulling a Shoe

When a horse shows evidence of a loose or shifted shoe, notify your superiors immediately. A farrier will be engaged to remove the shoe and replace it as soon as possible. The groom may be required

Fig. 5–33. The value of racehorses is such that they must receive the best of care and judgement from those entrusted with their well-being.

to remove the shoe in an emergency situation. These situations include when a farrier cannot be reached, or the shoe is dangling so that the horse cannot put its foot down without stepping on the protruding nails. Even then, a farrier should be summoned as soon as possible to fix the hoof and reshoe the horse.

Removing a loose shoe is a simple procedure and can be accomplished in just a few minutes. The equipment needed to properly pull a shoe is as follows:

- clinch cutters
- nailing hammer
- shoe pullers

The procedure to pull a loose shoe, shown in Figure 5–34, can be accomplished in four basic steps:

1. Begin by resting the horse's foot just above your knee. This position allows you to use both hands to pull the shoe.
2. Straighten the nail clinches on the outside of the hoof wall. To do this, hold the clinch cutter under each clinch and strike the clinch cutter with the nailing hammer using upward strokes.
3. Flex the foot so that you can see the bottom, and hold the foot between your knees. Grip one heel section of the shoe with the shoe pullers in the area between the wall and the shoe. Press the shoe pullers downward and toward the center of the foot. Repeat the process on the opposite side of the foot.
4. Continue to work the shoe pullers on alternate sides of the foot toward the toe until the shoe is completely removed from the foot.

Pulling a Shoe (Fig. 5–34.)

Step 1

Begin by resting the foot just above your knee.

Step 2

Straighten each clinch with the clinch cutter and nailing hammer.

Step 3

Loosen the heels of the shoe with the shoe pullers.

Step 4

Work the shoe pullers toward the toe until the entire shoe is removed.

FOOT CARE PRODUCTS

Different trainers have various hoof care supplements, dressings, or packs that a groom will be required to use in maintaining healthy hooves. A common product among most trainers is hoof dressing, or hoof oil. Hoof dressing is used to prevent or treat dry, brittle feet by containing moisture in the hoof wall to prevent cracking. Some trainers apply it before a horse goes out onto the track, and some apply it after the horse is bathed. It is available in ointment or liquid form and should be applied directly to the hoof wall with a brush or spray bottle after the hoof has been washed. Some of the ingredients found in hoof dressing are lanolin, pine tar, and oils of turpentine.

Fig. 5–35. Hoof dressing can be applied with a paintbrush.

Be careful not to apply hoof dressing too often. In wet weather or moist conditions, it is not usually necessary to use hoof dressing. Even in dry areas, using hoof dressings on a daily basis can make the hoof wall too soft. If the foot becomes too soft, the farrier will have difficulty clinching the horseshoe nails to the hoof wall. The nails rip through a

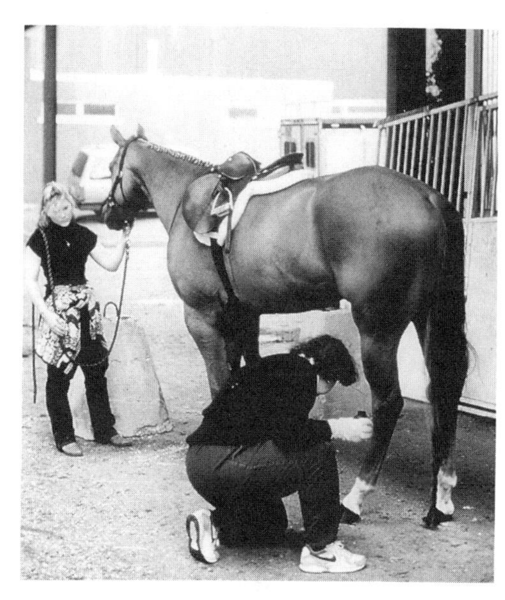

Fig. 5–36. Some trainers apply hoof dressing before exercise.

soft wall, making it difficult to affix the shoe. The farrier is the best person to ask about a horse's hoof texture and moisture content.

Another common foot care product is a commercial thrush remedy. Kopertox™ is one example of such a product. One of the most effective components of a thrush remedy is formaldehyde, which kills bacteria. If a horse has thrush, this medication (or a similar one) should be applied daily. Any treatment should be discontinued once the thrush problem has been eliminated, as this medication tends to dry the hoof.

SUMMARY

No one examines the horse's feet as often or as closely as the groom, which is why he or she must give careful daily attention to their condition. The feet must be cleaned before and after every race or workout. In most cases, the feet must be washed and packed with mud as well. It is also good to know any conformational faults of the horse's feet and any lameness the horse may be predisposed to. This way, the groom will be alerted early to any signs of stress in these areas. Given the amount of stress a racehorse's legs and feet undergo each day, the importance of proper preventive foot care cannot be overemphasized.

Chapter

6

Mucking Stalls

Racehorses spend most of their time confined to their stalls, so it is important to keep the stalls as clean and fresh as possible. A clean stall does much to keep a horse healthy and comfortable; daily mucking (cleaning) helps prevent many ailments, and fresh bedding encourages a horse to lie down and rest. (Many horses will not lie down in their own manure and urine.)

Fig. 6–1.

A lazy person can easily "parlay" a stall by removing the visible piles of manure and adding fresh bedding. The underlying urine and manure may go undetected for a few days. But sooner or later, the dirty stall will be discovered by the trainer or foreman, and the groom will be looking for a new job. Done the right way on a daily basis, mucking the stall is not a difficult task.

BEDDING MATERIALS

The two most popular types of bedding used at the racetrack that grooms should be aware of are straw and wood shavings. Some trainers feel that straw is a more comfortable bedding for the horse than wood shavings. On the other hand, wood shavings are more absorbent than straw. In situations where a horse eats straw bedding, it would probably be wise to switch the horse to wood shavings (only under the trainer's directions).

One important disadvantage of straw is that it can become moldy, and any horse that eats it may become sick. Upon breaking open each bale of straw, be alert to the following signs of bad straw:

- musty odor
- dampness
- straw that is dark or black with white patches

If any of these factors is present, the straw is moldy and the groom must discard it. There are advantages and disadvantages to both straw and wood shavings, and the decision to use either depends on availability and the trainer's preference.

MUCKING TOOLS & EQUIPMENT

The tools and equipment needed to properly muck a stall are:

- 4-prong fork for straw, *or*
- 8- or 10-prong fork for wood shavings
- wheelbarrow/muck sack/muck basket
- metal garden rake
- dehydrated garden lime
- broom

Forks

When mucking a straw stall, use a 4-prong fork to remove the manure and urine-soaked straw from the stall. Use the 8- or 10-prong fork when mucking a stall bedded with wood shavings. To carry the manure and wet bedding to the manure pit, you will need a wheelbarrow, muck sack, or muck basket.

Wheelbarrow

If you use a wheelbarrow, it should be placed outside the stall door where it can be filled with manure and wet straw or shavings. When the wheelbarrow is full, you can push it to the manure pit and empty it. A wheelbarrow is probably the easiest method of transporting dirty bedding from the stall to the manure pit.

Muck Sack

The muck sack is used primarily for the disposal of straw bedding. It has the advantage of holding a large amount of manure, and it usually requires only one trip to the manure pit.

Fig. 6–2. A muck sack holds a large amount of manure, usually requiring only one trip to the manure pit.

Making a Muck Sack

The muck sack is made of four burlap feed sacks sewn together into one large sheet. This sheet is then laid outside the stall and the manure is deposited in the middle of the sack. Once filled, each corner of the sack is tied together. You may lift the sack on your back or toss it on a wheelbarrow to take it to the manure pit.

To make a muck sack you need the following materials:

- scissors
- four burlap feed bags
- large sewing needle
- heavy duty twine

The muck sack may be re-used several times, but will not last as long as a wheelbarrow or muck basket. *(See Figure 6–3 for instructions on making a muck sack.)*

Making a Muck Sack (Fig. 6–3.)

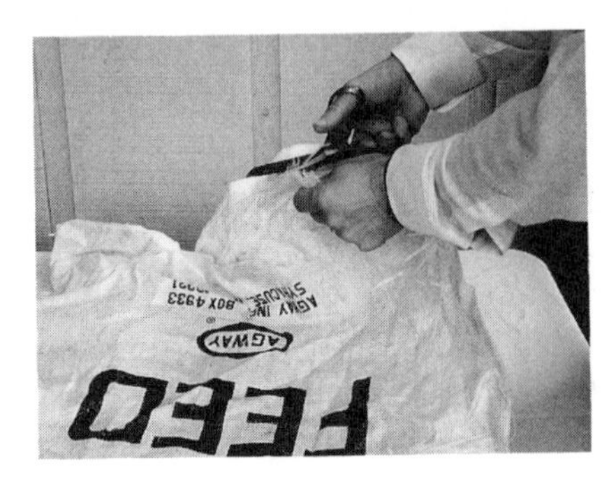

Step 1

Begin with scissors by cutting open the seam on one side and the bottom section of the feed bags.

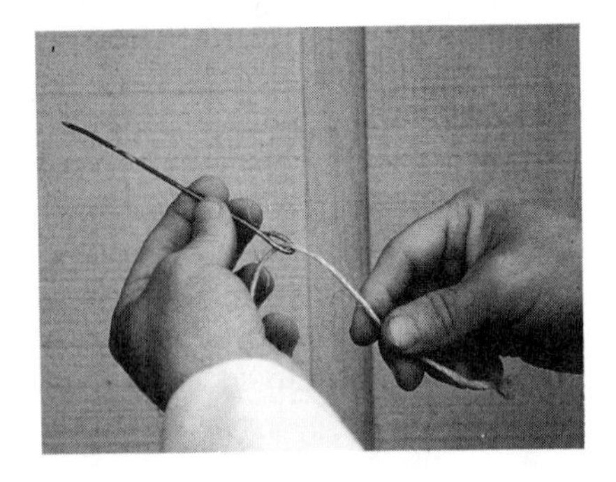

Step 2

Thread the large sewing needle with the twine.

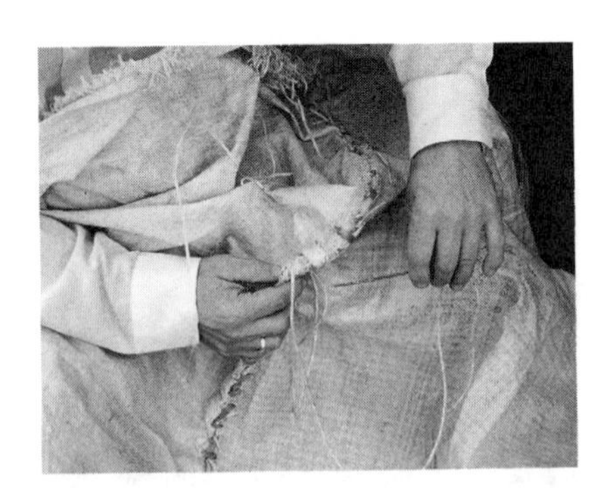

Step 3

Sew the four cut sheets of burlap together into one large square.

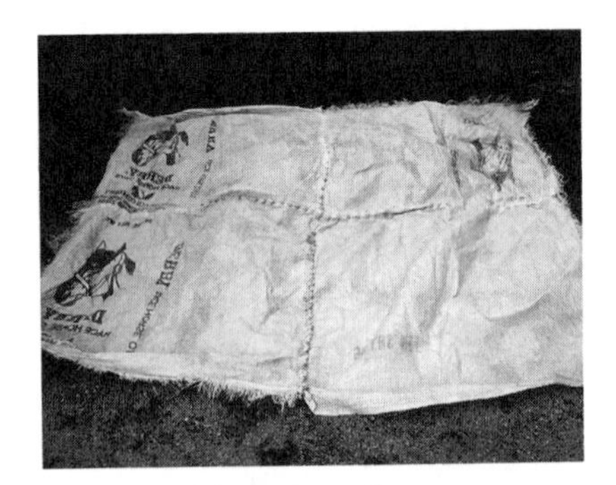

Step 4

The completed muck sack measures $5\frac{1}{2}$ x $6\frac{1}{2}$ feet.

Muck Basket

This basket is usually made of plastic with two nylon rope handles. It is best if a groom leaves the muck basket outside the stall when mucking a stall with the horse still inside. That way, there is less chance of the groom or the horse tripping over the muck basket and getting injured if something frightens the horse. Naturally, if mucking a stall when the horse is out, the muck basket can be taken inside the stall. Once the muck basket is filled with manure, it can be lifted and carried to the manure pit to be emptied.

Metal Garden Rake

This rake is used to remove small pieces of manure missed by the fork and to remove the chaff (small bits of hay, straw and/or shavings) which settles on the floor of the stall. After laying down fresh wood shavings, use the rake to rake up stray shavings from the dirt shedrow outside the stalls.

Dehydrated Garden Lime

As most stall floors are made of clay, rather than cement, wet spots develop from constant buildup of urine and manure. Lime is sprinkled on these wet spots to dry them up and help keep the stall smelling fresh. However, avoid using too much lime at one time (a few scoops is plenty), as a horse's eyes and nasal passages may become irritated.

Broom

After the stall is completely mucked, the shedrow area outside the stall should be raked if necessary, and swept clean. Also, any cobwebs inside or outside the stalls or in the shedrow should be swept away with the broom. *Appearance counts.*

MUCKING PROCEDURE

If possible, muck the stall while the horse is out being walked or exercised. If the horse must remain in the stall, it should be tied in some manner to prevent it from moving around and disturbing the cleaning process. If you need the horse to swing its hind end away from where you are cleaning, place the palm of your hand on the horse's hip and gently but firmly push the horse over. Encourage the horse with your voice as you do this.

When cleaning the stalls, feed tubs and water buckets must be taken out and cleaned. Use only water and a stiff brush to clean any buckets the horse eats or drinks out of. Soap may leave a residue that could cause the horse to stop eating or drinking.

Mucking a Wood Shavings Stall

Once the horse is out of the stall, or has been tied up in the stall, gather all equipment necessary to muck a wood shavings stall. Remove the feed tub, water bucket, and hay net from the stall.

1. With the 8- or 10-prong shavings fork, remove the visible piles of manure and discard them into the wheelbarrow, muck basket, or muck sack.
2. With the rake, lightly rake the surface of the shavings from the rear of the stall to the front. This will remove any small pieces of manure and hay. Discard this into the wheelbarrow, muck basket, or muck sack also.
3. With the shavings fork, begin turning over the shavings. Remove the wet shavings to the wheelbarrow, muck basket, or muck sack, and set aside the dry shavings.
4. Sprinkle lime on any wet spots, if necessary.
5. Dump fresh shavings in the middle of the stall and with the fork, mix the fresh shavings with the old. Depending on the trainer's preference and how much bedding was discarded, one to two bales of wood shavings (one or two wheelbarrow loads if the shavings are delivered by the truckload) should be sufficient.
6. Using the rake, spread the shavings evenly on the stall floor.

After cleaning the stall, rinse the feed tub and hang it up outside the stall door. Clean the water bucket using water and a stiff brush. Rinse it thoroughly and replace it in the stall with fresh water. Also, fill the hay net or rack with fresh hay along with any good hay left over. Finally, sweep the area outside of the stall to make it look neat.

Mucking a Wood Shavings Stall (Fig. 6–4.)

Step 1

With the fork, remove all visible piles of manure and wet shavings and discard them into the wheelbarrow, muck basket, or muck sack.

Step 2

Rake the surface of the shavings to remove any small pieces of manure and chaff. Discard them into the wheelbarrow, muck basket, or muck sack.

Step 3

With the fork, turn over the shavings, removing the soiled bedding to the wheelbarrow, muck basket, or muck sack, and moving the dry shavings into the corner of the stall.

Step 4

Sprinkle a small amount of lime on any wet spots.

Step 5

Mix one or two bales (or two wheel-barrows full) of new shavings in thoroughly with the old shavings.

Step 6

Using the rake, smooth the top layer of the shavings even.

Mucking a Straw Stall

Again, tie or cross-tie the horse in the stall when mucking the stall with the horse still inside. Then get all the equipment necessary to muck a straw stall, and remove the feed tub, water bucket, and hay net from the stall.

1. With the 4-prong straw fork, remove the visible piles of manure and discard them into the wheelbarrow, muck basket, or muck sack.
2. With the fork, sift through the stall and discard the remaining manure and wet straw into the wheelbarrow, muck basket, or muck sack. Pile up any dry straw in one corner or side of the stall. Continue until you have the entire floor of the stall exposed except for the dry straw piled in one corner or side of the stall.
3. With the metal garden rake, rake the entire floor from the rear to the front. This collects the chaff on the stall floor.
4. Sprinkle lime on any wet spots, if necessary.
5. Before putting down new straw, bed down the stall with the leftover dry straw you piled into one corner or side of the stall. Then add new straw by breaking open one or two bales of straw with your fork. (This entails cutting the baling wire or twisting one of the prongs of the fork around the baling wire until it breaks.) Once the bale is open, *discard the baling wire immediately,* as it may become entangled in the legs and injure a horse if left on the ground.
6. Take each square or "flake" of straw and shake it thoroughly in the stall with the fork. If you cannot break up the straw with the fork, use your hands. Spread the fresh straw evenly over the old, dry straw throughout the stall.
7. Now "bank" the three sides of the stall away from the door with a little extra straw.
8. Flatten out the straw in the center by tapping it with the tip of the fork until you see a neat, even appearance throughout the stall.

Remember to rinse out the feed tub and hang it up outside the stall. Clean the water bucket using water and a stiff brush to remove any film, hay, or grain. Rinse the water bucket thoroughly and replace it in the stall with fresh, clean water. Also, fill the hay net or rack with fresh hay along with any good hay saved from the day before. Then sweep the area outside the stall.

Mucking a Straw Stall (Fig. 6–5.)

Step 1

With the fork, remove all visible piles of manure and wet shavings and discard them into the wheelbarrow, muck basket, or muck sack.

Step 2

Sift through the straw, removing the soiled bedding. Pile the clean straw in one corner of the stall.

Step 3

Rake the stall floor clean of chaff.

Step 4

Sprinkle a small amount of lime on any wet spots.

Step 5

Use the fork to open a fresh bale of straw. Then discard the baling wire immediately.

Step 6

Spread new straw evenly over the old straw. Use your hands if necessary.

Step 7

Bank three sides of the stall with a little extra straw.

Step 8

Pat down the straw evenly in the center of the stall.

Fig. 6–6. When the stall is clean and bedded, it should look appealing to the horse.

Note: It is important to be conscious of where your tools are at all times. Tools should be organized and kept close at hand. Also, the wheelbarrow should be kept to the side of the aisle rather than in the middle to allow free passageway for other people and horses.

WHEN A HORSE BECOMES CAST

When a horse lies down close to a stall wall and rolls its legs up against the wall, it sometimes becomes trapped, or "cast," and is unable to stand up. Instead of rolling back over into the middle of the stall, some horses panic and paw at the wall with all four feet, trying to regain their footing and get up. This struggle to stand up may go on for only a few seconds or for several hours. In the majority of cases, the horse is able to

Fig. 6–7. A cast horse.

right itself and escape without injury. If not, however, continued struggling can result in injury to a horse's feet and legs as well as other parts of the body.

When you are alerted to this situation (a sudden, loud scrambling noise against the wall is a good indication that a horse is cast), notify the trainer immediately and try to assist the horse.

One method of assistance is to loop the leather portion of a lead shank over the pastern of the front leg and another over the hind leg that is closest to the ground. (You may need to ask another person to help.) Pull the legs upward and toward you *slowly* to help the horse roll over away from the wall. Or, if the horse has a halter on, it may be easier to carefully pull the horse's head toward the center of the stall, which moves the front feet away from the wall, thus allowing the horse to stand.

When attempting to free a horse from a cast position, be prepared to exit the stall once the horse is able to stand up. When released from such a position, horses have been known to jump, buck, shake, or rear up. To avoid possible injury, the groom should be out of the horse's way.

Banking a stall is a good way to prevent a cast horse; not only does banking provide good footing, but it discourages the horse from lying down too close to a wall. Many trainers have their grooms bank all of their stalls.

SAFETY FACTORS

Some safety factors to remember when mucking any stall are as follows:

- Discard all baling wire immediately after you remove it from the bales.
- Always check for moldy straw or hay when opening bales.
- Never leave mucking tools or equipment in the stall or shedrow where a horse can step on them.
- Before dropping bales of straw or hay from a storage loft to the shedrow, always alert everyone to clear the area below.
- When mucking a stall with a horse still inside, always hook the horse to a tie chain or put it on cross-ties.
- Never reach under a horse with a manure fork when mucking a stall; instead, move the horse over.
- Fill up any holes found in the stall floor, as they are a danger to the horse.

SUMMARY

A professional groom takes pride in keeping stalls fresh and comfortable for the horses. When mucked out daily (or several times a day), stalls are not difficult to keep clean. A groom should be familiar with how to clean a stall bedded with straw or wood shavings, and should know how to bank a stall to help prevent horses from becoming cast. Finally, before putting a horse back in a freshly cleaned stall, make sure the bedding is absent of mold and foreign materials, all mucking tools have been removed from the stall, and the water buckets have been cleaned and filled with fresh water.

Chapter

7

TACKING UP

Preparing a horse properly for exercise or a race requires a certain amount of skill and attention to detail. Not only must the caretaker know how to put on a horse's tack, he or she must be able to fit each article correctly to the horse, so that the equipment is effective and safe. It is also important that the caretaker pay attention to the daily exercise schedule, usually posted in the barn, which indicates the type of exercise that the trainer wishes each horse to perform that day.

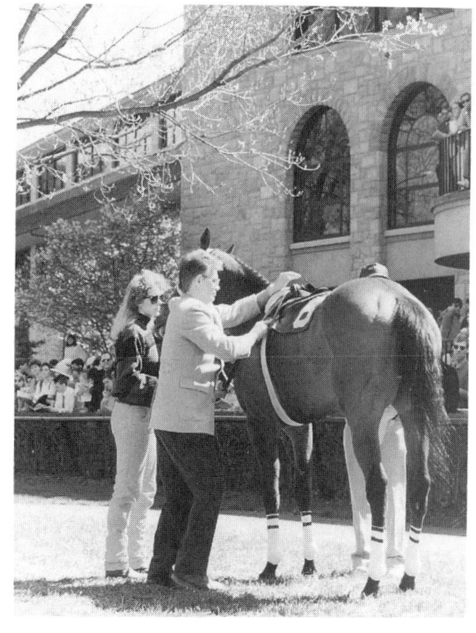

Fig. 7–1.

For the running horse, exercise might include:

- walking under the shedrow either by hand with a hotwalker or under tack with a rider
- going to the track with a pony for a slow gallop
- going to the track with a rider for a gallop
- going to the track with a rider for fast work

For the harness horse, exercise might include:

- walking
- jogging three to six miles
- training miles in heats

TACK ROOM

The tack room is the area (often an extra stall) designated for the storage of all equipment. It is usually kept locked to discourage thieves. Bridles and martingales are usually hung on nails or bridle hooks on the wall. (The names of the horses may or may not be affixed to the wall above each bridle.) Each saddle is set on a saddle rack which is attached to the wall. Or, two or more saddles may be placed on each wooden saw horse inside the tack room. Harnesses may be hung from the wall and may be covered with a harness bag. Large storage trunks are usually arranged on the floor of the tack room to provide storage for miscellaneous tools and equipment. Under no condition should bridles, saddles, or harnesses be dumped or dragged on the ground. Always treat tack and equipment with care.

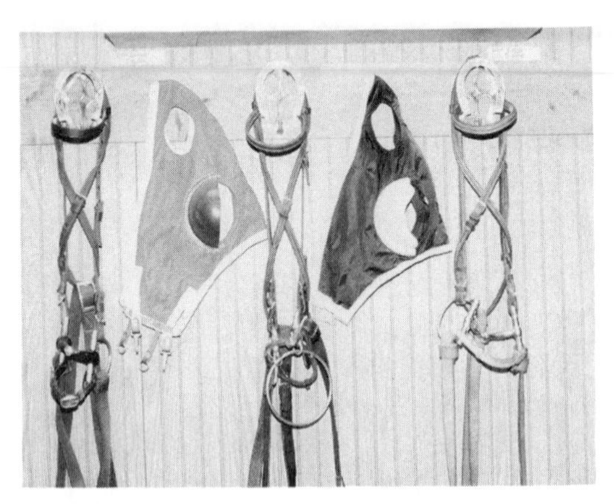

Fig. 7–2. Bridles are hung on nails or bridle hooks on the wall.

The discussion of running horse tack begins in the next section. The discussion of harness horse tack begins on page 186.

TACKING UP THE RUNNING HORSE

The Thoroughbred or Quarter Horse groom who knows the exercise schedule knows what equipment each horse needs. While it is normally the trainer's responsibility to make sure the horse is properly turned out for a race, it is the groom's job to see that the horse is correctly equipped for exercise.

The stable supplies its employees with the necessary saddle and bridle. It is unusual to have a saddle and bridle for each horse in the stable. In a stable that has a large number of horses it is not uncommon for horses to go out to the track in sets. Each set may consist of 2 – 10 horses depending on the number of horses in the stable. Therefore it is not uncommon to wait and use the tack from a horse in an earlier set to tack up a horse in a later set.

Tack varies depending on the type of exercise a horse will be performing. For example, if a horse is scheduled for a fast breeze, a running martingale may be eliminated from the usual equipment and a tongue tie added. *(See Chapter 8.)*

Preparing for a Workout

The groom is in charge of tacking up the horse when the trainer has scheduled the horse to go to the track for exercise. Tack up the horse with all the equipment prescribed by the trainer, including any track bandages which may be required. *(See Chapter 10 for instructions on applying bandages.)* Then, if the rider is not ready, walk the horse around the shedrow until he or she is ready.

The equipment used in exercise may vary greatly among trainers, but the following sections describe some of the basic pieces of equipment with which a groom needs to be familiar.

Exercise Saddle

Racehorses are never exercised in a racing saddle; running horse trainers use a heavier saddle that is designed strictly for exercise. (Exercise saddles weigh approximately 5 – 8 pounds, depending on the saddle manufacturer.)

The parts of the exercise saddle include:

- pommel
- panel
- flaps
- cantle
- tree (not shown)
- stirrup leathers
- seat
- billets
- stirrup irons

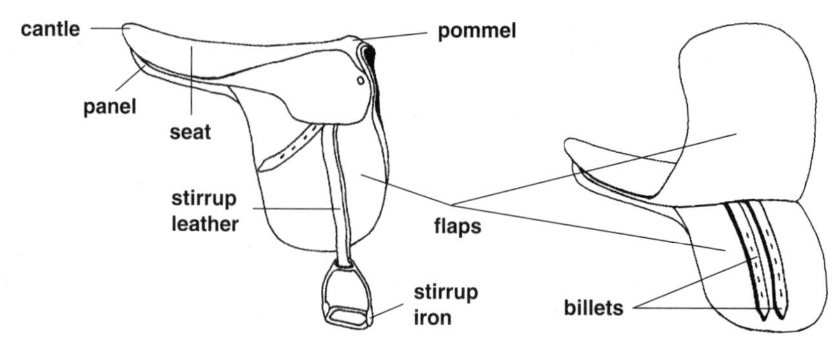

Fig. 7–3. Parts of the exercise saddle. The tree is the frame of the saddle and cannot be seen here.

The seat of the exercise saddle may vary in length from 16½ to 17½ inches. The tree (the frame of the saddle) may be 6½ – 8 inches wide. A shaped elastic-ended leather girth holds the exercise saddle on the horse, and it attaches to two billets on either side underneath the flaps of the saddle. Exercise stirrups are 4¼ – 4½ inches wide, made of stainless steel, and weigh approximately 6 – 8 ounces. (They are heavier than racing stirrups.)

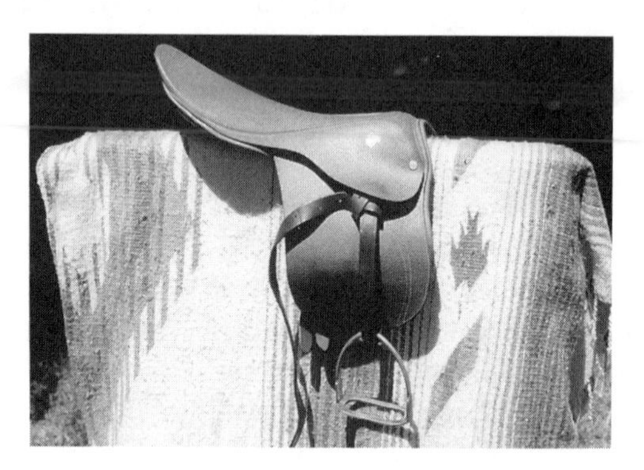

Fig. 7–4. An exercise saddle should always be stored on a saw horse or saddle rack.

Saddle Accessories

Saddle Cloth

The saddle cloth is usually made of cotton. It is placed on the horse's back to absorb sweat and prevent chafing on the back and withers. The saddle cloth is approximately 34 inches long and 38 inches wide. *(See Figure 7–5.)*

Pommel Pad

The pommel pad keeps the horse's withers well-protected from the saddle during exercise. Pommel pads are usually made of wool and are 10½ inches wide and 19 inches long. *(See Figure 7–5.)*

Fig. 7–5. On the left, a saddle cloth prevents chafing on the horse's back and withers. On the right, a pommel pad further protects the withers.

Saddle Pad

The saddle pad protects the horse's back from the exercise saddle, distributing weight more evenly and absorbing concussion so the horse's back will not become sore or tender. It is made of felt, sheep's wool, sponge rubber, or foam rubber.

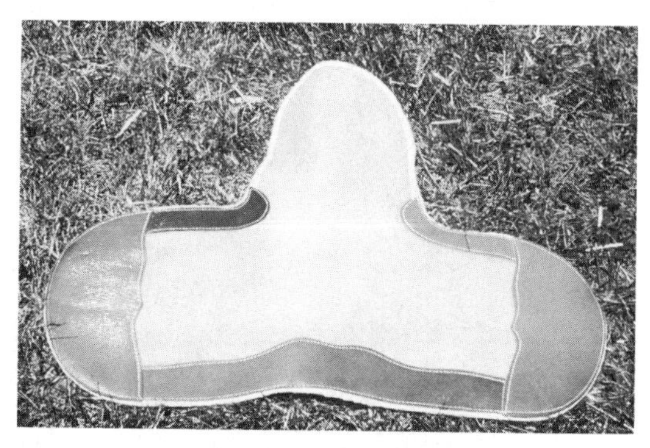

Fig. 7–6. A saddle pad distributes the rider's weight more evenly and absorbs concussion.

Exercise Girth

The exercise girth is usually made of leather and is shaped at the elbows to prevent chafing. Girths vary in length from 36 to 56 inches. There are two buckles at each end of the girth which fasten it to the two billets on the exercise saddle. One end of the girth is usually made of

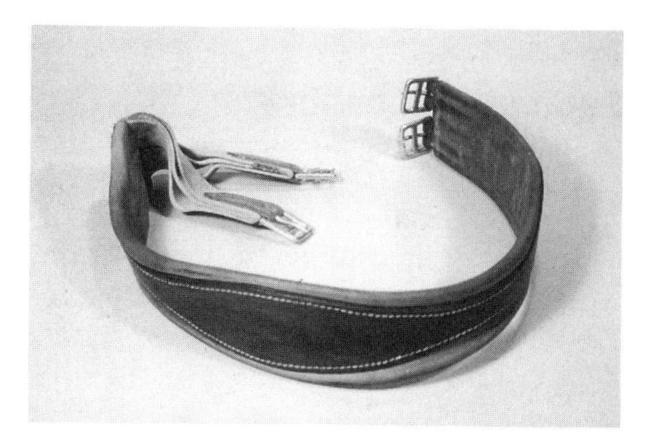

Fig. 7–7. The girth has two buckles at each end which fasten it to the exercise saddle.

strong elastic, which makes it easier to tighten. (This end is always buckled on the near side.)

Girth Cover

Girth covers are made of either flannel or fleece and prevent chafing of the horse's skin from the girth. Another advantage of using girth covers is that they prevent the spread of skin diseases from one horse to another if the actual girth is used on more than one horse. Girth covers are used primarily during training sessions.

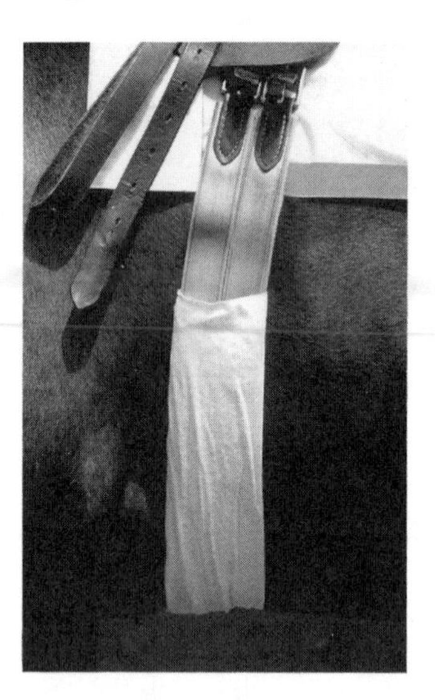

Fig. 7–8. On the left, a flannel girth cover. On the right, a fleece girth cover.

Saddling Procedure

The following instructions, illustrated in Figure 7–9, outline a good procedure to follow when tacking up a racehorse for exercise. Note that the saddle is put on before the bridle so that the horse can remain tied up (without the interference of the bridle and reins) during saddling. When tacking up for a race, the opposite is true; the bridle is put on before the saddle.

1. Tie or cross-tie the horse to the wall by clipping the tie chain(s) to the ring(s) of the halter.
2. After grooming the horse, lay the saddle cloth over the withers and back. Do not pull the saddle cloth forward, because it is against the direction of hair growth, and can irritate the horse.
3. Gently lay the saddle pad over the saddle cloth so that it covers the withers and back. Fold the front of the saddle cloth back over the saddle pad about six inches.
4. Fold the pommel pad in half and lay it on top of the saddle pad and saddle cloth, over the horse's withers.
5. Set the saddle on the horse's back right behind the withers. With your left hand, crease the saddle cloth, saddle pad, and pommel pad up underneath the pommel of the saddle. This prevents chafing of the withers, and allows the saddle to fit snugly with no slipping.
6. Go around the horse to the off side. Buckle the non-elastic end of the girth to the off saddle billets. Return to the near side. With your right foot, pull the elastic end toward you before reaching for it. This keeps your head out from under the horse.
7. Buckle the elastic end of the girth onto the near saddle billets. Make sure the girth is tight enough so that the saddle will not slip. Also, make sure the buckles rest on the leather flap, and are not rubbing against the skin.
8. Standing to the side, gently lift each of the horse's front legs forward to stretch the skin smooth. This is done so that the skin does not get pinched underneath the girth.

The horse is properly saddled and ready for bridling.

Saddling the Running Horse for a Workout (Fig. 7–9.)

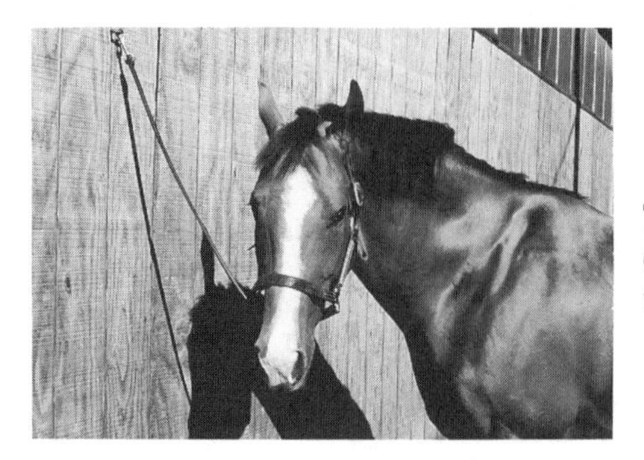

Step 1

Tie the horse to the wall by clipping the rubber tie chain to the halter.

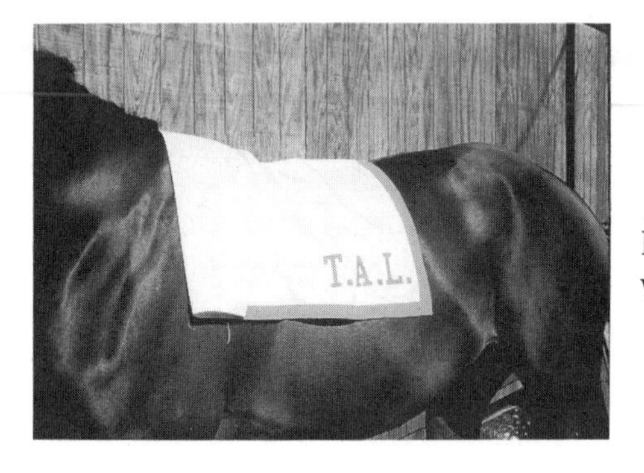

Step 2

Lay the saddle cloth over the withers and back.

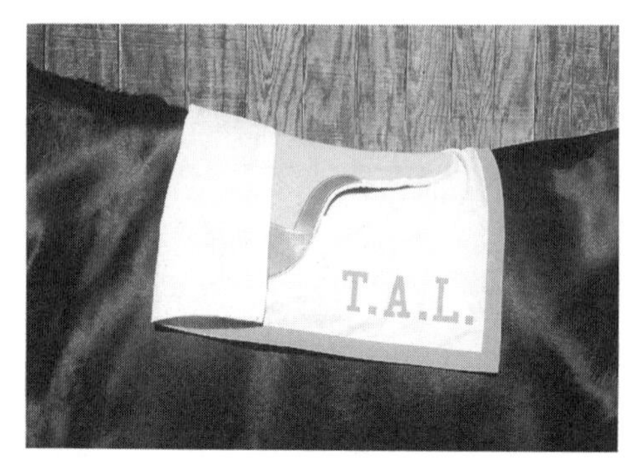

Step 3

Lay the saddle pad over the cloth. Fold the front of the saddle cloth back over the saddle pad about six inches.

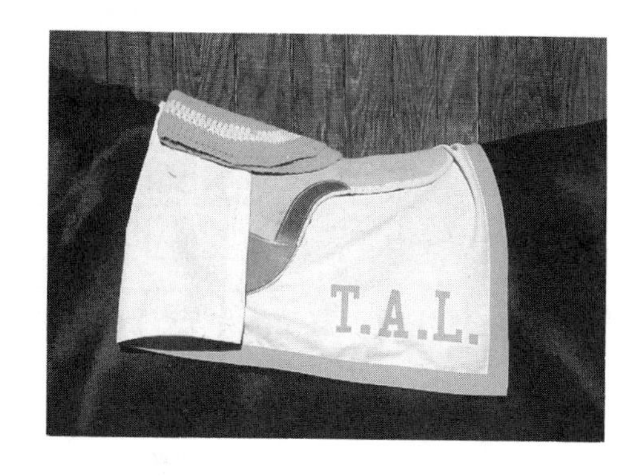

Step 4

Fold the pommel pad in half and lay it on top of the saddle pad and saddle cloth, over the horse's withers.

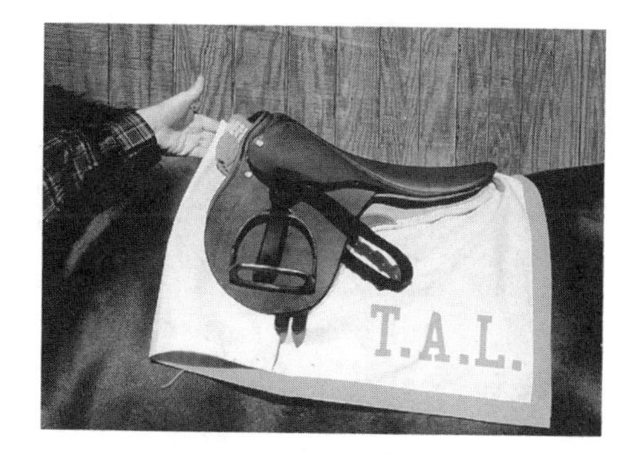

Step 5

Place the saddle right behind the withers. With your left hand, crease the saddle cloth, saddle pad, and pommel pad up underneath the pommel of the saddle.

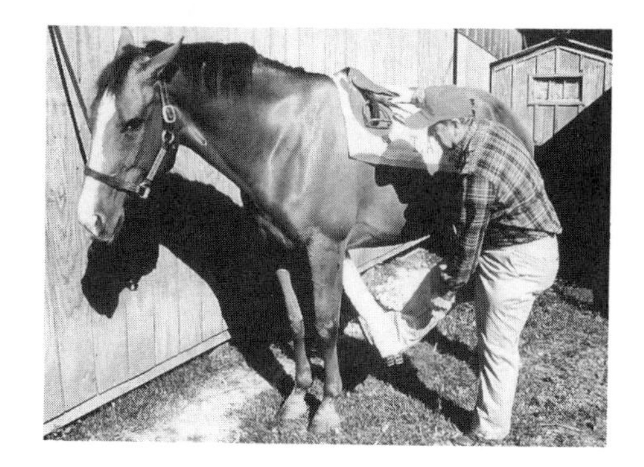

Step 6

Buckle the girth to the off billets. Return to the near side. With your right foot, pull the elastic end toward you before reaching for it.

Step 7

Buckle the elastic end of the girth onto the near saddle billets.

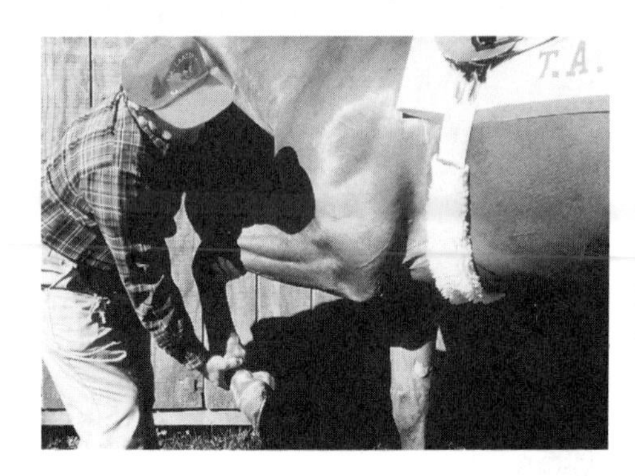

Step 8

Standing to the side, gently lift each of the horse's front legs forward to stretch the skin smooth under the girth.

Step 9

This horse is properly saddled and ready for bridling.

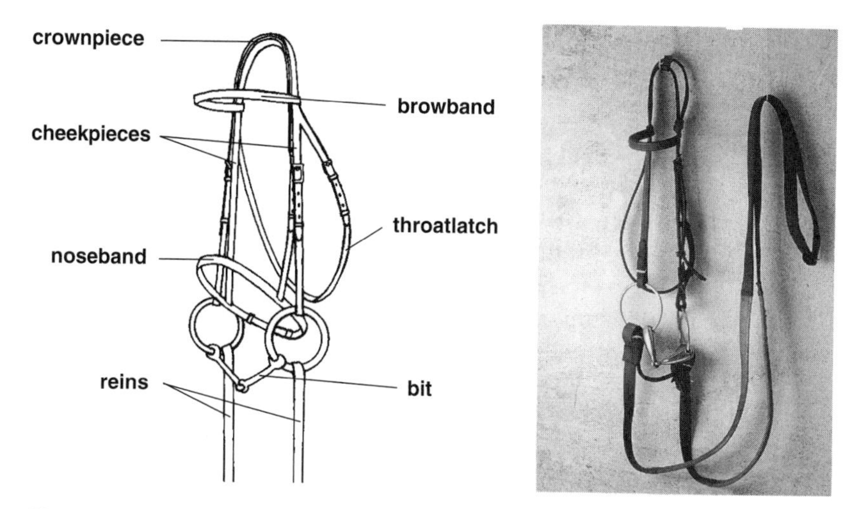

Fig. 7–10. The exercise rider controls the horse using the bridle.

Exercise Bridle

The bridle is a very important piece of exercise equipment. It is the means by which the rider controls the horse. The bridle is made of either leather or nylon and includes the crownpiece, browband, cheekpieces, and throatlatch. Attached to the bridle are the bit and reins, and perhaps a noseband. The material is sturdy, usually about ½ – ⅝ inches wide. The reins vary in length from 60 to 66 inches and are usually 1 inch wide. Most reins have 18-inch rubber positions to provide a better grip for the rider.

Basically, there is no difference between an exercise bridle and a racing bridle. The trainer may race the horse in a different bit, but will probably use the same bridle.

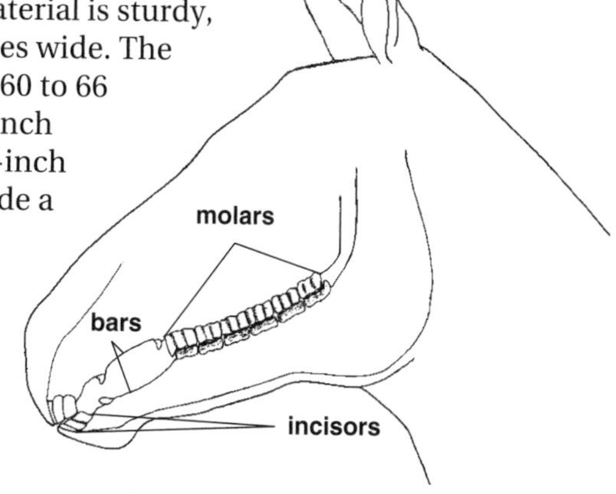

Fig. 7–11. The bit rests on the bars of the mouth, between the incisors and the molars.

Bridling Procedure

The following is the procedure for bridling the running horse:

1. Standing on the near side, place the reins over the neck.
2. In your right hand, hold the cheekpieces of the bridle together, just beneath the browband. Curl your right arm underneath the horse's jaw, and place your right hand, with the bridle, over the nose. This position prevents the horse from throwing its head about during bridling.
3. Hold the bit in your left hand with your fingers spread and place your left thumb through the large ring. Align the bit with the mouth and insert your thumb in the horse's mouth at the bar. When the horse opens its mouth, slip the bit in. (Avoid letting the bit hit the teeth, as this can hurt the horse.) The bit should rest on the bars of the horse's mouth, where there are no teeth.
4. When the bit is in the mouth, keep enough tension on the upper part of the bridle with your right hand so that the bit does not slide back out. Now you should have both hands free to gently fit the crownpiece over the horse's ears.
5. Tilting the bridle as necessary, slowly and cautiously pull the crownpiece over the off ear. Then gently fold the near ear forward just far enough to slip the crownpiece over it. Do not force the crownpiece over the head too quickly, as this pinches the ears back and is uncomfortable for the horse. Also, do not fold the off ear, because you cannot see the ear and you might accidently hurt the horse by folding the ear too much. When completed, make sure the crownpiece rests on the bridle path (the clipped area just behind the ears).
6. Once the bridle is on, buckle the throatlatch. The throatlatch should rest lightly against the cheekbone, not be tight up against the throat, as this restricts the horse's breathing.
7. When the throatlatch is fitted properly, you should be able to place four fingers from the jawbone to the throatlatch. Be sure all leather straps are run through their keepers so no bridle parts fly about while the horse is running.
8. Make sure that the horse's forelock is pulled free of the browband and lays flat on the forehead, and that the bit is fitted correctly.

If the bridle was last used on a different horse, chances are it will need adjustment. There are buckles on both cheekpieces to make the bridle looser or tighter. Most trainers place the bit where the horse is

Bridling the Racehorse (Fig. 7–12.)

Step 1

Place the reins of the bridle over the horse's neck.

Step 2

In your right hand, hold the cheekpieces of the bridle together, just beneath the browband. Curl your right arm underneath the horse's jaw, and place your right hand, with the bridle, over the nose.

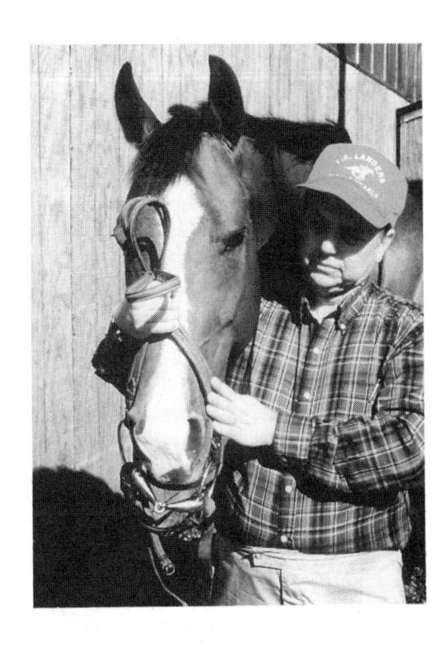

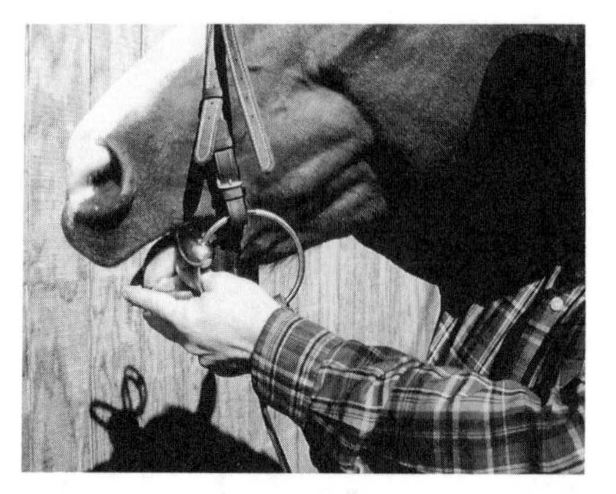

Step 3

Insert your thumb in the bar of the mouth. Wiggle your thumb, and when the horse opens its mouth, slip the bit in.

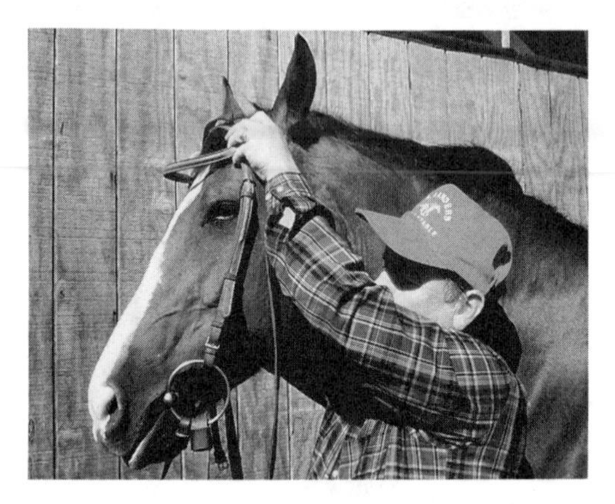

Step 4

Keep enough tension on the upper part of the bridle so that the bit does not slide back out. Gently fit the crownpiece over the horse's ears.

Step 5

Place the crownpiece over the off ear. Then fold the near ear forward to slip the crownpiece over it, until the crownpiece rests on the bridle path.

Step 6

Buckle the throatlatch so that it rests lightly against the cheekbone.

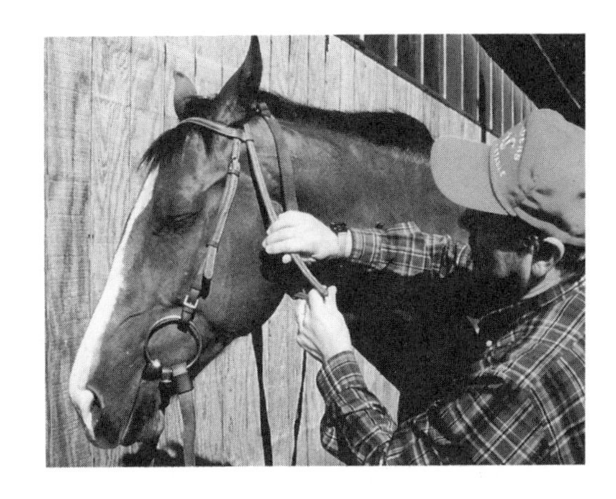

Step 7

When the throatlatch is fitted properly, you should be able to fit four fingers between the cheekbone and the throatlatch.

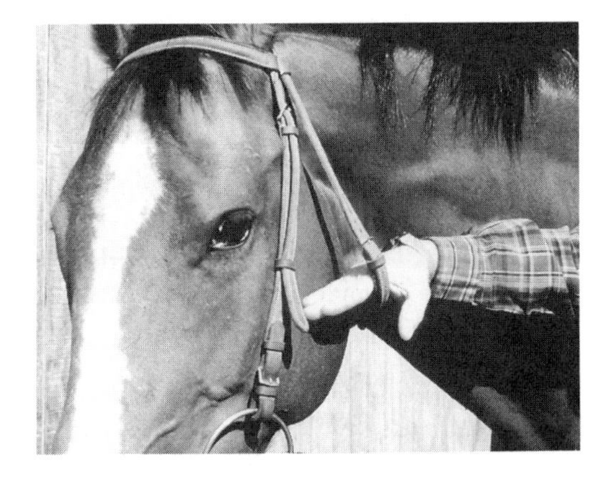

Step 8

Make sure the bit is fitted correctly, with two wrinkles at the corner of the mouth.

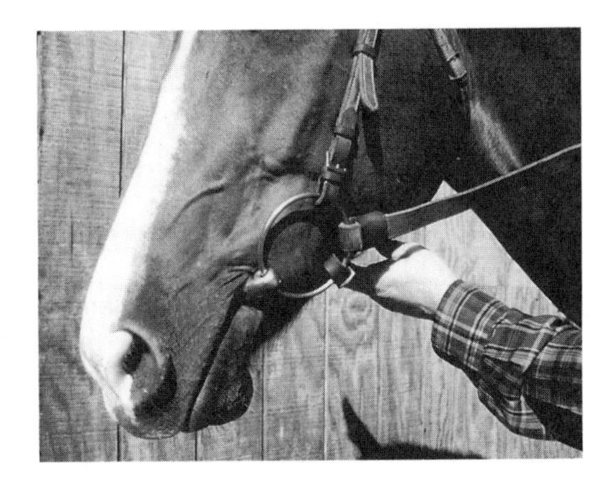

"smiling" with two wrinkles at the corners of the mouth. The bit is too tight if there are several wrinkles at the corners of the mouth. The horse will probably let you know the bit is too tight by opening its mouth excessively the whole time the bridle is on. The bit is too loose if it is low in the mouth and the horse is playing with it. Also, be sure that the horse's tongue is under the bit. If the tongue is over the bit, the rider will have little control, and the horse will be distracted.

If the rider is not ready, begin walking the horse around the shedrow. If for some reason you cannot do this, tie the horse to the stall wall with a halter on over the bridle. Make sure the stirrups are run up; they should be, because the saddle is stored that way. Never leave a tacked horse alone unless it is tied to the stall wall.

A horse should not be tied up for a long time after being tacked. When a horse is tied to the wall for a long time, there is always the chance that the horse will become irritable or frightened, which may lead to injury. After a horse is tacked up, it anticipates going to the track and becomes excited. Fifteen minutes is the maximum amount of time for any horse to be tied to the wall after being tacked. It is better to walk the horse under the shedrow with its tack for the amount of time needed to await an exercise rider.

Once the rider is ready, remove the halter. Clip the lead shank onto the ring of the bit on the near side and walk the horse around the shedrow at least one time before giving the rider a leg up.

Giving the Rider a "Leg Up"

The exercise rider must be assisted by the groom to properly mount the horse. This is a common skill which is easily mastered. The usual method to accomplish this task is to give the rider a "leg up" using one hand. (It is not normally the groom's responsibility to give the jockey a leg up before a race, only the exercise rider.)

1. Stand on the horse's near side and hold the reins in your left hand to control the horse.
2. With your right hand, grip the rider's left leg just above the ankle.
3. Coordinate your efforts with the rider and gently lift the rider up into the saddle.
4. Once the rider is in the saddle, walk with the horse until the rider has adjusted the girth, stirrups, and reins. Upon the command of the rider, release the horse and walk away from it to your left.

Giving the Rider a Leg Up (Fig. 7–13.)

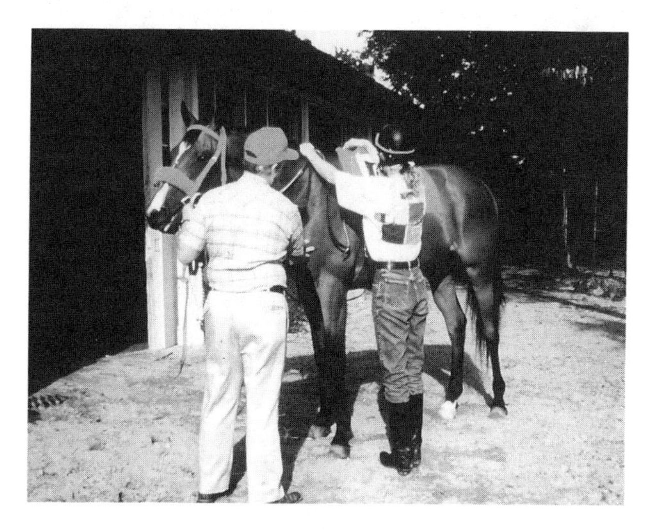

Step 1

Stand on the horse's near side and hold the reins in your left hand to control the horse.

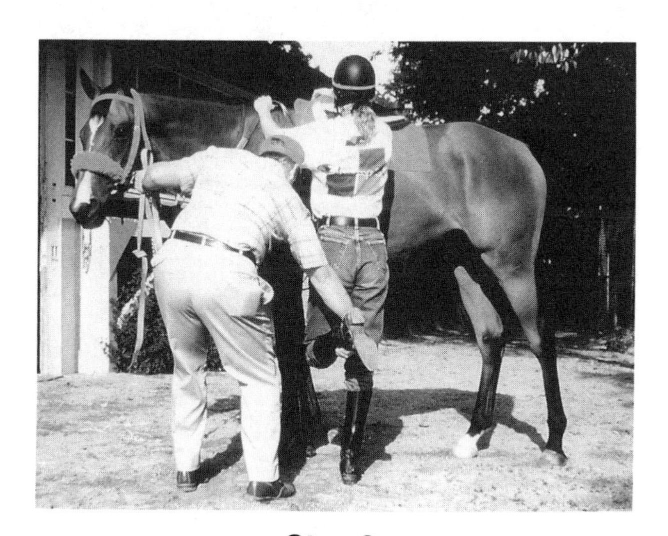

Step 2

With your right hand, grip the rider's left leg just above the ankle.

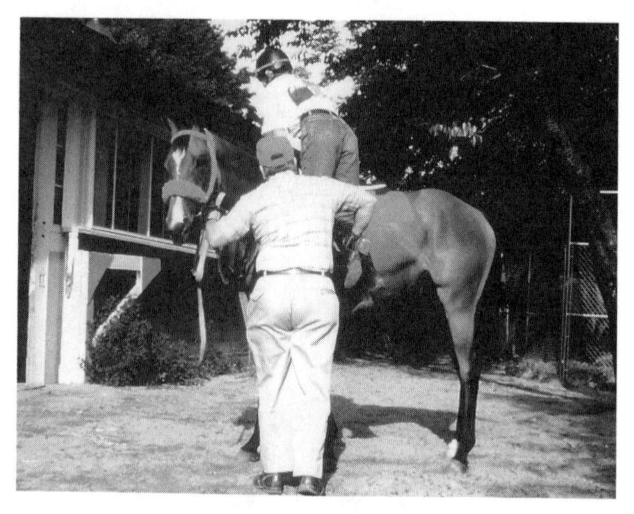

Step 3

Coordinate your efforts with the rider and gently lift the rider up into the saddle.

Step 4

Walk with the horse until the rider has adjusted the girth, stirrups, and reins.

If you are in charge of more than one horse, tack the second horse up now with its prescribed equipment.

Preparing for a Race

Before discussing tacking up for a race, it should be understood that the equipment used for exercising a racehorse may not necessarily be used for racing. For example, a martingale may be used to train a running horse, but it is too restrictive for racing. Any change of racing equipment must be requested by the trainer and approved by the racetrack officials before the horse races. Immediately before a race (in the saddling paddock), a racing official checks each horse to be sure it is racing with the equipment listed in the official's records.

The groom's duty is to hold the horse while it is being tacked up and perhaps to walk it around the saddling paddock. (If you hope to become a trainer then you should learn how to tack a horse up for a race. In fact, it might be part of the practical test that is required for a trainer's license.) Keep the horse as quiet and calm as possible during pre-race tacking. If you are aware of the steps involved in pre-race tacking, particularly the tightening of the girths, you can be prepared to control the horse if it becomes unruly. The saddling paddock at the racetrack can become quite congested with both people and horses before a race. You must be alert to prevent any dangerous situation which might cause injury.

Racing Saddle

Jockeys provide their own saddles. Most jockeys have several different custom-made saddles of different weights. All racing saddles fit all horses; it is the elastic girths which are different lengths. While the racing saddle has the same basic parts as the exercise saddle, the parts are lighter

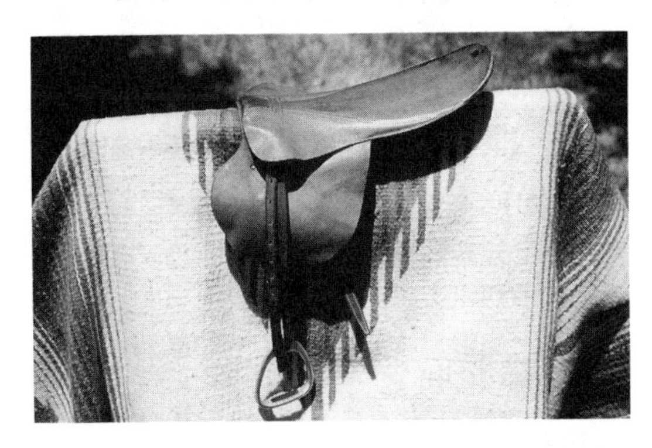

Fig. 7–14. A racing saddle has the same basic parts as an exercise saddle, but it is lighter.

and designed specifically for racing. The weight of the average racing saddle varies from one to three pounds. The tree (and seat) is about 16 inches long and 6 inches wide. The saddle has a single billet, which is secured to an elastic undergirth. A second girth—an elastic overgirth—is also used for additional stability. The stirrups are usually lighter on a racing saddle than those used with an exercise saddle. Racing stirrups are made of lightweight aluminum and weigh approximately four ounces each. The average width of the racing stirrup varies from 3¾ inches to 4½ inches.

Saddle Accessories

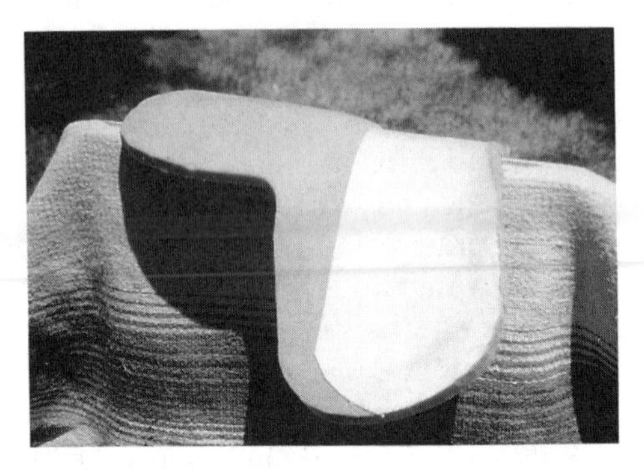

Fig. 7–15. A foam saddle pad protects a horse's back from the racing saddle.

Foam Saddle Pad

This type of saddle pad is usually made of foam rubber about one inch thick that is cut in the shape of the saddle. This pad absorbs concussion and protects the horse's back from the saddle. This protection may be especially important for horses with tender backs.

Fig. 7–16. The numbered saddle cloth is visible to the public.

Numbered Saddle Cloth

Each racehorse is assigned a number which correlates with the program for that race. The number is visible to the public and appears on both sides of the saddle cloth. The saddle cloth is usually made of cotton and prevents chafing on the horse's back and withers from the saddle.

Girth Channel

A girth channel prevents the girth from slipping and from chafing the horse. It is usually made of soft rubber or light foam material that is easily washed after each use. The girth channel is placed under the elastic undergirth. It may also be used under a leather girth to prevent chafing and slippage.

Fig. 7–17. The girth channel prevents slipping and chafing.

Elastic Racing Girths

The racing saddle requires two elastic girths to keep it on the horse during a race. These two girths are called the undergirth and the overgirth. The undergirth is attached to the racing saddle like a regular exercise girth, while the overgirth goes over the seat of the saddle and buckles together behind the front legs over the undergirth.

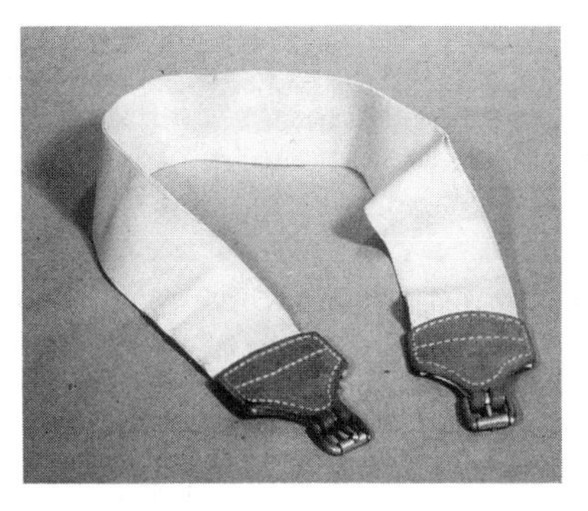

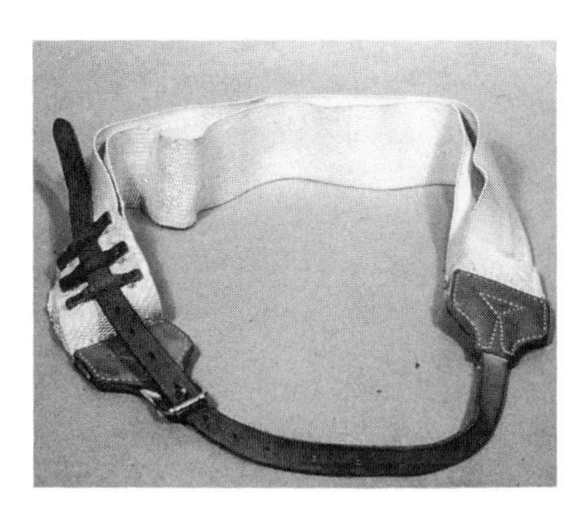

Fig. 7–18. On the left, the undergirth is buckled to the saddle billets. On the right, the overgirth goes over the saddle and buckles together behind the horse's front legs.

Racing Bridle

As mentioned earlier, the same bridle may be used for racing and exercise. However, there may be a different bit that the horse wears only on race day. The groom may have to change bits out or add equipment to the racing bridle, such as a special noseband or blinkers. *(See Chapter 8 for more information on racing accessories.)*

Bridling and Saddling Procedure

This procedure is reserved for a licensed trainer or the trainer's assistant. The trainer usually bridles the racehorse while it is still in the stall. This is done just before the horse is led to the saddling paddock before a race. There, the jockey valet aids the trainer in the tacking procedure.

1. Once in the saddling paddock, the trainer places the foam saddle pad over the horse's back. Some trainers put a damp chamois cloth over the horse's back before putting on the saddle pad, to prevent the saddle from slipping.
2. The trainer then places the numbered saddle cloth over the foam pad and folds the front of the cloth back about six inches.
3. The trainer places the saddle gently on the horse's back over the foam pad and numbered saddle cloth. Before securing the undergirth, the trainer slides the foam girth channel directly under the undergirth.
4. Then the trainer fastens the undergirth to the saddle billet, which holds the girth channel in place.
5. The trainer drapes the overgirth over the saddle just behind the pommel and pulls downward. The jockey valet then hands the trainer the end of the girth for the trainer to buckle as tightly as he/she feels fit. The excess portion of the strap is fastened under the leather keepers on the off side.
6. The trainer then lifts each front leg forward to stretch the skin smooth so that it does not get pinched under the elastic girth.

Saddling the Running Horse for a Race (Fig. 7–19.)

Step 1

Place the foam saddle pad over the horse's back.

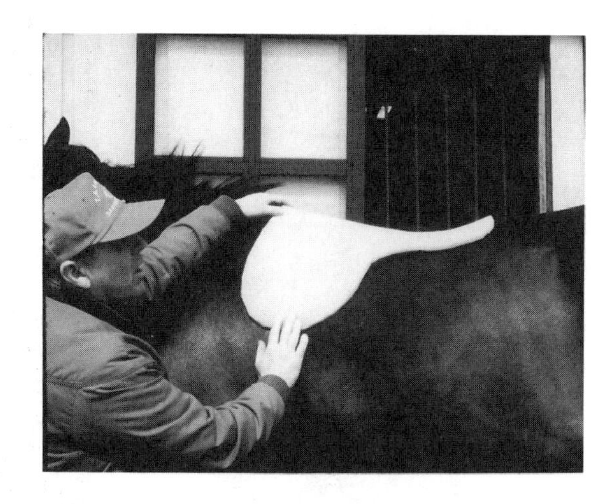

Step 2

Then place the numbered saddle cloth over the foam pad and fold the front of the cloth back about six inches.

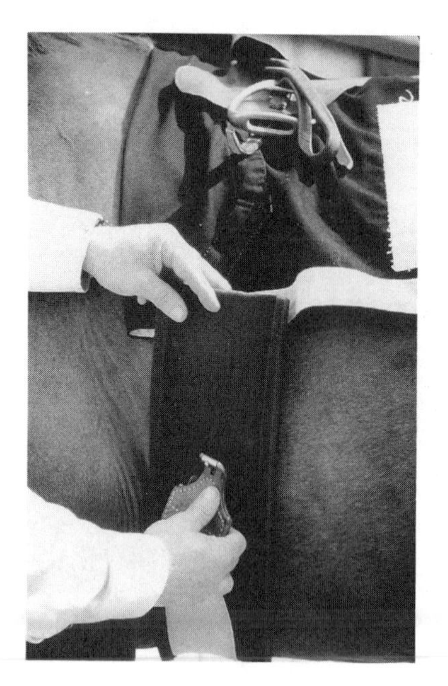

Step 3

Place the saddle gently on the horse's back over the foam pad and numbered saddle cloth. Before securing the undergirth, slide the foam girth channel directly under the undergirth.

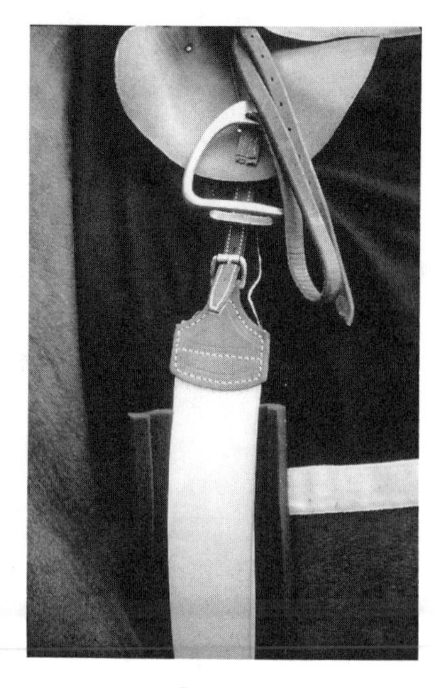

Step 4

Fasten the undergirth to the saddle billet to hold the girth channel in place.

Step 5

Buckle the overgirth, fastening the excess portion of the strap under the leather keepers on the off side.

Step 6

Lift each front leg forward to stretch the skin smooth so that it does not get pinched under the elastic girths.

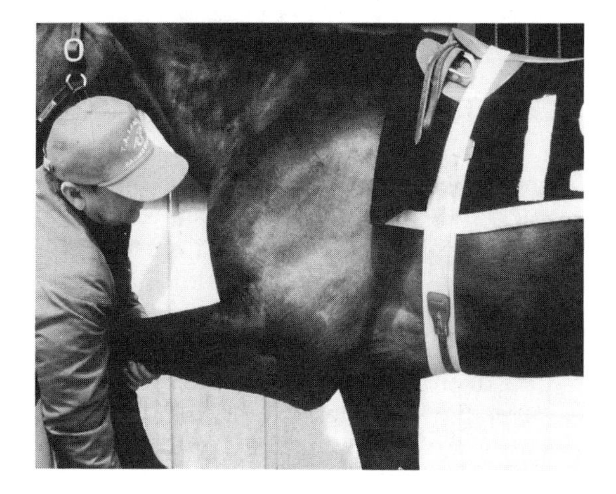

HARNESSING THE STANDARDBRED

The equipment used on the Standardbred can look quite complex. It is considered by most trainers as the cornerstone of harness racing. For a harness horse to trot or pace to its maximum physical ability, various types of harness equipment must be applied. Some of the basic functions of this equipment are:

- control
- protection
- balance
- correcting a faulty gait

Most Standardbred trainers maintain very accurate records on the type of equipment that is worn by each horse in the stable. The reason for this record-keeping is that it allows the trainer to keep up with what equipment best suits each horse to achieve its maximum potential as a trotter or pacer. Accurate records also help a new employee determine which equipment is to be used on each of the horses in his or her care.

Preparing for a Workout or Race

In general, the same equipment used for training is also used for racing. Some exceptions to this rule are the racing sulky, head number, and numbered saddle pad. A racing bike, or sulky, is used exclusively for racing. The jog cart is used only for training. Likewise, head numbers and numbered saddle pads are used for identification during a race. Finally, the overcheck is often omitted during a workout, to allow the horse to relax and to strengthen the neck muscles.

Not all of the equipment is used on both trotters and pacers, however. It is essential that the Standardbred groom become familiar with all of the equipment used on the Standardbred, including its function and proper adjustment.

After the Standardbred is completely groomed, the harness and bridle are placed on the horse while it is still in the stall. The horse is then led outside and hitched to the cart. The hitching procedure usually requires two people; one to hold the horse and the other to position and attach the cart to the horse. The groom then hooks up the overcheck rein, mounts the cart, and drives the horse to the track to exercise his/her horse according to the instructions of the trainer.

Unlike the running horse groom, the harness horse groom is responsible for harnessing and hitching the horse up for a race. Depending on the horse's temperament, it may be harnessed and/or hitched in the barn or in the paddock. If the horse is even-tempered and reliable, it is safe to harness and hitch it up in the paddock. When the horse is ready, the trainer mounts the sulky and takes the horse out for its pre-race warm-up. At post time, the catch-driver, who is sometimes engaged by the trainer to drive the Standardbred horse in the race, mounts the sulky.

Harness

The harness is the piece of equipment which attaches the race-horse to the jog cart or sulky. Each horse usually has its own harness. It is made of leather or synthetic material and includes a saddle, crupper (also spelled "crouper"), girth, and driving lines.

Saddle

The saddle rests on the horse's back just behind the withers. On each side of the saddle are terrets, or rings, through which the driving lines pass. A check hook, or water hook, is attached to the top of the saddle. A saddle pad or a folded towel placed between the horse and the saddle protects the horse's back from chafing. **Note:** Folded towels are not allowed during racing—only saddle pads.

Saddle pads are made of fleece and are often in the stable colors. They fasten to the saddle with strong Velcro straps.

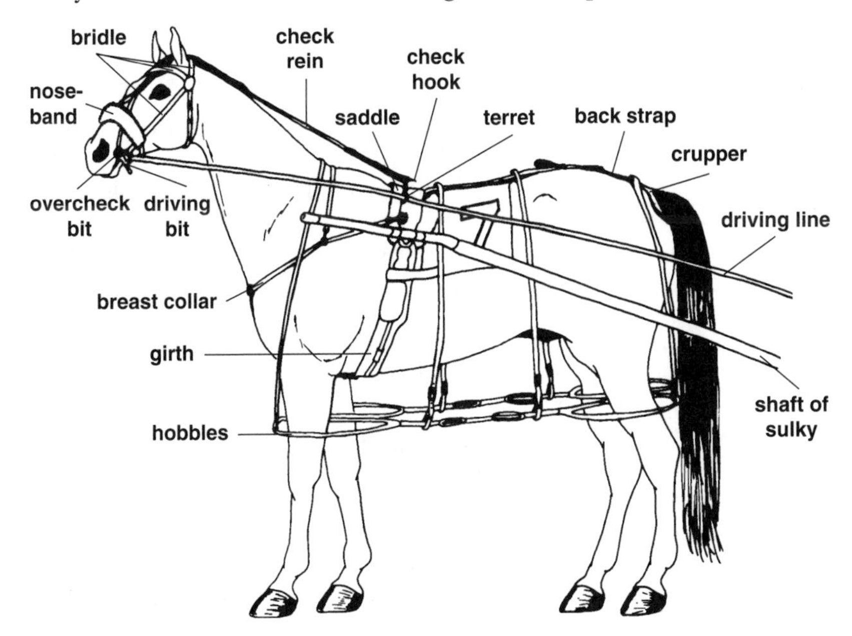

Fig. 7–20. Basic equipment used on the Standardbred.

Crupper

A leather strap attaches the back of the saddle to the crupper. The crupper is a leather loop through which the horse's tail is pulled. It prevents the harness from slipping forward.

Girth

The girth fastens the harness to the horse's body. It is attached to the off side of the saddle, encircles the horse just behind the front legs, and is buckled to a billet on the saddle on the near side. Most girths are made of leather or rubber, and may be shaped or have elastic ends for a snug fit.

Driving Lines

The driving lines are the means of communication from the driver to the horse. They run from hand holds in the driver's hands, through the terrets on the saddle, and buckle into the driving bit in the horse's mouth. If a ring martingale is used, they pass from the terrets, to the martingale, and then to the driving bit. To store the harness and to make the harnessing procedure easier, the driving lines are fastened to the check hook until the driver is ready to mount the cart. Driving lines are made of leather, nylon, or a combination of the two.

Hobbles

This unique piece of equipment is used exclusively on the legs of the pacer. Unless the horse is "free-legged," that is, can pace without hobbles (sometimes called "hopples"), it is a mandatory piece of equipment which prevents the horse from changing its gait.

Hobbles are made of leather, plastic, or nylon. The weight of the hobbles varies from two to three pounds. The two forward oval loops encircle the middle of the forearms, and the two rearward oval loops encircle the middle of the gaskins. The loops are supported by a total of eight adjustable upright straps called hangers. The front and rear loops are connected by an adjustable center strap on each side, thus connecting the front leg and the hind leg on the same side.

The two front hangers are placed over the neck just in front of the withers. The two rear hangers are attached to the back strap.

The four center hangers include forward and rear hangers. The center forward hanger is attached to the back strap just behind the saddle. The center rear hanger is also attached to the back strap near the loin and hangs down just in front of the hip.

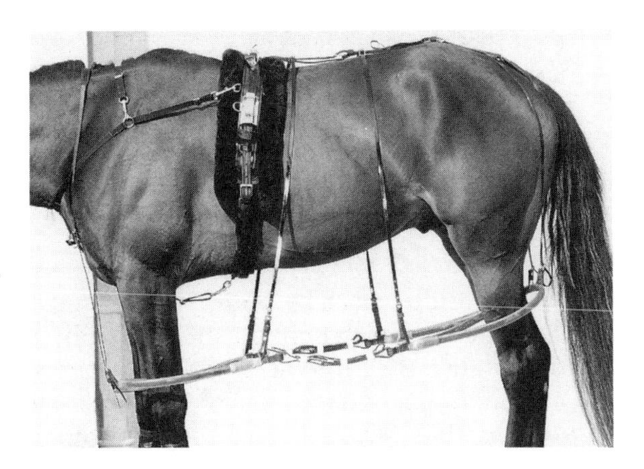

Fig. 7–21. There are eight hangers which support the hobbles.

Adjustment of the hobbles varies with the size and gait of the horse. If the front hobble loops are too low, the horse may hit them with the bottom of its feet; if the rear hobble loops are too low, they ride up and down the horse's hind legs with every stride and "burn" or chafe the horse. On the other hand, if the front hobble loops are too high, the center straps slap the horse on the belly with every stride; if the rear hobble loops are too high, they restrict the horse's ability to pace and can cause it to make a break. If the center straps are too long or too short, the horse may also make a break. Adjustment considerations may also include racing conditions—such factors as the circumference of the racetrack and the condition of the track surface.

Trotting Hobbles (Half Hobbles)

This type of hobble is rarely used, but it is worth mentioning. Trotting hobbles are basically used on trotters to help them maintain the trotting gait. They are placed on the front legs in the same manner as the pacing hobbles. The loops encircling

Fig. 7–22. Trotting hobbles are occasionally used on trotters.

the forearms are joined by a connecting strap running through a pulley. The pulley is supported by a strap which is attached to the back strap in the center of the horse's back. Another strap runs from the back of the pulley, between the horse's hind legs, and attaches to the cart.

Standardbred Bridles

Most bridles worn by trotters and pacers consist of two cheekpieces which hold the driving bit, browband, crownpiece, throatlatch, the front piece of the overcheck which holds the overcheck bit, and the check rein. Sometimes a horse is exercised without the overcheck. To remove it from the bridle, unbuckle the overcheck, grasp the front piece, and pull it and the check rein free from the bridle. When not in use, the overcheck is stored in the tack room on a hook.

There are several different types of bridles used on Standardbreds. Some of the most popular harness racing bridles are discussed in the following sections.

Fig. 7–23. The horse on the left (number 4) is wearing an open bridle. The horse on the right (number 3) is wearing a closed bridle.

Open Bridle

The open bridle allows the horse to see all around. It is generally used on relaxed, quiet, or seasoned horses. Occasionally, it is used on a young, inexperienced horse during the initial training stages to allow the horse to become familiar with everything around it.

Closed Bridle

This type of bridle is similar to the open bridle except for the presence of blinds on each side of the bridle. The blinds limit the horse's vision so it is unable to see things to the side or directly behind it. (The horse is unable to see the driver, cart, or other horses.) A closed bridle encourages a horse to focus its vision on what is ahead. Adjust each blind so that it is over the middle of the horse's eyeball.

"Kant C Bak" Bridle

As the name indicates, this bridle is designed so that the horse is unable to see directly behind it. It is useful on horses that become nervous or frightened when other horses are driven up behind them during a race or training session.

Fig. 7–24. "Kant C Bak" bridle.

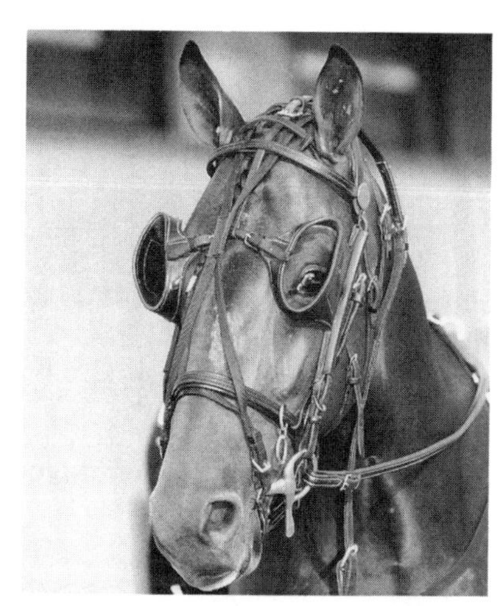

Fig. 7–25. "Peek-A-Boo" bridle.

"Peek-A-Boo" Bridle

This type of bridle, also called a Telescopic bridle, is very restrictive. The rounded blinds limit the horse's forward and downward vision as well as its rearward vision.

Head Number

Each horse is assigned a number which correlates with the program for that race. Once in the paddock, the groom will find the plastic head number in the stall assigned to the horse. The groom clips it to the crownpiece of the horse's bridle.

Numbered Saddle Pad

The saddle pad used in a race has an extension which has the horse's number on both sides. It is usually made of an oilskin material. Although it does go on under the harness, it does not prevent chafing on the horse's back and withers—it is for identification only.

Training or Jog Cart

This type of cart is used exclusively for training the Standardbred trotter or pacer. The parts of a jog cart include:

- shafts
- mud apron
- wheels
- stirrups
- seat
- wheel axle

A training jog cart is generally made of either wood, aluminum, and/or steel, depending on the personal preference of the Standardbred trainer. The shafts of the modern jog cart are usually attached to the saddle with the quick hitch device (discussed later) and secured with safety straps. A mud apron is only attached to the cart (or sulky) when the track is muddy. Its purpose is to prevent mud from flying into the driver's face. It fits under the horse's tail and is tied to each side of the cart.

Fig. 7–26. A jog cart is used for exercising the Standardbred racehorse.

Fig. 7–27. A racing sulky is lighter than a jog cart.

Racing Bike or Sulky

The racing sulky is used exclusively for racing. It has the same basic parts as the jog cart, and it is attached to the saddle with the quick hitch device.

Most Standardbred trainers are very particular about construction of their racing sulky. The racing sulky can be custom-made to meet the individual tastes of the trainer. In fact, some racing sulkies are custom-built to suit a particular horse. They can be specially designed to meet the needs of a horse which may be above average in size, or one that has an unusual way of going.

There are some notable differences between the racing sulky and the jog cart which should be mentioned at this point. A sulky is lighter than a jog cart, even though it has a larger wheel base, wider wheels, and is taller. This is because sulkies are made of chrome alloy steel with no wooden parts (jog carts often have wooden shafts). Also, the stirrups on the jog cart are usually located directly in front of the driver on the cross bar section. The stirrups on the racing sulky are attached to the inside of the shafts. The position of the driver is much closer to the rear end of the horse in a racing sulky than a jog cart. Finally, sulkies are more expensive than jog carts.

Harnessing and Bridling Procedure

Before harnessing the trotter or pacer, groom the horse thoroughly and apply the necessary boots or bandages. The harness is usually hanging from the tack room wall. With the horse tied or cross-tied in its stall, or being held by an assistant, put the harness on the horse.

Note: Bridling the Standardbred racehorse is much the same as bridling the running horse. For this reason, only an abbreviated version of bridling the Standardbred is given here. *(For a more detailed description of the bridling procedure, see Figure 7–12.)*

Use the following steps to correctly harness and bridle the Standardbred racehorse for a workout or race. This procedure is illustrated in Figure 7–28.

1. Standing on the near side, snap the buxton breast collar together around the horse's neck. If a buxton breast collar is not used, skip to step 2.
2. Place the pad or folded towel over the horse's back just behind the withers.
3. Set the saddle in the middle of the horse's back, behind the pad or towel. This should leave enough slack in the back strap to slip the crupper under the tail.
4. Slide the loop of the crupper over your right wrist, and with

your right hand, grasp the tail near the end of the bone. Gently pull the entire tail through the loop.

5. Pull the back strap and crupper forward as far as they will comfortably go.
6. After the crupper is properly set under the tail, pick up the saddle and move it forward so that it rests just behind the withers on the pad or towel. Do not drag the saddle forward.
7. Buckle the girth onto the saddle billets. Tighten the girth from the near side. The saddle should fit snugly on the horse's back—not so tightly that it restricts the horse and makes it uncomfortable, but not so loosely that it slips backward or sideways. Then snap the standard or buxton breast collar into place on the saddle and girth.
8. Put the hobbles on the horse by lifting one foot at a time and placing it through the appropriate loop. Adjust the hobbles according to the instructions of the trainer. If hobbles are not used, skip to step 9.
9. Grasp the top or cheekpieces of the bridle in your right hand over the horse's forehead. Holding the bit in your left hand, slip your thumb into the horse's mouth. When the horse opens its mouth, slip the bit inside, keeping tension on the crownpiece with your right hand so the bit does not slide back out when you let go. (If the overcheck is used, you must slip both bits into the horse's mouth at the same time.) With your left hand now free, slip the crownpiece over the off ear first and then fold the near ear slowly and cautiously forward to slip it under the crownpiece.
10. After the bridle is in place, buckle the throatlatch loosely under the throat.
11. Fasten the driving lines to the rings of the driving bit.

Hold both ends of the driving lines and lead the horse to the hitching area. The horse is ready to be hitched up.

Harnessing and Bridling the Standardbred (Fig. 7–28.)

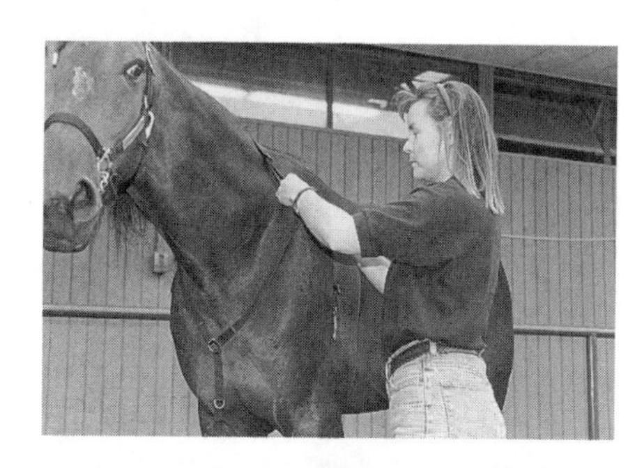

Step 1

Snap the breast collar around the neck, if applicable.

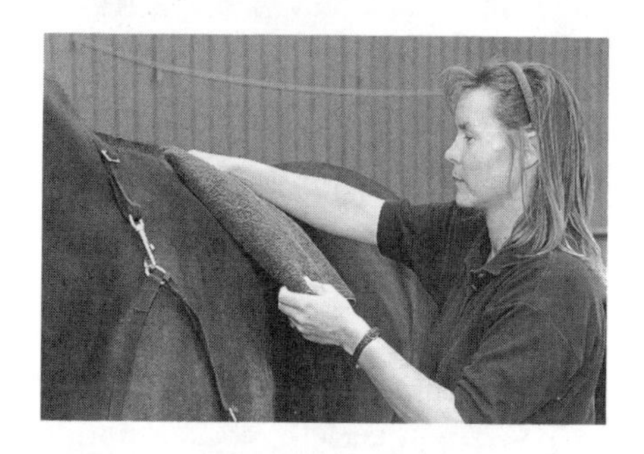

Step 2

Place the pad or folded towel just behind the horse's withers.

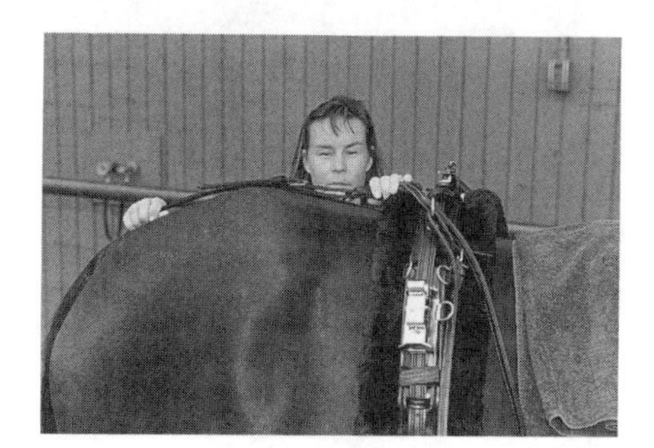

Step 3

Set the saddle in the middle of the back, behind the pad or towel.

Step 4

Pull the tail through the crupper.

Step 5

Pull the back strap forward as far as it will comfortably go.

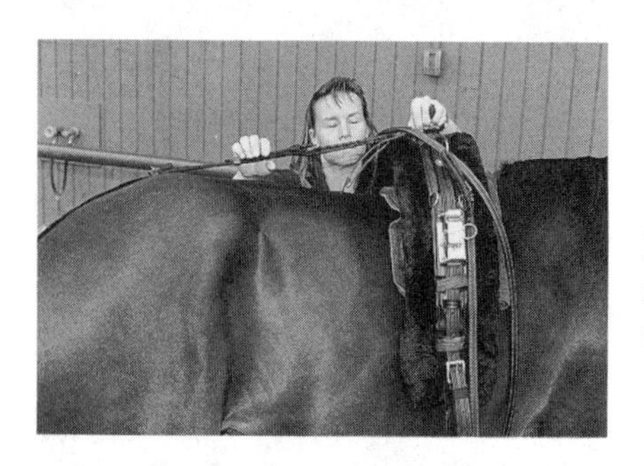

Step 6

Pick up the saddle and move it forward to rest on the pad or towel.

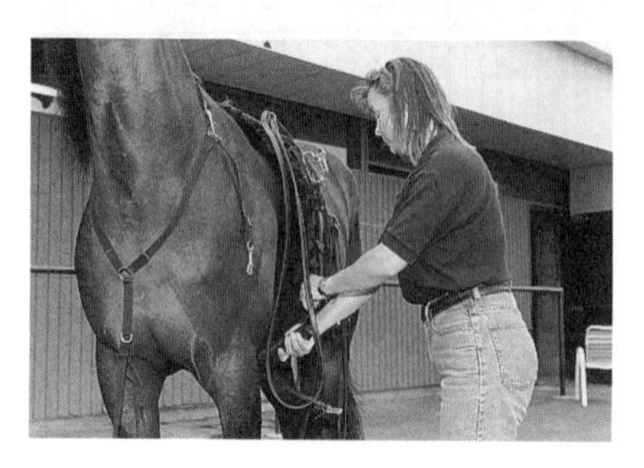

Step 7

Adjust the girth from the near side.

Step 8

Gently lift each foot in turn and place it through the appropriate loop of the hobble.

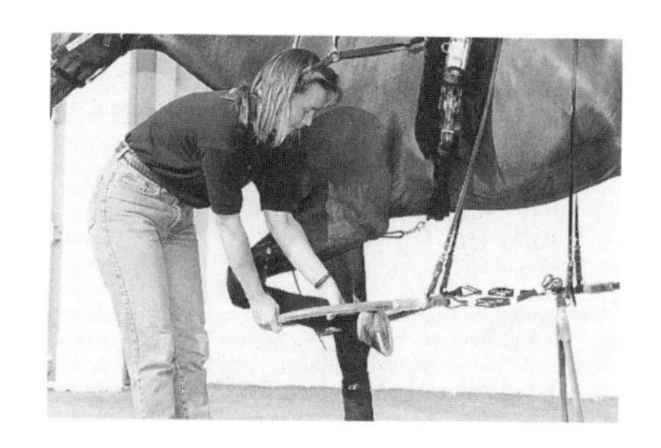

Step 9

Holding the bit in your left hand, and the top (or cheekpieces) in your right, slip the bit into the mouth.

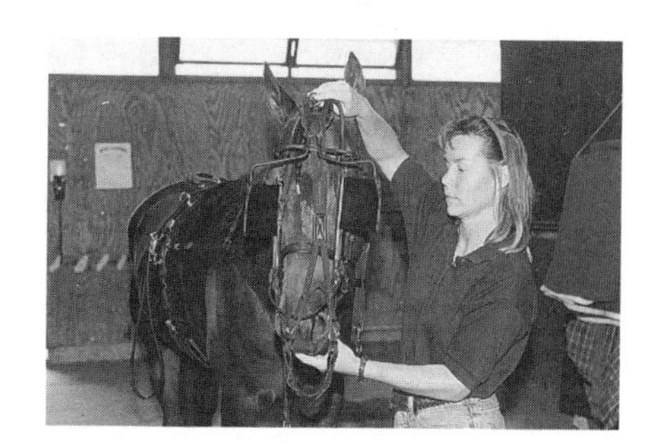

Step 10

Buckle the throatlatch loosely.

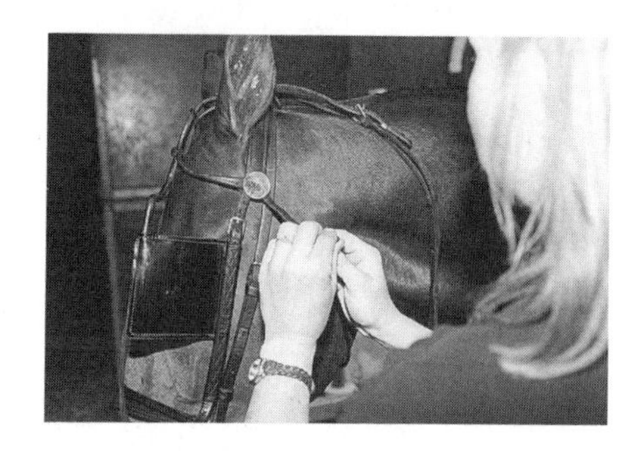

Step 11

Fasten the driving lines to the rings of the driving bit.

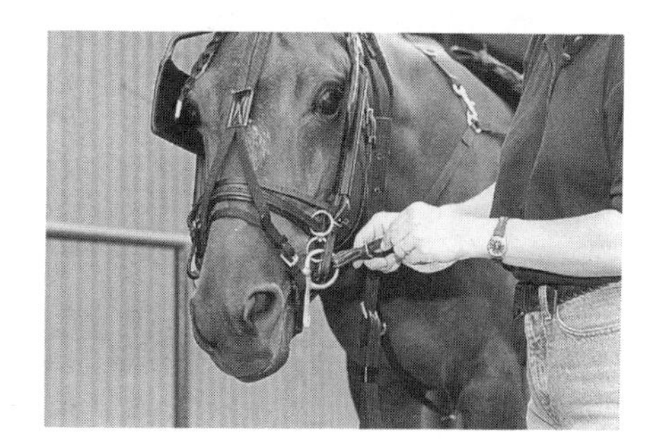

Hitching Procedure

While it is possible for one person to hitch a horse up, it is much safer to work with an assistant. One person stands on the near side of the horse, by its head, holding the driving lines. The other pulls the jog cart or sulky up to the horse.

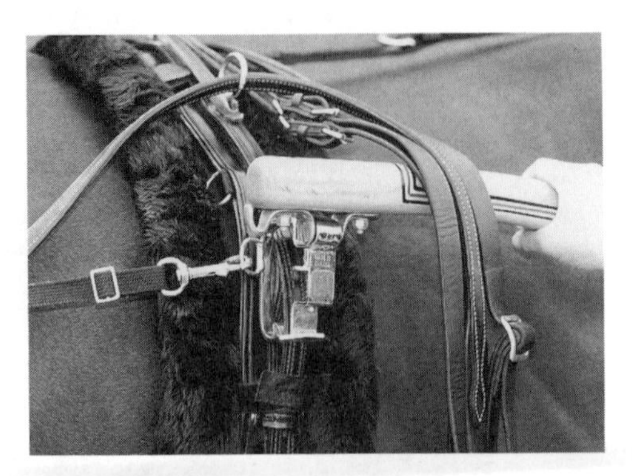

Fig. 7–29. The quick hitch device is a system of brackets and spring locks.

The most popular hitching method used today (for both the jog cart and sulky) is the "quick hitch" system. This system consists of a metal bracket with a spring lock permanently attached to each side of the saddle. The second part of the quick hitch device consists of a metal bracket with one or two setting prongs permanently affixed to the tips of the shafts of the jog cart or sulky.

One advantage of this system is that it allows the shafts of the cart to be attached to the saddle in the same position day after day. It is also useful to the trainer and driver when they must release the shafts of the cart quickly in the event of an emergency.

To attach the cart to the horse's saddle, merely line the shafts up and lock them into place on each side of the saddle. To do this, use the following steps as illustrated in Figure 7–30:

1. One person holds the horse from the near side while the other person slowly pulls the cart toward the horse from the rear. Never back a horse into the cart.
2. When the shafts are properly lined up on each side of the horse, they are locked into place on the saddle by means of the quick hitch method described above.
3. When the driver's ends of the driving lines are released from the check hook, the horse is harnessed and hitched up for a training session or race. At this point, the driver may decide to braid the tail with a tail rope. *(See Chapter 4.)*

If it is used, the check rein should always be attached to the saddle before the groom or trainer mounts the cart. Gently lift the horse's chin with your left hand and slip the check rein onto the check hook

Hitching Procedure (Fig. 7–30.)

Step 1

Have an assistant pull the cart up to the horse.

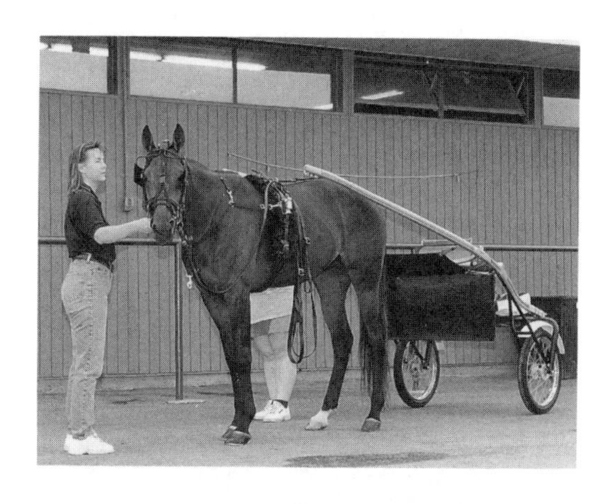

Step 2

Lock the shafts of the jog cart or sulky in place on the saddle.

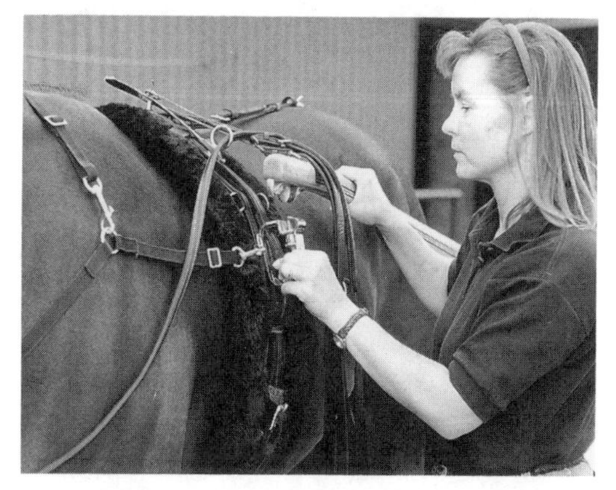

Step 3

Release the driving lines and the horse is ready to go to the track.

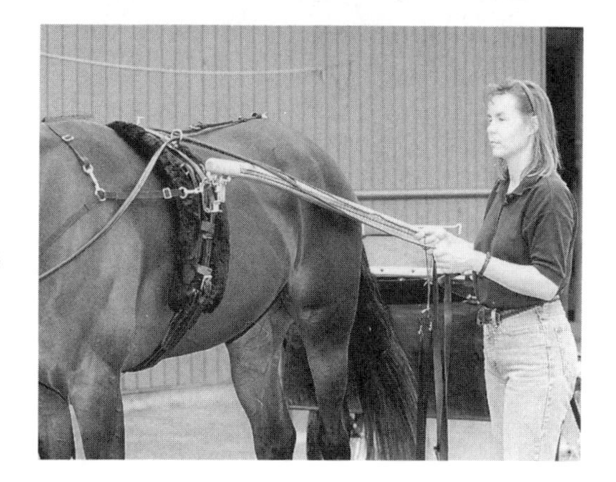

Fig. 7–31. Sit down on the seat with both legs hanging on the left side.

Fig. 7–32. Swing both legs around to the right, toward the front of the cart.

Fig. 7–33. Insert your middle two fingers into the hand holds.

with your right hand. Do not lift the horse's head by pulling on the check rein as it may hurt the horse. Most harness horse trainers check the horse's head up until the nose is even with the withers.

Mounting Procedure

Once the horse is properly harnessed and hitched to the cart, with the check rein attached, the groom or trainer can mount the cart. While holding the driving lines for control, the driver sits down on the seat of the cart or sulky with both legs hanging on the left side of the cart behind the left wheel. The driver swings both legs around to the right (toward the front of the cart), putting one foot in each stirrup. Then the driver inserts his/her middle two fingers into the hand holds of the driving lines. To dismount from the cart, the driver simply reverses this procedure.

Training Procedure

The average harness horse may race one day each week with one day off after racing. A typical weekly training pattern for a Standardbred racehorse may consist of five days of jogging, one day of training, and one day of rest.

On those days scheduled for jogging it is the duty of the groom to exercise the horse on the track and then return to the barn to cool the horse out. On those days designated for training both groom and trainer participate jointly in exercising the horse. The groom may be responsible for warming the horse up at a two-mile jog. Then he or she turns the horse over to the trainer for the first of several one-mile training sessions, or "heats" for the day. Between heats, the trainer returns the horse to the barn. Then the trainer dismounts from the cart and unhooks the check rein. With the aid of the groom, the cart is unhitched and is safely backed off the horse.

Lead the horse to its stall and remove the harness and bridle. Place the halter on the head and allow the horse to drink a few swallows of water at this time. Sponge the horse off with a damp sponge and then rub it thoroughly with several rub rags. If it is a cool day, blanket the horse to prevent a chill. Then allow the horse to drink a little more water. Remove the boots and hobbles and wipe them thoroughly, and perhaps dust them with talcum powder. This is very important as wet boots can slip down on the horse's legs and chafe the horse. (The harness, bridle, and all other equipment are also wiped with rags if the horse is sweating alot.) Standardbreds have a much greater problem with chafing than running horses because of the extensive equipment and the fact that they train in heats. To minimize chafing, it is important to keep the horse and equipment as clean and dry as possible.

After about 45 minutes, harness and hitch the horse up again and return to the track for another mile heat with the trainer. One training session may include three heats.

Switching Control of the Horse

On a training day, after the horse has been jogged, the trainer takes the horse back out for a training heat. As the groom pulls the horse up to a stop, the trainer approaches the horse and holds its head. The groom then dismounts from the cart on the left side (using the reverse procedure of mounting). With the driving lines in hand, the groom walks toward the horse's head. At this point, the trainer takes the lines from the groom, and the groom holds the horse's head. The trainer then walks toward the horse's rear with the driving lines in hand and mounts the cart. In this way, the horse's head and driving lines are always under control.

SUMMARY

Tacking up correctly for every workout or race is an important skill. The groom must make sure the appropriate tack is put on and that it fits the horse correctly and comfortably. Poorly-fitted tack can cause a horse to perform badly. The following chapter discusses many of the accessories with which a groom must also be familiar when tacking up a racehorse.

Chapter

8

Racing & Training Accessories

Racing and training accessories play an important role in obtaining top performance from a racehorse. Most of the accessories that are discussed in this chapter fulfill at least one of the following three functions:

- improving a horse's racing ability
- preventing injury
- correcting bad habits

The groom has several responsibilities when it comes to equipment: knowing the correct article to use, fitting the article correctly to the

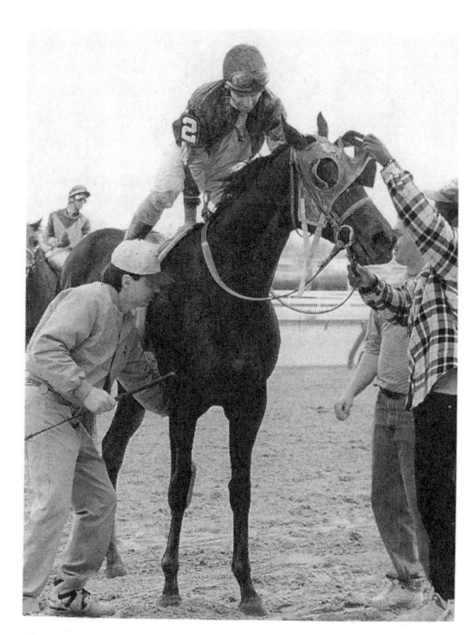

Fig. 8–1.

horse, and removing and cleaning the article after exercise or a race. Sometimes, the groom may not be responsible for cleaning or maintaining the equipment, but it should still be treated with utmost care. In this chapter, the groom will become familiar with the most common accessories and their roles as correction or training aids.

MARTINGALES

The basic function of a martingale is to prevent the horse from throwing its head up. (This bad habit makes it difficult for the rider or driver to control the horse.) The martingale keeps the horse's head in a normal position and allows the rider or driver more leverage, resulting in more control. Many running horse trainers use a martingale as a training device, but not for racing, as it is too restrictive when a horse is fully extended. A martingale is more likely to be used as racing equipment on a harness horse.

Martingales are adjustable to different sized horses. Many are put on the horse before the girth is tightened, as the girth is often run through the martingale strap before being buckled. However, some martingales just snap on to an already tightened girth. The following sections discuss the most common martingales used on racehorses.

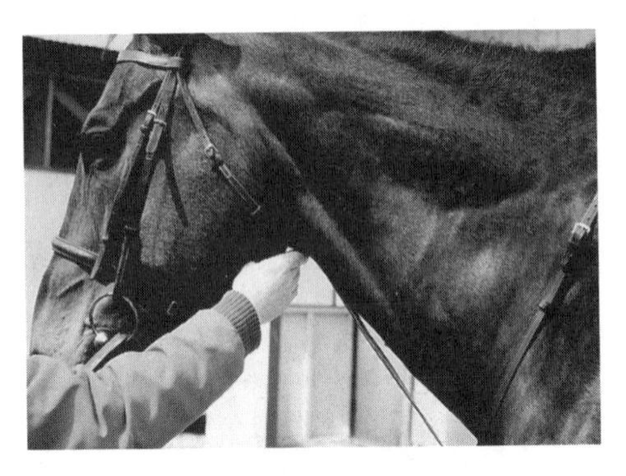

Fig. 8–2. The standing martingale should be adjusted loose enough at the girth so that the leather strap will reach the horse's throatlatch when pulled tight. After it is properly adjusted, it may be fastened to the bottom of the noseband.

Running Horses

Standing Martingale

This martingale consists of a neck strap and a strap that runs between the legs and attaches to the leather exercise girth. On the other end, this single leather strap attaches to the bottom of the noseband. The standing martingale is the most restrictive type of martingale, as it can be adjusted to hold the horse's head down below its normal position. To prevent this restriction, it is important to adjust the martingale loose enough at the girth so that the leather strap will reach the horse's throatlatch when pulled tight. After it is properly adjusted, it may then be fastened to the bottom of the noseband.

Running Martingale

This type of martingale consists of a strap that runs between the horse's front legs and attaches to the girth, and a split leather strap (separate from the neck strap) with a ring on the end of each strap. The reins are passed through these rings.

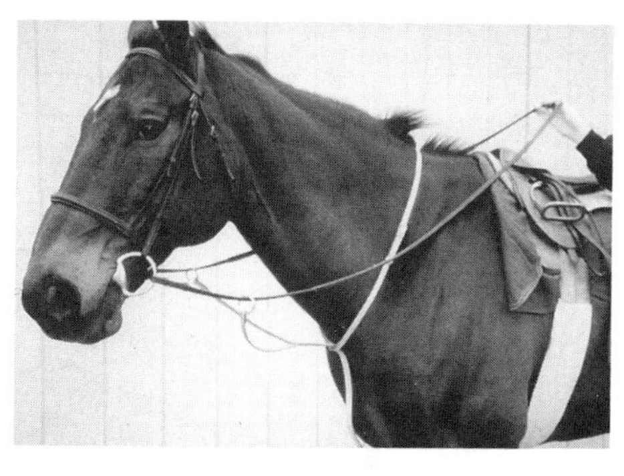

Fig. 8–3. A running martingale.

Solid Yoke Running Martingale

This type of running martingale has a strap attached to the middle of the girth that goes up between the front legs, attaching to the neck strap, which rests on the withers. However, instead of a split leather strap like the traditional running martingale, this martingale has a solid V-shaped yoke. There are two rings (one on each side of the "V") which the reins are passed through.

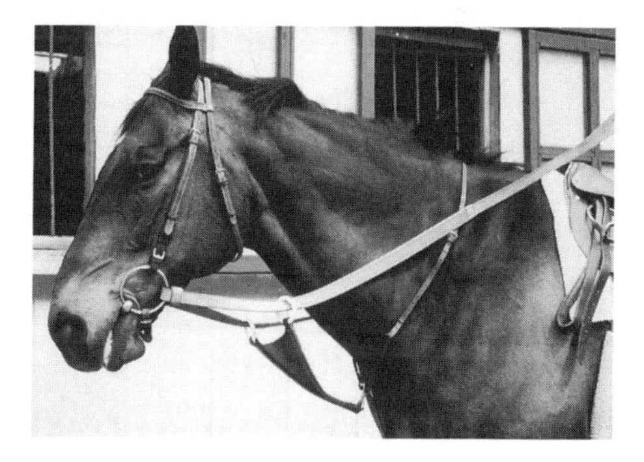

Fig. 8–4. A solid yoke running martingale.

Harness Horses

As with martingales used on runners, the basic function of a martingale for harness horses is to prevent the horse from carrying its head too high. The martingale maintains the racehorse's head in the proper position and prevents the horse from tossing its head up,

Fig. 8–5. Top left: a standing martingale. Bottom left: a running or ring martingale. Right: a split martingale.

which often occurs at the starting gate. There are three basic types of martingales used on the harness horse—the running or ring martingale, the standing martingale, and the split martingale.

Standing Martingale

The standing martingale consists of a single strap running between the front legs, which is snapped to the girth or breast collar. The other end of the strap is attached to the center ring of the head halter. Again, the standing martingale is the most restrictive aid for holding the horse's head down and in position.

Running or Ring Martingale

This type of martingale is used in conjunction with the driving lines to create a downward pull on the horse's mouth. It has a single strap running between the front legs, which is connected to the girth. This strap divides into two straps, each with a ring on the end. The driving lines are passed from the bit, through the rings on each side of the neck, through the terrets, and back to the driver's hands. A second strap is placed over the neck for support. This strap ends with a metal snap on each side of the neck, each snap being attached to one of the rings as well.

Split Martingale

This type of martingale has a single strap attached to the girth in the same manner as the running martingale. The last few inches of the other end of the strap is split into two—one section is attached to a metal ring on the left side of the head halter and the other is attached to a metal ring on the right side. Because of its configuration, the split martingale helps to keep a horse's head straight and down at the same time.

NOSEBANDS & HEAD HALTERS

The noseband and the head halter are similar pieces of equipment, but nosebands are primarily used on running horses and head halters are used specifically on harness horses. The purpose of both the noseband (sometimes called a caveson) and head halter is to keep the racehorse's mouth closed. By keeping the mouth closed, the rider or driver has better control over the horse. The noseband or head halter also prevents the horse from playing with the bit(s).

The running horse groom places the noseband on the horse's head as part of the bridle, but leaves it unbuckled until the bit is in the mouth and the bridle is fitted properly on the head. Then the noseband is buckled underneath the jaw and/or chin. (It should also be unbuckled before the bridle is removed.) The harness horse groom puts the head halter on and fastens the nosepiece before putting the bridle on. Both the noseband and head halter are properly placed on the head when they do not, in any way, interfere with the effectiveness of the bit(s).

Not all racehorses need nosebands or head halters, and there is no sense in using this equipment if it is not necessary. However, many racehorses require the use of this equipment for top performance.

Running Horses

There are various types of nosebands that are used on running horses. Some of the most common types are standard, figure-eight, dropped, flash or hinged dropped, and Kineton nosebands.

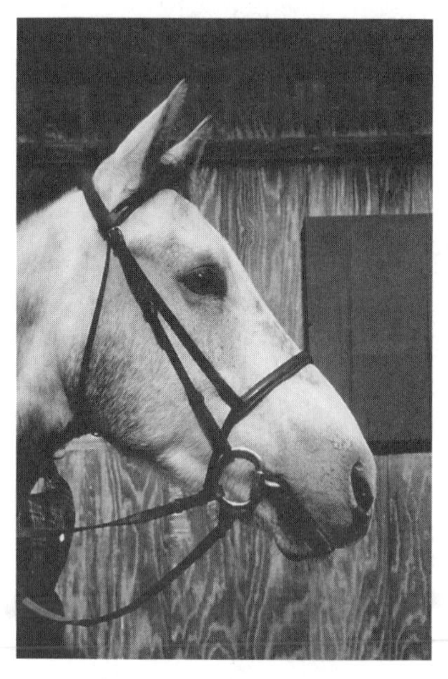

Fig. 8–6. Standard Noseband.

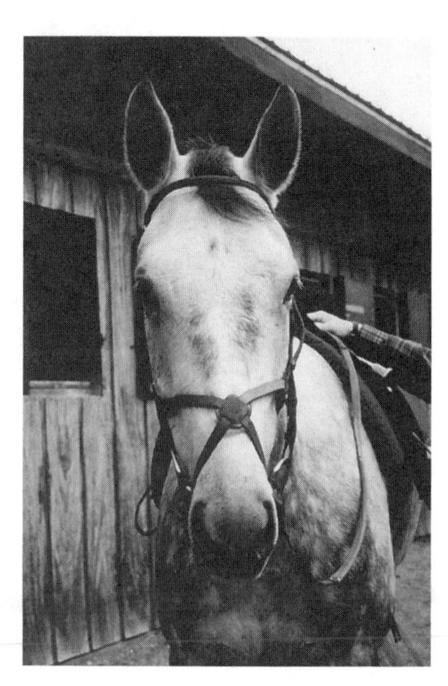

Fig. 8–7. Figure-Eight Noseband.

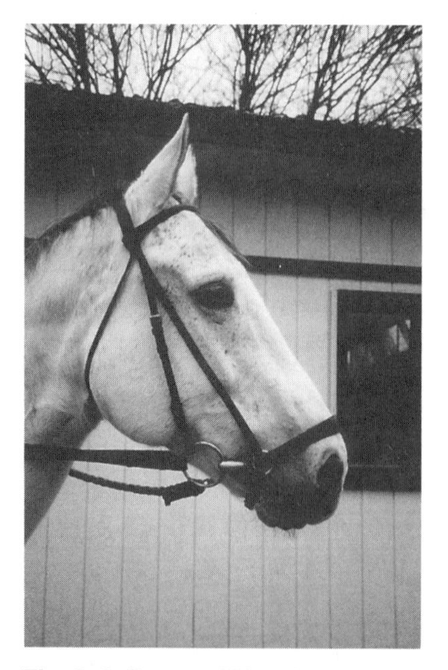

Fig. 8–8. Dropped Noseband.

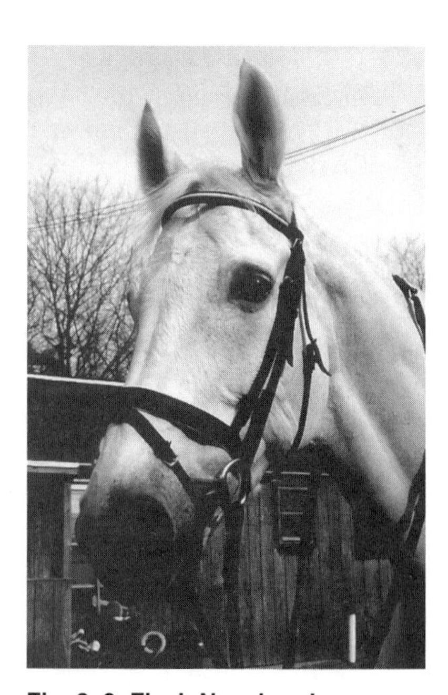

Fig. 8–9. Flash Noseband.

Standard Noseband

This type of noseband may be adjusted high or low by the buckle on the left cheek strap of the noseband. The standard noseband should be fitted about two inches down from the horse's protruding cheekbones.

Figure-Eight Noseband

The figure-eight noseband is so named because it resembles the number eight (8). A figure-eight noseband is useful on a horse that constantly plays with its tongue and gets its tongue over the bit. It is fastened by one strap buckled under the jaw and by another strap buckled in front of the bit under the chin. A third strap is fastened behind the ears, and a fleece-lined leather disc lies on the bridge of the nose to protect it from the pressure of the adjustment straps.

Dropped Noseband

This type of noseband exerts pressure on the sensitive end of the horse's nose, allowing the rider more control. This type of noseband is often used on "pullers," or horses that are difficult to slow down or stop. By fastening in front of the bit, the dropped noseband also keeps the horse's mouth closed and prevents the horse from playing with its tongue during racing or training.

Flash (Hinged Dropped) Noseband

This noseband is a variation of the figure-eight noseband. A "flash," or second strap, is attached at the top of the regular noseband, which rests on the bridge of the nose. The strap runs in front of the bit and fastens under the chin, keeping the horse's mouth closed and preventing the tongue from sliding over the bit.

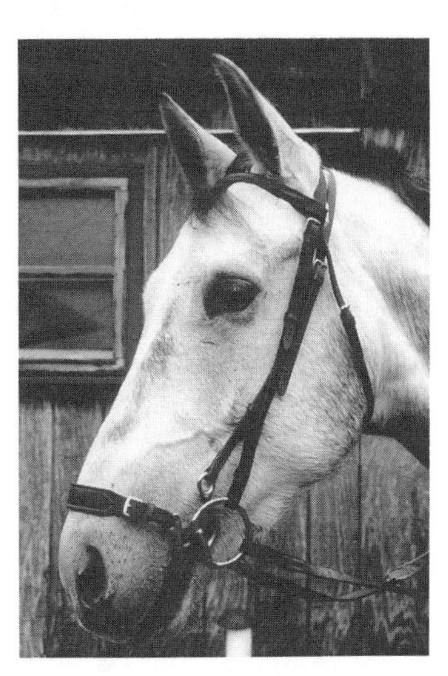

Fig. 8–10. Kineton Noseband.

Kineton Noseband

Like the dropped noseband, this type of noseband is used to control horses that are hard "pullers." When the rider pulls on the reins, the nosepiece places pressure just above the nostrils, which causes pain and affects the horse's breathing. However, the Kineton noseband is even more severe because it also places pressure on the horse's poll, forcing the head down. This renders even the most difficult horse controllable. For proper fitting, the steel loop fittings of the Kineton noseband should be placed directly behind the bit. (This noseband is used primarily for training, as it would be too restrictive if used in a race.)

Harness Horses

The head halter is a piece of racing equipment used on the harness horse to keep the horse's mouth closed during racing or training. The head halter rests underneath the bridle and buckles under the chin. There are two common types of head halters: standard and figure-eight.

Fig. 8–11. A standard head halter should be adjusted high enough on the nose so it does not interfere with the driving bit or the lines.

Standard Head Halter

The standard head halter has a noseband which encircles the horse's muzzle, keeping the mouth closed so the driver can control and steer the harness horse more effectively. It should not be confused with the regular halter, which is used on all types of racehorses around the stable and while walking or grazing. The noseband buckles under the jaw and should be adjusted high enough on the horse's nose so that it does not interfere with the driving bit or the lines. This type of head halter can be worn by both the trotter and pacer.

Figure-Eight Head Halter

The figure-eight head halter is much the same as the figure-eight noseband, which was discussed earlier regarding its application to runners. It buckles under the jaw and in front of the driving bit. The figure-eight head halter is effective in keeping the horse's mouth closed, but its use is not as widespread among harness horse trainers as the standard head halter.

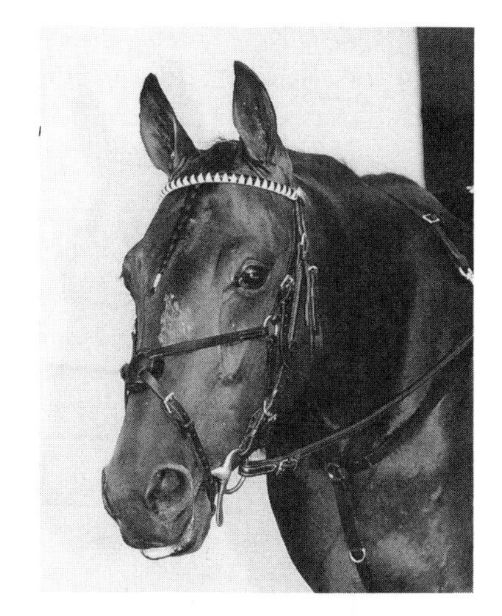

Fig. 8–12. A figure-eight head halter.

SHADOW ROLLS

While a shadow roll serves to keep the horse's mouth closed like a noseband or head halter, its most important function is to prevent the horse from spooking at "shadows" during a race or workout. Shadow rolls prevent the horse from seeing shadows, wet spots, dirt clods, etc. directly in its path or to the sides and rear. For instance, shadow rolls are used on most pacers, as they have a tendency to spook due to the hobbles they wear. Horses that spook like this or attempt to dodge or jump over shadows during a race are likely to lose valuable strides, possibly even injuring themselves.

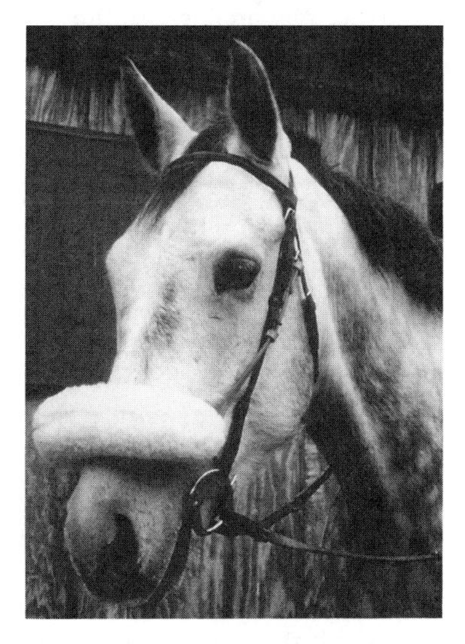

Fig. 8–13. A standard shadow roll fitted to a running horse.

Fleece Shadow Roll

The standard shadow roll is made of fleece and is used frequently on horses running on dirt or turf (grass) courses. This shadow roll restricts what the horse can see directly ahead of it.

The standard shadow roll is also a vital piece of equipment to the trotter and pacer. Like that used on the running horse, its main function is to restrict the vision in such a way that the horse is unable to see shadows, debris, or any other distraction directly in its path that may cause it to spook or break gait.

Brush Shadow Roll

This type of shadow roll is used only on harness horses. It is similar to the standard fleece shadow roll, but instead of a fleece noseband, a "brush" or series of upright, stiff fibers extend over the nose.

Fig. 8–14. A brush shadow roll is used only on harness horses.

A shadow roll should be attached to the bridle before the bridle is put on. The noseband portion of the shadow roll fits over the nose (just under the eyes) and fastens under the jaw. **Note:** It is important to adjust the shadow roll after the head is checked up on a harness horse. This ensures that the shadow roll is not blocking the horse's vision straight ahead.

BLINKERS

Blinkers are special hoods with cups over part of the horse's eyes that restrict what the horse can see to the sides and rear. They are commonly used to correct a variety of problems affecting performance. For example, some horses shy or react negatively to other horses that approach them from behind in a race. Others prefer to stay *with* a pack of horses, instead of moving out in front of them. Neither practice wins a race.

Blinkers are usually made of a nylon hood with two holes for the ears and either leather or plastic cups around the eye holes. On running horses, they are placed on the head over the bridle, but the side flaps are slipped under the cheekpieces before they are fastened under the jaw with Velcro tabs. For harness horses, caretakers must put the blinkers on under the bridle to avoid interfering with the check rein.

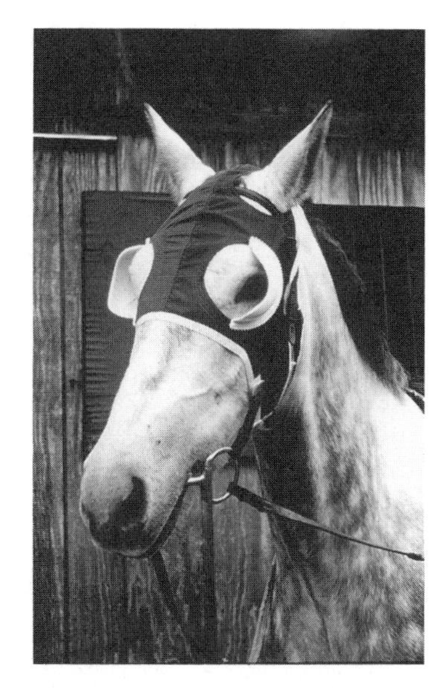

Fig. 8–15. Some trainers trim back the blinker cups to suit a particular horse.

The trainer decides if a horse needs blinkers and what type of blinker cups should be used. As a general rule, the trainer should use the least restrictive blinkers that correct the performance problem.

Some trainers will buy blinkers and trim one or both cups back themselves to suit a particular horse. Or, a trainer might cut a small opening in the back of both blinker cups to enable the racehorse to see horses coming up behind it on a limited basis. The theory behind this practice is that seeing another horse coming up behind provokes the horse to dig in and go faster.

It is important to be familiar with the different types of blinkers used on racehorses, as you will be responsible for putting them on the horse before exercise, and possibly before a race. Most running horses are walked to the saddling paddock with their bridles and blinkers already on before they are saddled.

Running Horses

Some of the most common types of blinkers used on running horses are standard cup, duck bill cup, and closed cup blinkers.

Standard Cup Blinkers

On standard cup blinkers, both cups are of equal size. They are available in various cup sizes (quarter cup, half cup, and full cup) according to how much vision is limited. For instance, a full cup would restrict the horse's vision more than a quarter cup.

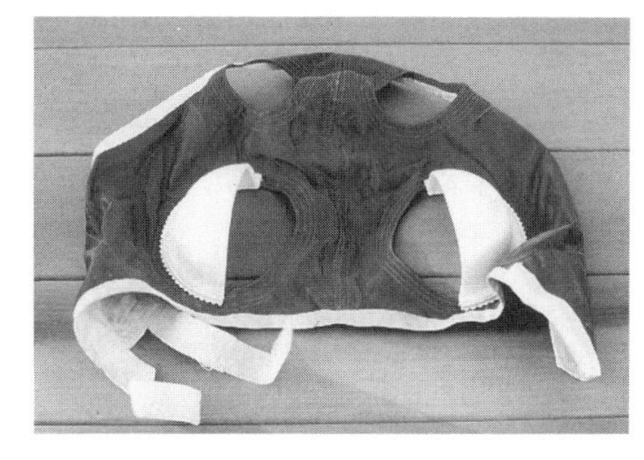

Fig. 8–16. Standard full cup blinkers.

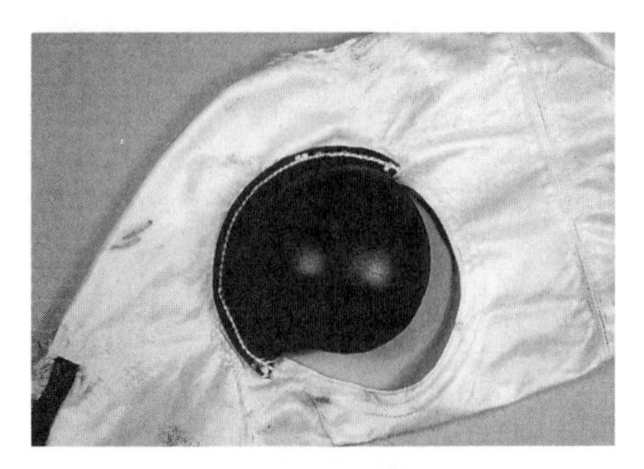

Fig. 8–17. Duck bill cup blinkers.

Duck Bill Cup Blinkers

This blinker has one full cup and one "extended," or duck bill cup. The duck bill cup is used on horses that drift in or out. For example, a trainer may use a duck bill cup on a horse that has a habit of drifting wide toward the outside rail around the final turn of the track. The extended cup helps prevent the problem. The duck bill cup limits the horse's sight even more than the full cup blinker.

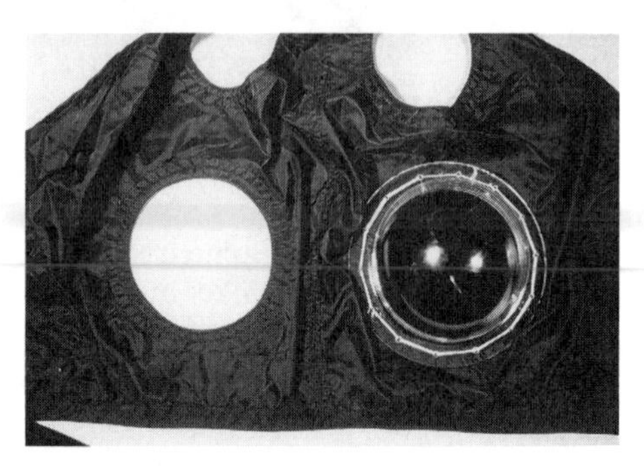

Fig. 8–18. Closed cup blinkers.

Closed Cup Blinkers

This type of blinker is used on racehorses that are blind or partially blind in one eye. The good eye is open with no cup at all. The closed cup protects the blind eye from dirt clods.

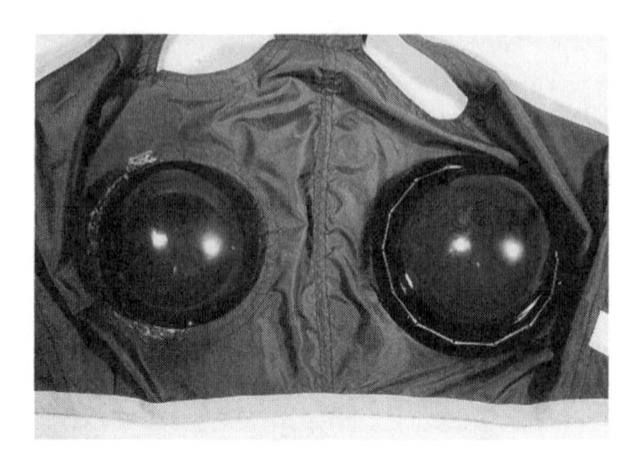

Fig. 8–19. Plastic goggles are good for racehorses that are sensitive to dirt clods hitting their eyes.

Plastic Goggles

Some racehorses are very sensitive to dirt clods hitting their eyes during a race and, as a result, refuse to extend themselves. Plastic goggles protect the horse's eyes from flying dirt clods. They are made of a hard, clear plastic and are placed over the horse's eyes. The plastic goggles may be placed on the horse in the saddling paddock or the horse may come to the paddock with the goggles already on.

For the horse to become accustomed to blinkers or goggles, it is best to first put them on during the day while the horse is confined to its stall. The next step might involve the horse being walked with the blinkers or goggles on and finally, allowing the horse to go to the racetrack for exercise with them on. The time frame for this adjustment period varies with each horse, but it should take approximately two weeks.

Harness Horses

Blinkers

Blinkers are more common with running horses than with harness horses, although it is not uncommon to see a trotter or pacer wearing blinkers. As with those used on runners, blinkers used on trotters and pacers limit the horses' rear and side vision. Some harness horse trainers may use an adjustable blinker cup to increase the effectiveness of the blinkers by placing the cups under the eye, thus limiting the horse's vision downward. In the case of a horse that has a tendency to jump and shy over shadows on the ground, blinkers with the cups adjusted under the eye may be used to correct the problem. **Note:** Blinkers should only be worn under an open bridle.

Fig. 8–20. Blinkers limit the harness horse's rear and side vision.

Fig. 8–21. One fleece roll should fit around each cheekpiece.

Cheek Rolls

Instead of using blinkers or one of the various types of closed bridles (discussed in Chapter 7) to prevent a harness horse from seeing behind itself, a trainer often uses fleece cheek rolls. One roll should be placed on each cheekpiece of the open bridle, just behind the eyes. The fleece cheek roll restricts what the horse can see to the sides and rear, encouraging the racehorse to focus only on what is ahead.

BREASTPLATE OR COLLAR

The breastplate prevents the running horse's saddle from slipping backward during training or racing. On a trotter or pacer, the breast collar prevents the saddle section of the harness from slipping back from its position directly behind the withers. These two pieces of equipment are used on horses with flat, round withers that are prone to having the saddle slip during a workout or race. They are usually applied first, before the saddle or harness.

Running Horses

Breastplate

Breastplates are used on running horses when necessary to keep the exercise or racing saddle from slipping back. The wide strap of the breastplate crosses the front of the chest and attaches to the girth on either side by buckled straps. The breastplate is supported up on the chest by a narrow strap that lies in front of the withers and connects to each side of the chest strap.

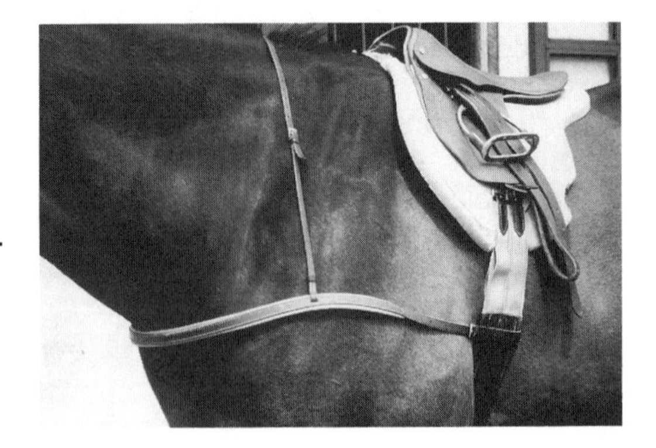

Fig. 8–22. A breastplate prevents the running horse's saddle from slipping backward.

Harness Horses

There are two main types of breast collars used in harness racing—standard and buxton.

Standard Breast Collar

This type of breast collar consists of a straight strap running horizontally across the horse's chest and snapping to rings on each side of the saddle. There is also a second strap running over the neck and snapping to both sides of the first strap for support.

Fig. 8–23. This horse is wearing a standard breast collar across its chest.

Buxton Breast Collar

This type of breast collar is "V" shaped. A single strap snaps to the girth and runs up between the front legs. In the middle of the chest, it splits into two straps, which snap onto either side of the saddle. The buxton breast collar is supported by another strap which runs over the neck.

Fig. 8–24. This horse is wearing a buxton breast collar.

HEAD POLE

The purpose of the head pole is to help the harness horse to keep its head straight while exercising or racing. It may be used on either

Fig. 8–25. The head pole keeps the horse's head straight. The head pole burr (in the middle of the head pole) keeps the horse from leaning into the head pole.

the trotter or pacer. The head pole (because it is a tube within a tube) is adjustable to the length of the horse's neck and can be positioned on the near or off side. The smaller end of the head pole is inserted into a ring on the head halter. The larger end has a slot. A leather strap, which is attached to the saddle, is slipped through this slot and then hooked to the check hook.

If the horse is still leaning against the pole after it is properly adjusted, a head pole burr may be used. This device is a strip of leather or plastic with sharp rivets fixed around the head pole alongside the horse's neck. The rivets on the head pole burr prevent the horse from leaning against the head pole, thus increasing its effectiveness.

MURPHY BLIND

This accessory is a stiff piece of leather (approximately eight inches square) that snaps on the cheekpiece of the bridle behind the horse's eye. Generally, the Murphy blind is only used on one side. The blind is cupped around the eye to restrict the vision of a horse that lugs in or out on that side.

GAITING STRAP OR POLE

This device involves a leather strap or a wooden pole which runs on one or both sides of the horse's hindquarters to keep the back end straight with the rest of the body. The strap or pole runs from the tip of the shaft to the cross bar of the jog cart or racing sulky.

The gaiting strap or pole is usually effective in straightening a horse that has a bad habit of carrying its hindquarters off to one side in a race or workout. However, make sure that the sideways travel is merely a bad habit and not the result of a hind leg lameness before trying to correct it with this device.

EAR ACCESSORIES

Scrim Ear Net

This piece of equipment is placed over the horse's ears during the summer months when flies are a problem. It is made of fine net material, which protects the ears from annoying flies and gnats. The Scrim ear net is generally used during training sessions only. It is placed over the ears (and over the bridle) and the tapes are tied in a normal shoestring bow underneath the jaw.

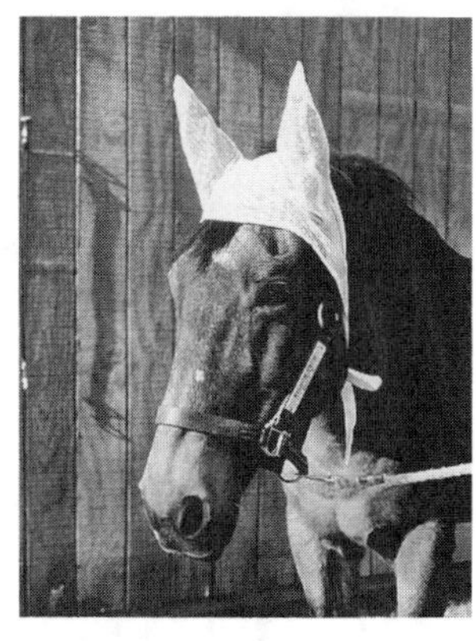

Fig. 8–26. A Scrim ear net.

Earplugs and Hoods With Cones

Harness horse trainers use earplugs to prevent a horse from being frightened or distracted by noises during a workout or race. Earplugs for horses may be purchased, or you can make your own. If you buy them, earplugs are usually available in foam or cotton, with or without strings.

To make an earplug, roll a wad of cotton into a ball, then insert the cotton into the toe of a knee-high pantyhose. Often, the trainer will ask you to sew a long string from one earplug to the other. This string must be long enough so that the driver can tug on it, which pulls the earplugs out during a race or workout. Then the horse can hear another horse coming up behind it, which may encourage it to go faster.

Hoods with hard plastic cones for the ears are also popular. Like the Scrim ear net, the hood is placed over the horse's ears and fastened under the jaw. The purpose is to deaden the noise during exercise. Both the hood and earplugs may be used on a particularly nervous horse.

RACING BAT/WHIP

The jockey or driver may use the bat or whip during a race, while the exercise rider or driver may use it during a training session. The purpose of the bat or whip is to get the horse to go faster, or to reprimand the horse for behaving badly. The racing bat or whip is also used to keep a horse on a "straight course" during a race.

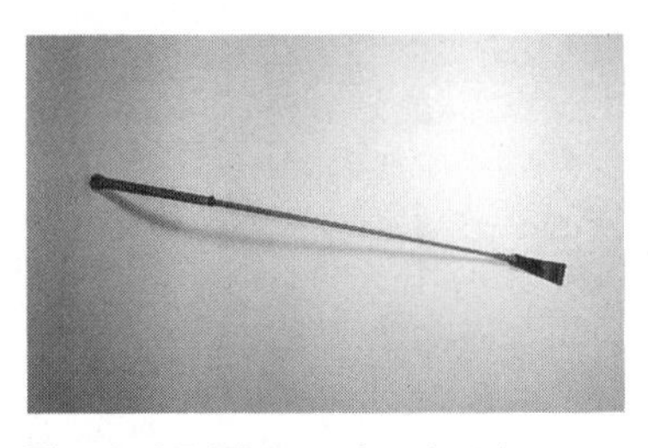

Fig. 8–27. The racing bat is used during racing or training.

Sometimes a jockey or driver abuses a horse with a whip during a race. As a result, that person is fined or suspended by racing stewards for those actions. An abused horse will likely have large welts on both sides of its rump when it returns from a race or workout.

TOE PROTECTOR

This piece of equipment is actually misnamed, as its purpose is not to protect the exercise rider's toe, but rather to protect the skin on the horse's side from abrasion by the toe of the exercise rider's boot. The toe protector is usually made of fleece. It is designed to slip over the toe of the rider's boot and is held in place by an elastic strap behind the heel.

Fig. 8–28. A toe protector.

BITS

The bit is the part of the bridle which goes in the horse's mouth, over the tongue. It is usually made of non-rusting metal or rubber. The purpose of the bit is to allow the rider or driver to control the horse. There are many types of bits—it would take an entire book to adequately discuss all the different types of bits ever used on racehorses. Therefore, this section will familiarize grooms only with the most commonly used bits.

There are several reasons why a groom should learn about bits and their functions. If a trainer asks you to go to the tack room to retrieve a bridle with a specific bit, you should be able to select the correct bridle and bit. Or, the trainer may ask you to replace a bit on a bridle with a different type of bit. Finally, if you wish to become a trainer someday, you will find this information invaluable.

Running Horses

Racing Dee Bit

On the racing dee bit, there is a jointed mouthpiece attached to a movable "D" shaped ring at each end. The mouthpiece and rings are usually made of stainless steel and aluminum. The mouthpiece varies in size but the average length is five inches. This standard racing bit may be covered with rubber for horses with sensitive mouths.

Dexter Snaffle Ring Bit

The ring on this bit prevents the horse from grabbing either side of the bit with its teeth. The ring also exerts pressure on the jaw and cheek when the rider pulls on the reins. This bit is generally used on racehorses that lug. (The horse pulls to the inside or outside and wastes costly strides instead of staying in a straight line.)

Springsteen Bit

This is a very severe bit. When pressure is applied by pulling on the opposite rein, the spoon-shaped prongs jab into the corner of the horse's mouth. Trainers use the Springsteen bit on racehorses that are hard to control.

Regulator or Sidelining Bit

This bit prevents a racehorse from lugging in or out, particularly on the turns during a race or workout. The regulator or sidelining bit extends six inches outside the mouth on one side, giving the rider a great deal of leverage when going around the turns.

Slide Pipe Run Out Bit

The trainer may also use this bit on racehorses that have a habit of lugging in or out during a race or workout. The sliding bar allows for additional leverage. For example, when the horse begins to lug out,

Types of Bits Used on Running Horses (Fig. 8–29.)

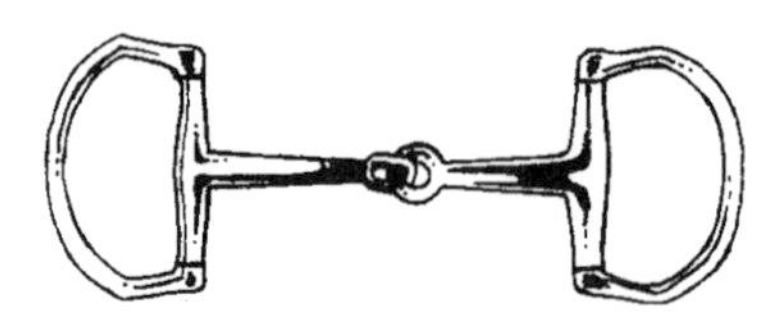

Racing Dee.

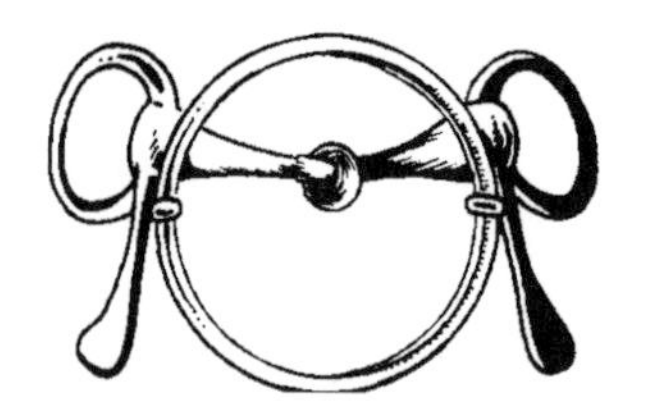

Dexter Snaffle Ring.

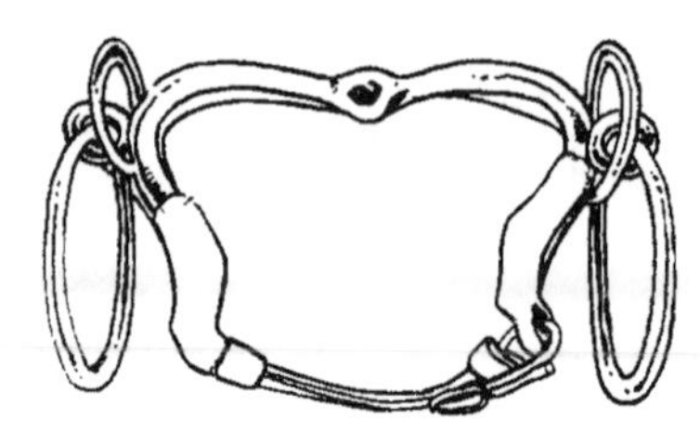

Springsteen.

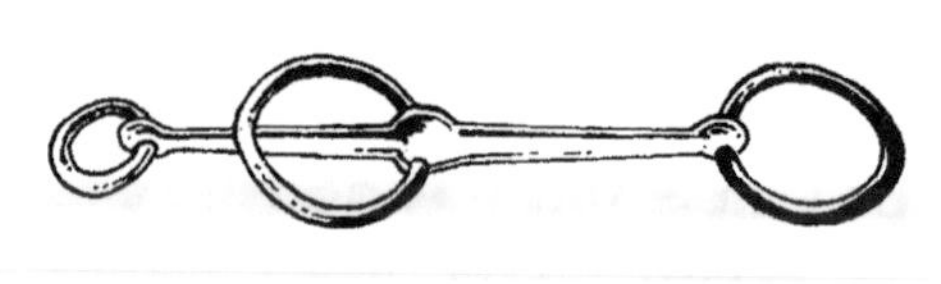

Regulator or Sidelining.

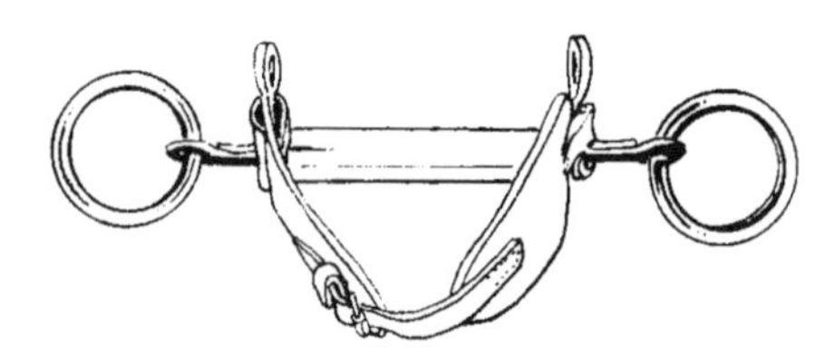

Slide Pipe Run Out.

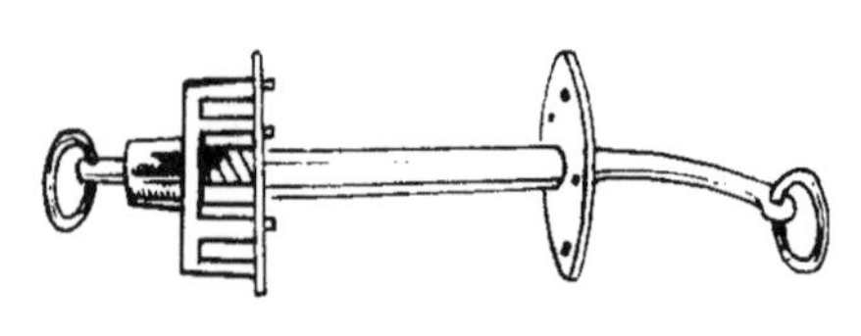

Belmont Run Out.

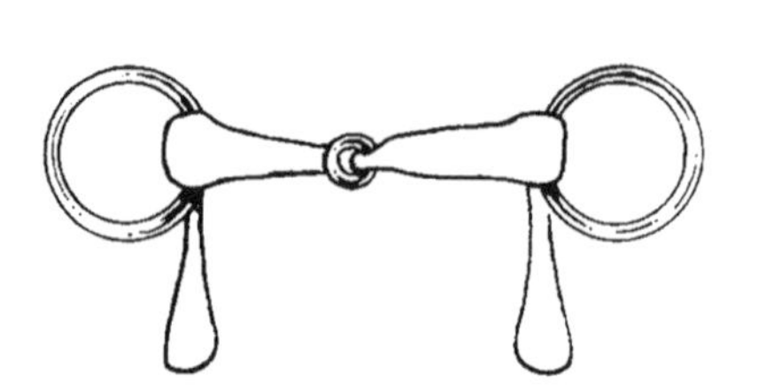

Half Cheek Rubber Snaffle.

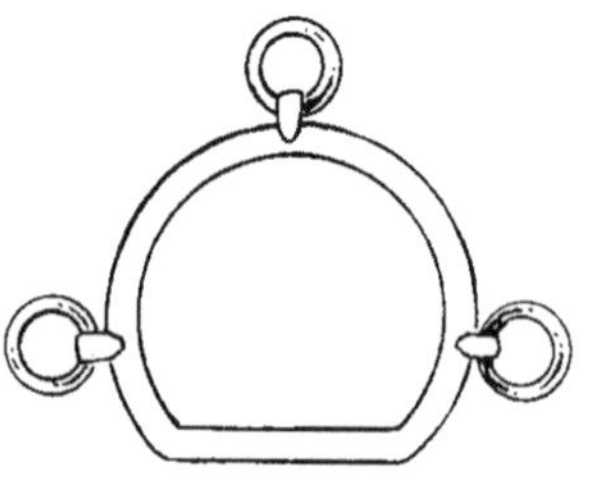

Chifney.

the rider pulls on one rein to allow the bar to slide to the same side of pull, thus keeping the horse straight during a race or workout.

Belmont Run Out Bit

This is the most severe bit used to correct the problem of lugging in or out during a race or workout. When the rider pulls on the opposite rein, metal prongs on the bit jab the corner of the horse's mouth. When the rider releases the tension on the rein, the spring allows the prongs to retract. Some trainers recommend using a standard noseband with this bit to prevent the rider from pulling the bit through the horse's mouth.

Half Cheek Rubber Snaffle Bit

This bit is used on a horse that has a tender or light mouth. It is also a good bit for a horse that has a tendency to play with its tongue during a race or workout.

Chifney Bit

The Chifney bit is not considered a racing bit. It is designed to be used with a halter and lead shank. The two snaps on the bit hook onto the lower rings of the halter and the shank is passed through the center ring of the halter. This bit is popular because it can be used to make any horse more controllable from the ground.

Harness Horses

There are two categories of bits used on harness horses—driving bits and overcheck bits. Both types are used at once.

Driving Bits

Driving bits are part of the bridle for trotters and pacers. Most of these bits are made of stainless steel and measure 4¼ – 5 inches long. The bit may be covered with rubber or leather for added comfort in the horse's mouth. Once the bit is placed in the horse's mouth, the driving lines are buckled to the rings of the bit, sliding through the terrets found on each side of the saddle, and then running back to the driver's hands. The following are some of the most common driving bits used on harness horses *(see Figure 8–30)*.

Standard Unjointed Straight Bit

This bit is used on horses that do not require a severe bit for control. It is also available in rubber, which is easy on the horse's mouth.

Half Cheek Snaffle Bit

As with any snaffle bit, control comes from a "pinching" pressure exerted on the lower jaw. When the driver pulls on the driving lines, the snaffle bit puts pressure on both sides of the horse's mouth. Some horses react unfavorably to this pressure. However, for the most part a snaffle bit is a good bit for control.

Sidelining Bit

This sidelining bit is the same type that is used on runners for increased lateral control and leverage. It is designed to correct a horse that has a tendency to "pull" to one side. For example, if the horse pulls to the right, then the extension is placed on the right side.

Slip Mouth Sidelining Bit

This type of bit (also used on runners) is designed to correct a harness horse that tends to lug in or out during a race or workout. The bar portion of the bit slides through a hole in the center of the bit from one side to the other. In this manner, the bar extends to the appropriate side to allow extra leverage for the driver.

Dr. Bristol Bit

This type of driving bit is quite effective for trotters or pacers that are considered "pullers" or that tend to play with the bit. The Dr. Bristol bit is similar to a snaffle bit except that the center jointed portion is replaced by a small metal plate which is linked to the mouthpiece. This center plate allows the driver more leverage in controlling the horse with less pressure being exerted on the horse's mouth.

Eggbutt "Barrel-Mouth" Bit

This type of driving bit is characterized by a larger diameter at the ring joint than at the center joint of the mouthpiece. The wider end of the bit lays snug against the corners of the horse's mouth, which reduces the amount of pinching at the corners of the mouth. The eggbutt design increases lateral control for the driver by pressing against the sides of the horse's mouth.

Driving Bits Used on Harness Horses (Fig. 8–30.)

Standard Unjointed Straight.

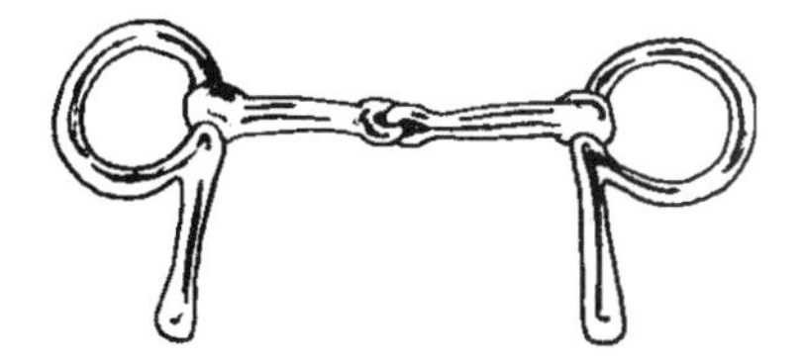

Half Cheek Snaffle.

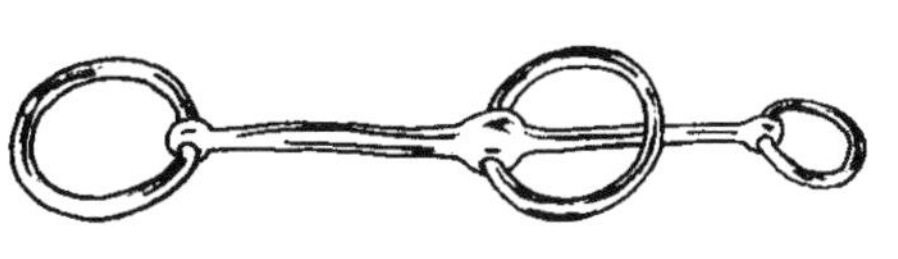

Sidelining.

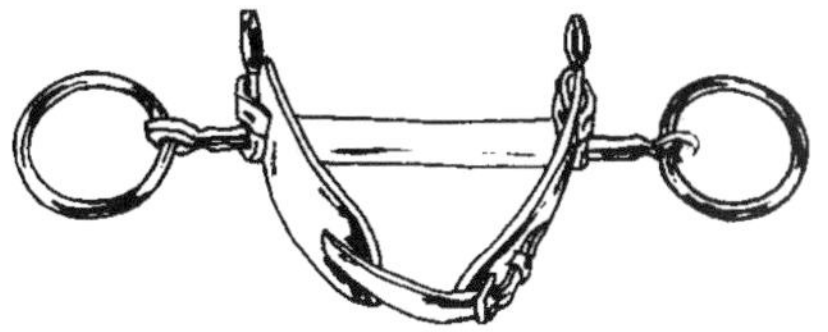

Slip Mouth Sidelining.

Dr. Bristol.

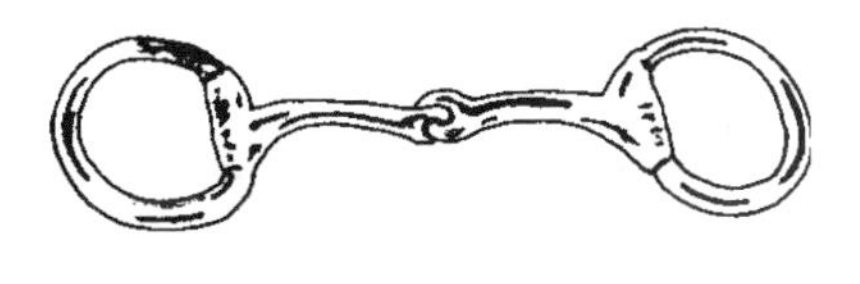

Eggbutt "Barrel-Mouth."

Overcheck Bits

The check rein is looped over a hook at the top and center of the saddle and runs straight up the horse's neck to the crownpiece of the bridle. At the crownpiece, the check rein is divided into two straps which extend down the face to attach on each side of the overcheck bit. All overcheck bits are placed in the mouth along with the driving bit during the bridling process.

The function of the overcheck is to hold the horse's head so the nose is level with the withers—a position that maximizes performance. The horse is unable to lower its head after it is "checked up."

There are several types of overcheck bits with which a groom should be familiar. They include the Plain, Speedway, and Burch overcheck bits *(see Figure 8–32)*.

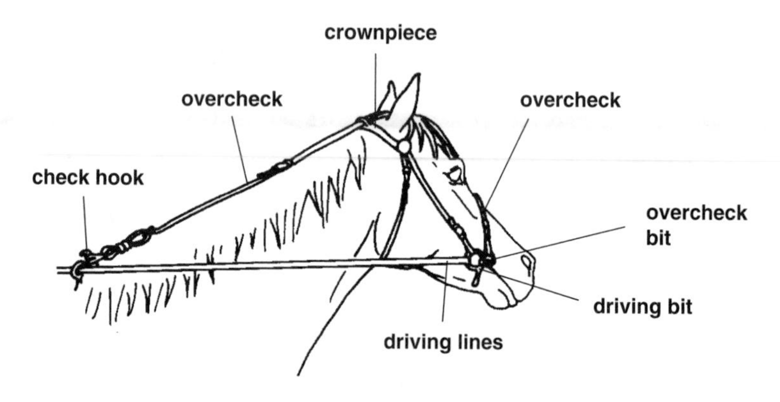

Fig. 8–31. The overcheck extends from the check hook up to the crownpiece, then down the center of the horse's face, where it splits and buckles to each side of the (smaller) overcheck bit. The driving lines are buckled to the (larger) driving bit.

Plain Overcheck Bit

This type of overcheck bit is nothing more than a straight metal bar with rings at each end for the overcheck reins.

Speedway Overcheck Bit

This type of overcheck bit must be used with a chin strap. The openings on the ends of the bit are designed to attach to the chin strap, which keeps the horse's mouth closed.

Overcheck Bits Used on Harness Horses (Fig. 8–32.)

Plain.

Speedway.

Burch.

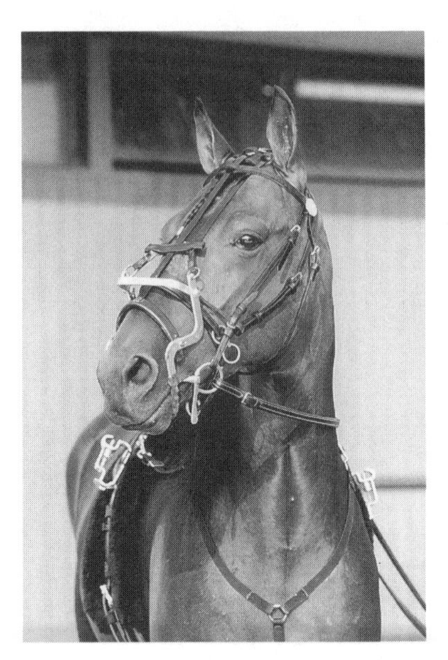

Fig. 8–33. A Z-Guide prevents a horse from hogging down without interfering with its mouth.

Burch Overcheck Bit

When a horse requires a more severe overcheck bit to counteract the problem of "hogging down" onto the overcheck bit (when the horse pulls its head down against the action of the overcheck), then the Burch overcheck bit may be used. This bit lacks the leverage needed to hold a horse's head up. But when a horse hogs down, the raised center plate of the Burch overcheck bit presses against the roof of the mouth.

Z-Guide Overcheck

The Z-Guide overcheck functions like an overcheck bit without interfering with the horse's mouth (no overcheck bit is used). This cumbersome-looking device allows more leverage than an overcheck bit and is very effective in preventing a horse from "hogging down." The fact that no overcheck bit is used with this device makes it especially good for horses with sore mouths.

BIT ACCESSORIES

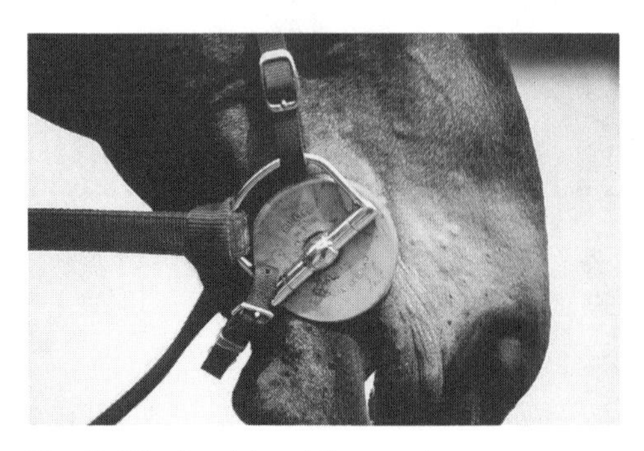

Fig. 8–34. A rubber bit guard.

Rubber Bit Guards

These rubber discs are designed to be used on any snaffle bit. They are used primarily on racehorses of any breed that have tender mouths. They prevent the bit from pinching the corners of the horse's mouth.

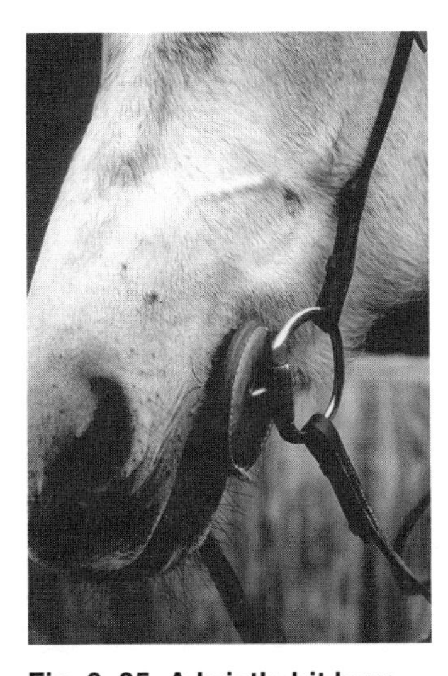

Fig. 8–35. A bristle bit burr prevents lugging in.

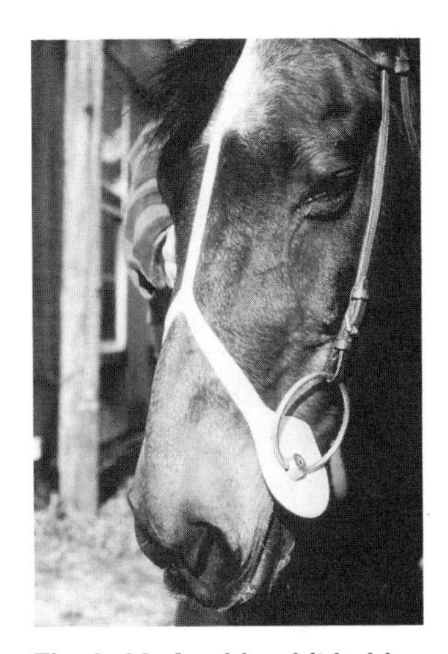

Fig. 8–36. A rubber bit holder prevents a horse from getting its tongue over the bit.

Bristle Bit Burr

This piece of equipment is a leather disc with stiff bristles set in a circular pattern on one side. It is useful on a horse that has a habit of lugging in. The bit burr is placed on the same side of the mouth that the horse usually lugs in. When the problem occurs, the rider or driver simply pulls on the opposite rein. This action presses the bristles of the bit burr against the cheek, forcing the horse to keep its head straight.

Rubber Bit Holder

The bit holder is designed to prevent a horse from getting its tongue over the bit. It is adjusted with a buckle attached to the crownpiece of the bridle. The high tension noseband keeps the bit on the roof of the mouth. Using the rubber bit holder eliminates the need for tying the tongue.

Tongue Tie

Sometimes during exercise or a race, a horse rolls its tongue far back into its mouth and then attempts to swallow. If a horse does this, it could choke on its own tongue. The purpose of tying the tongue is to prevent the horse from playing with its tongue and getting it over the bit during a race or workout.

A tongue tie may be made from rubberbands, leather straps, or a cloth strip. Most trainers believe that a clean cloth strip about 1 inch wide and about 24 inches long is the best type of tongue tie.

If the trainer instructs you to, you are responsible for tying the tongue of the horse for a workout in the morning training hours. The trainer may tie the tongue for a race, but it would not be unusual for you to be asked to tie the tongue then also.

How to Tie the Tongue

It is important that the tongue tie be placed correctly on the horse's tongue. Be sure that the tongue is lying flat under the bit. It is a good practice to wet the cloth to make it more comfortable in the horse's mouth.

The correct procedure for applying a tongue tie is illustrated in Figure 8–37.

1. Stand on the near side of the horse. To get hold of the horse's tongue, insert the fingers of your left hand into the side of the mouth. When the horse opens its mouth, grasp and pull the tongue to the outside of the mouth on the horse's near side. Gently (but firmly) hold the tongue. Drape the cloth over the tongue and loop it around the tongue so that you have an equal amount of ends hanging from both sides of the lower jaw. The tongue tie is placed at the bars of the mouth (the gum area between the molars and incisors).
2. Pull both ends together under the lower jaw and tie a shoestring bow. Be sure that the tongue does not curl when you tighten the ends, as this irritates the tongue and defeats the purpose of the tongue tie. **Do not make the tie too tight as it will cut off the circulation in the tongue** (which will eventually turn black).
3. If fitted correctly, the tongue tie should prevent the horse from swallowing its tongue.

Use common sense when placing a tongue tie on a horse. In between heats or immediately after a race or workout, gently remove the tongue tie.

Tying the Horse's Tongue (Fig. 8–37.)

Step 1

Grasp the horse's tongue gently but firmly. Drape the cloth over the tongue and loop it around.

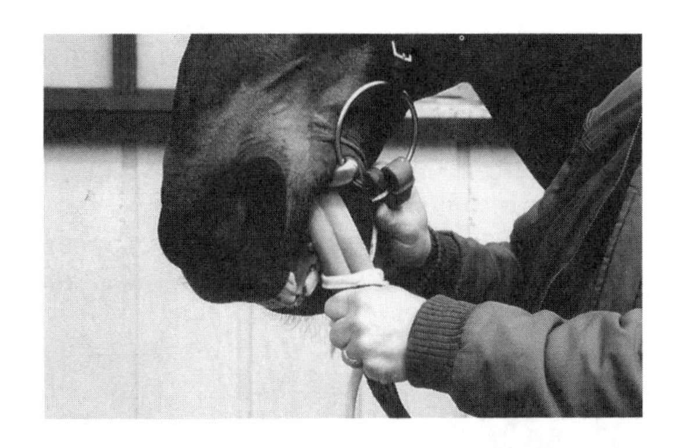

Step 2

Pull both ends together under the lower jaw and tie a shoestring bow.

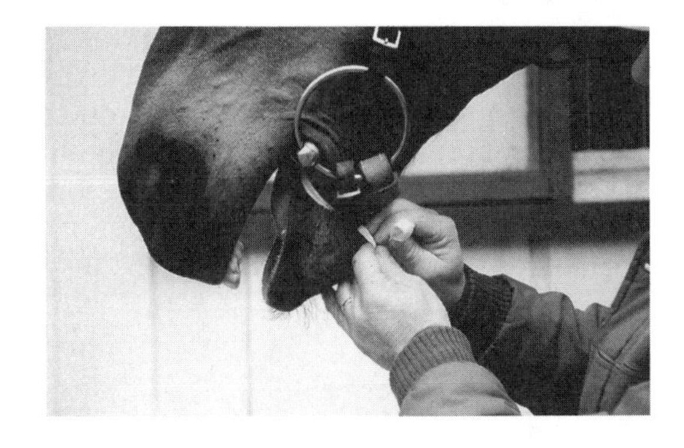

Step 3

A tongue tie, properly applied, prevents the horse from swallowing its tongue.

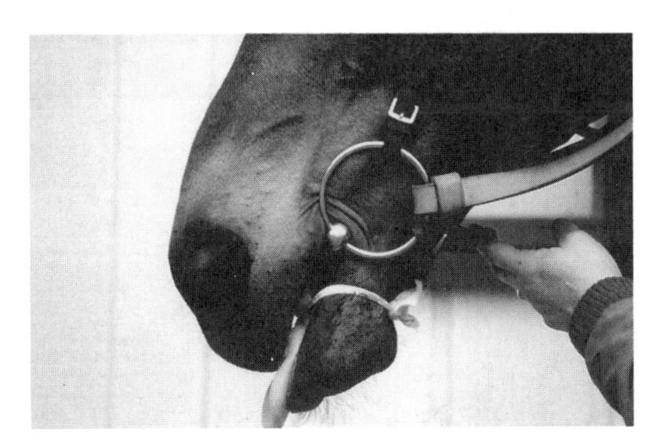

EQUIPMENT REGULATIONS

Each racing jurisdiction varies in its rules for allowing training accessories to be used in races, but in most cases, the stewards require that a trainer request permission for the use of or the removal of blinkers before entry in a race. The trainer must advise stewards if the horse is wearing shoes other than the standard shoes, including calks or bar shoes. Tongue ties, nosebands, bits, and bandages may or may not be required to be reported before a race, depending on the jurisdiction. *(See Chapter 10 for more information on boots and bandages used for a race or workout.)*

SUMMARY

Properly applied and used, all of the accessories in this chapter can be effective racing or training aids. Many accessories are used strictly for training, as they would be too restrictive in a racing situation. Also, some accessories are actually prohibited for use in races by the State Racing Commission. The accessories that are permitted for use in races, such as blinkers and shadow rolls, are generally designed to limit distractions to the racehorse, thus improving its performance.

Chapter
9

Untacking & Cooling Out

One of the first experiences a groom may have at the racetrack is walking a horse after a race or workout. This is probably the most basic, yet most time-consuming job there is, which is why this duty is often passed along to the newer stable hands. As simple as it may seem, cooling a horse down is serious business. There can be severe complications if a horse is put back in its stall

Fig. 9–1.

too soon by a lazy or careless person. If improperly cooled, a horse's muscles may tie up, which can result in both muscle and kidney damage. Or, the horse may colic or even founder if it is allowed to drink too much water too soon. Thus, the groom should not take this duty lightly.

AFTER A WORKOUT

After a workout, the running horse is usually brought back to the barn by the exercise rider. The groom holds the horse while the rider dismounts, and together they remove the bridle and saddle.

The Standardbred racehorse is driven back to the barn by the groom. An assistant holds the horse's head while the groom dismounts from the cart. Then the horse is unchecked, the driving lines are fastened to the check hook, and the horse is unhitched. The cart is backed away from the horse and the horse is led into the stall for untacking.

Removing the Bridle

It is important to learn the correct procedure for removing the bridle, even though it will become second nature in a short time. Do not forget to unbuckle the noseband or head halter before removing the bridle. Standardbred grooms should also unbuckle the driving lines from the driving bit and loop them through the terrets before removing the bridle. The following steps are illustrated in Figure 9–2.

1. Stand on the near side of the horse facing the head. Unfasten the throatlatch buckle.
2. Running horse grooms should bring the reins forward to rest just behind the ears.
3. Using both hands, lift the crownpiece (and reins) over the horse's ears. Allow the bit(s) to drop out of the mouth gently; avoid letting the bit(s) hit the teeth, as this hurts and frightens the horse.

After you remove the bridle (and noseband/head halter), put the halter on the horse and tie or cross-tie it to the stall wall.

Removing the Bridle (Fig. 9–2.)

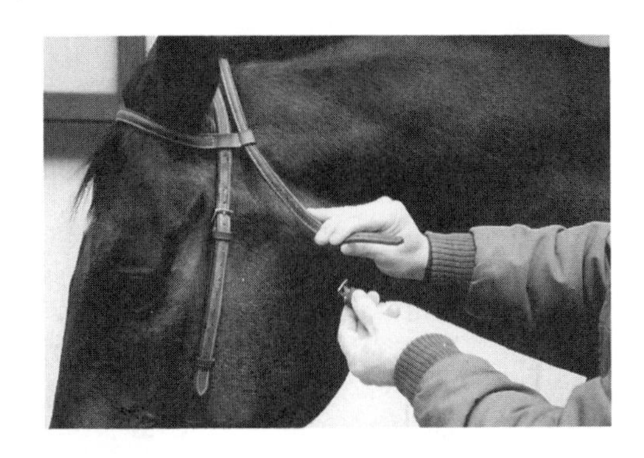

Step 1

Standing on the near side of the horse facing the head, unfasten the throatlatch buckle.

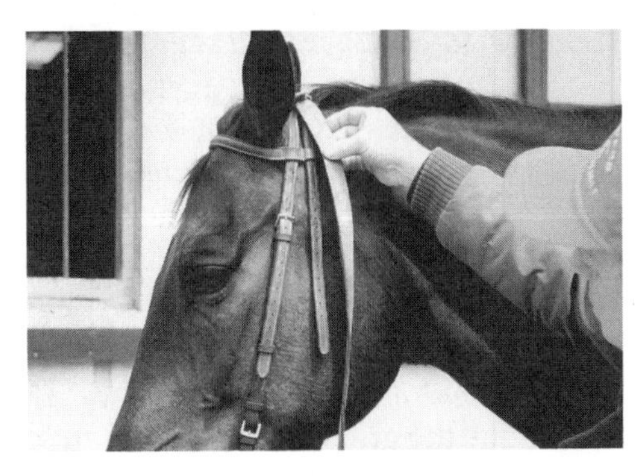

Step 2

If applicable, bring the reins forward and rest them just behind the ears.

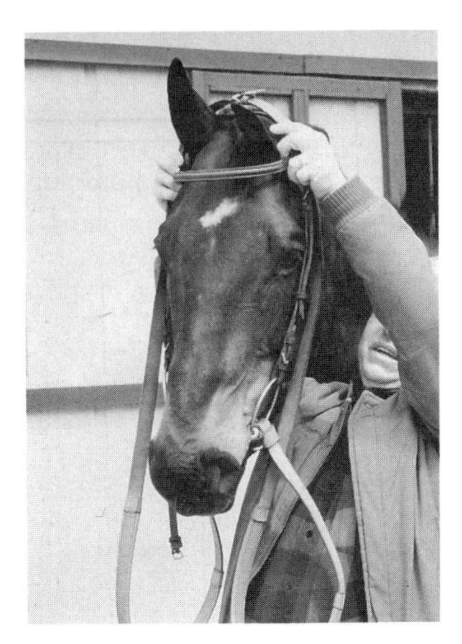

Step 3

Using both hands, lift the crownpiece (and reins) slowly over the horse's ears allowing the bit(s) to drop out of the mouth gently.

Removing the Saddle of the Running Horse

Before you remove the saddle, the stirrups must be "run up" to keep them from banging against the horse's sides or legs while the saddle is being removed. This process also makes the saddle easier to carry and store. Use the following guidelines to properly run the stirrups up and remove the saddle, as illustrated in Figure 9–3.

1. Standing on the near side, grasp the stirrup iron in your right hand. Holding the top strap of the stirrup leather in your left hand, slide the iron all the way up the back stirrup leather so that it rests just under the top flap.
2. Tuck the excess leather through the stirrup iron.
3. Move to the off side and repeat this procedure.
4. Moving back to the near side, unbuckle the girth from the saddle billets. Do not let the girth swing to the off side because it will hit the horse's front legs and frighten or hurt the horse; let it down gently.
5. Gently lift the saddle, saddle pad, pommel pad, and saddle cloth together off the horse's back. Pull the saddle toward you on the near side. Catch the girth as it comes toward you and drape it over the seat of the saddle.

If the weather is warm, the groom meets the horse and rider outside the shedrow to untack there. After removing the bridle and saddle, the groom and an assistant then bathe the horse outside. *(See Chapter 4 for more information.)* Once the horse has been bathed, it is walked for 45 minutes to one hour, or longer if necessary.

If it is cold or raining outside, the groom meets the horse and rider outside the shedrow and leads them both in. After the rider dismounts, the groom untacks and sponges the horse down. Then the horse is walked in the shedrow until it is cool and dry.

Removing the Saddle of the Running Horse (Fig. 9–3.)

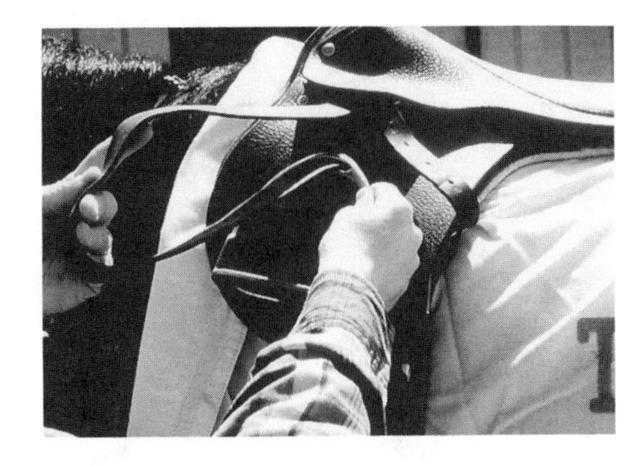

Step 1

Slide the iron up the back stirrup leather so it rests just under the top flap.

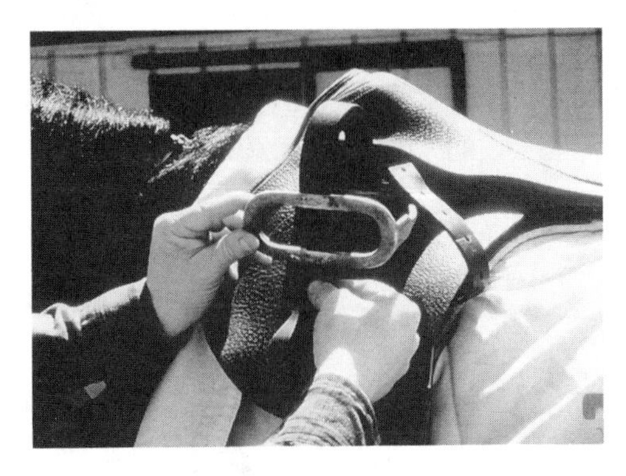

Step 2

Tuck the excess leather through the stirrup iron.

Step 3

Repeat this procedure on the off side. This holds the irons still and keeps them from banging against the horse's sides.

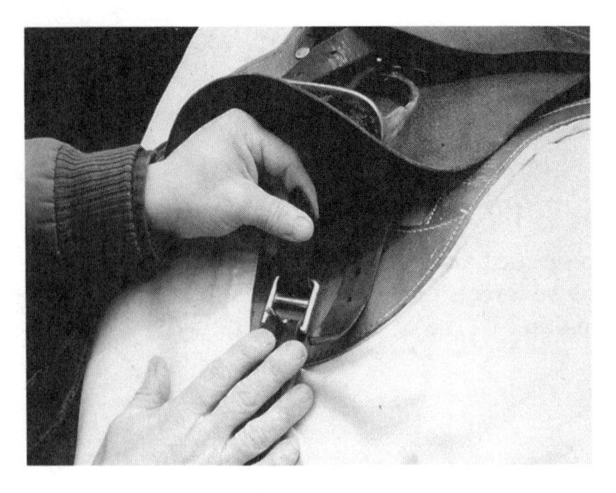

Step 4

Unbuckle the girth on the near side. Do not let the girth swing to the off side because it will hit the horse's front legs; let it down gently.

Step 5

Pull the saddle toward you on the near side. Catch the girth as it comes toward you and drape it over the seat of the saddle.

Removing the Harness of the Standardbred

There are actually several times on a training day when a trotter or pacer may be unharnessed. Most Standardbreds are trained in heats, and between heats, the horse is taken back to the stall for about 45 minutes and unharnessed. To efficiently unharness the Standardbred, use the following procedure as illustrated in Figure 9–4:

1. Unsnap the breast collar from the saddle on the near side. Then unsnap it from the girth. If it is a buxton breast collar, it may be removed from the horse last. Until then, it is allowed to hang around the horse's neck. If it is a standard breast collar, you may unsnap the neck strap and drape the breast collar backwards over the saddle.
2. Loosen the hobbles. Lift each foot in turn and pull it free of the hobble loops. Unbuckle the hobbles and remove them from the horse. If the horse does not wear hobbles, skip to step 3.
3. Unbuckle the girth from the billets on the near side. Do not let the girth swing to the off side because it will hit the front legs and frighten or hurt the horse; let it down gently.
4. Slide the saddle back until it rests in the middle of the horse's back.
5. Now you have enough slack in the back strap to lift the tail from the crupper.

Gently lift the harness, breast collar, and saddle pad or towel together off the horse's back, and then pull them toward you on the near side. Catch the girth as it comes toward you and drape it over the saddle. Hang the tack on the harness hook until you can clean it.

Unharnessing Procedure (Fig. 9–4.)

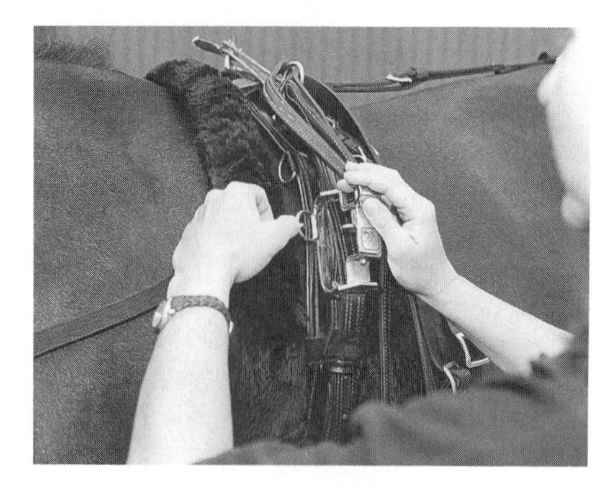

Step 1

Unsnap the breast collar from the saddle (on the near side) and from the girth.

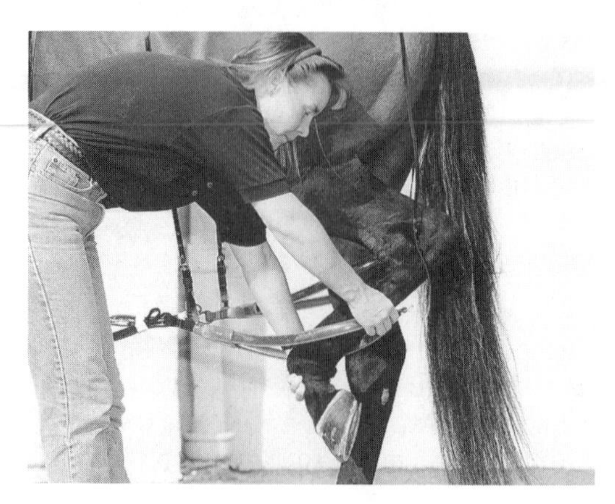

Step 2

Lift each foot in turn and pull it free of the hobble loops. Remove the hobbles from the horse.

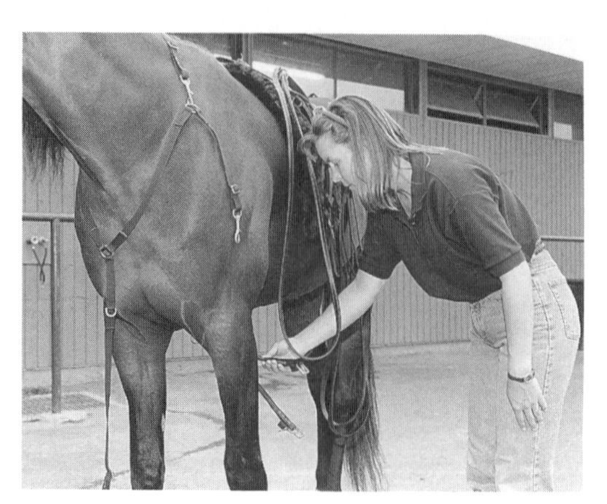

Step 3

Unbuckle the girth from the near billets. Do not let the girth hit the front legs; let it down gently.

Step 4

Slide the saddle back until it rests in the middle of the horse's back.

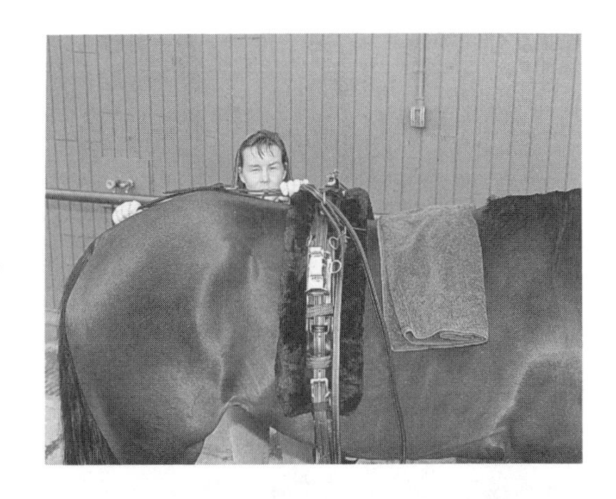

Step 5

Now you have enough slack in the back strap to lift the tail from the crupper.

AFTER A RACE

When the horse wins a race, the groom may escort the horse (and jockey or driver) to the winner's circle. Here, photographs are taken of the winning horse, owner, trainer, jockey/driver, and groom. The horse remains saddled or harnessed for the photos.

Fig. 9–5. When the horse wins a race, the groom may escort it to the winner's circle for photographs.

The running horse is then unsaddled by the jockey valet and the jockey, but the bridle is left on. In most cases, the horses placing first, second, and third must be taken to the state testing barn for drug testing. (Here, the Standardbred is unhitched and unharnessed.) The bridle may be replaced with the halter and lead shank. Now the groom can begin washing and cooling the horse out while waiting for the horse to urinate (so a urine sample may be taken). If your horse does not finish first, second, or third, and it is not being spot-tested, it is led back to the barn for cooling out.

WASHING & COOLING OUT

Washing the horse is a common routine on both the farm and racetrack. In the winter it is difficult to wash a horse, so cooling out must be done by rubbing down and walking. But in warmer weather, especially after a race, the horse is washed as part of the cooling out process.

There is generally no designated washrack at a racetrack, but each barn usually has an area where the horses are washed. However, the state testing barn does have a specific area where all horses are washed when they are brought in for testing after a race.

There should always be a stable hand available to hold the horse while the groom bathes it. If there is a shortage of help, it is not uncommon for the trainer, assistant trainer, or stable foreman to pitch in and hold the horse during bathing. Sometimes, the groom may be forced to hold the horse with one hand and wash the horse to the best of his or her ability with the other hand. Naturally, under these conditions, he or she is not expected to do a perfect job of bathing.

The following are some safety tips to remember whether cooling out in warm or cold weather:

- Never put cold water on the back or hindquarters of a hot horse. Doing so could cause a condition called "tying-up" which is a painful muscle cramping. *(See Chapter 13 for more information on tying-up.)*
- Never put the horse up when it is still hot to the touch or breathing hard. Improper cooling could cause the horse to founder. *(See Chapter 5 for more information about founder.)*
- If walking a wet horse on a windy day, put a surcingle (a leather or webbing belt) around the cooler to prevent drafts.

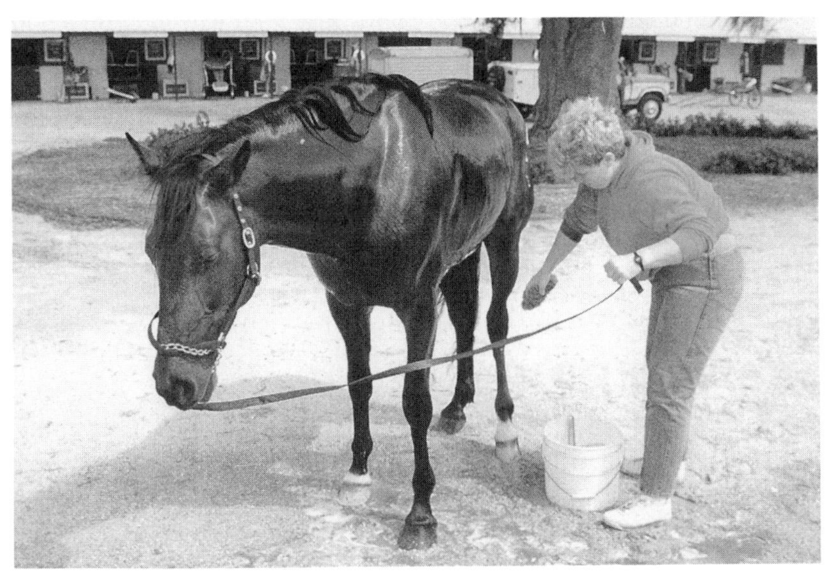

Fig. 9–6. Sometimes, the groom must hold the horse with one hand and wash it with the other hand.

In Cold Weather

When the horse has returned from the race or exercise, lead it into the stall and remove the tack, boots, and bandages. Place a halter on the horse, and tie or cross-tie the horse to the stall wall.

Dip a large body sponge in a bucket of warm water and wring the sponge out until damp. Merely sponge off the head, saddle area, girth area, and legs, removing all saddle marks and harness marks. Place a heavy woolen cooler on the horse and walk the horse under the shedrow to properly cool it out and allow it to dry. *(See Chapter 16 for more information on blanketing the horse.)*

In Warm Weather

A similar washing procedure as described in Chapter 4 should be applied in warm weather, except that shampoo is rarely used after a race. The primary objective after a race in warm weather is not to get the horse clean, but to cool the horse out as quickly as possible. This is usually accomplished by pouring rubbing alcohol in the wash water. (For each 12-quart bucket of warm water, one 16-ounce bottle of alcohol may be added to aid in the drying and cooling out process.) After rinsing the face with clear water (without alcohol), massage the alcohol mixture over the entire body (except the head) with a body sponge. Then scrape the body with a sweat scraper. The alcohol helps the horse cool off and dry out quicker.

You may also put an anti-sweat sheet on the horse to help it cool down more quickly. Still, in hot, humid weather a horse may not dry properly after a race and might actually break out in a sweat again while walking after the first washing. Under this circumstance, it is sometimes necessary to wash a horse a second time with warm water and alcohol.

HOTWALKING

A horse usually needs to be walked by a "hotwalker" for 45 minutes to an hour after a hard workout or race. The hotwalker can be a stable employee or an independent contractor who "hotwalks" horses for many trainers.

The hotwalker must not allow the horse to drink water too fast. After a race or workout, the horse should only be allowed to drink a few gulps of water before bathing, and then only if the horse is not "blowing," or breathing hard. (In this case, bathe the horse before allowing any water.) After the horse is washed and begins walking, it should be allowed to take another small drink of water. The horse may drink a few swallows about every two or three times around the shedrow or walking ring. When the horse is completely finished drinking and refuses to drink any more water, it is said to be "watered off." In some cases, a horse may require more than one bucket of water during the cooling out period. Each horse is different in its drinking habits; some horses drink at a slower pace than others. It is generally up to the trainer to set guidelines for watering off horses during the cooling out process.

After 45 minutes, the horse is usually put back in its stall, and the shank is removed. At this point, the hotwalker should watch and wait for the horse to urinate. The urine should be clear and not milky or dark. If the urine is discolored, it may be an indication that the horse has a kidney or muscle disorder. The hotwalker should report to the trainer if the urine is discolored. If the horse is straining and fails to urinate, the trainer should be notified, as this is also an indication of a kidney disorder.

There are several techniques used by stable personnel to induce a horse to urinate. Whistling, rustling the straw, or closing the stall door and darkening the stall may encourage a horse to urinate. These are some techniques that have been used with success, but do not expect them to work on all horses.

After the horse has urinated without any problems, take the horse from the stall and walk it for another 15 minutes.

Hotwalking Machines

The hotwalking machine can be found at racetracks and training centers all over the country. Most hotwalking machines have a capacity of four horses. These machines save labor by reducing the number of human hotwalkers required to properly cool down the horses.

However, a hotwalking machine does not allow the horses to stop and drink. Therefore, a stable employee should be assigned the duty of periodically stopping the machine and properly watering the horses. The horses' behavior should also be monitored while on the hotwalking machine. Horses that try to bite or kick other horses, or horses that are frightened of the machine, would be better off with a human hotwalker. Most properly trained horses, however, have no trouble with the hotwalking machine.

Fig. 9–7. A hotwalking machine does the work of several people, but horses should not be left on one unattended.

To put a horse on a hotwalking machine, make sure the machine is absolutely still. Then walk the horse toward the outer circle of the walking ring, and with your left hand snap the lead hanging from the overhead arm of the machine to the center ring of the halter. Once all the horses that need to be on the machine are hooked up, the machine may be turned on. Always be aware of the sex of the horses being placed on a machine. Never put a mare in front of a "studish" colt, especially if the mare is in heat.

To remove a horse from a hotwalking machine, simply reverse the previous procedure. Turn the machine off, unhook the horse, and lead it back to its stall.

Whether cooling a horse out after a workout or race, or just getting the horse out of the stall for awhile, walking the horse is a good form of exercise. It relieves the boredom of being confined in the stall and allows the horse to relax and "stretch its legs."

CLEANING THE TACK

When all horses have completed their daily exercise, all the saddles, harnesses, bridles, girths, breastplates, breast collars, and martingales are placed on saddle racks and hooks for cleaning prior to storage in the tack room. Cleaning of tack may take place inside the tack room or in some other designated area.

Especially in larger running horse stables, the exercise rider is usually responsible for cleaning the tack; this is not the groom's responsibility unless he or she is also the exercise rider. (The jockey valet cleans the racing saddle after a race.) The Standardbred groom is responsible for cleaning all of the horse's tack, including the harness and bridle. After the harness is cleaned, it is usually hung in the tack room on a hook. *(See Chapter 2 for more information on how to properly care for leather equipment.)* The bits are merely dipped in water to clean them, and the bridle is sponged off with leather cleaner, if it is made of leather. If it is not leather, then the bridle is sponged off with plain water.

Fig. 9–8. All of the tack is cleaned before being put away.

Wrapping a Bridle

Some trainers like the way a neat row of wrapped bridles looks on the tack room wall. In those stables, wrapping the bridle is a skill which separates the mediocre groom from the high-caliber groom. The task is simple and includes the following steps, as illustrated in Figure 9–9.

1. Take the buckle end of the throatlatch and pass it around to the front of the bridle. This is the first pass around the bridle.
2. When the end of the throatlatch reaches the back of the bridle, feed it through the reins at the point where they are buckled together. Standardbred grooms should merely wrap the throatlatch around the back of the bridle.
3. Continue passing the end of the throatlatch to the front of the bridle (the second pass), forming a figure-eight pattern. Buckle the throatlatch at the upper left side of the bridle.

Wrapping a Bridle (Fig. 9–9.)

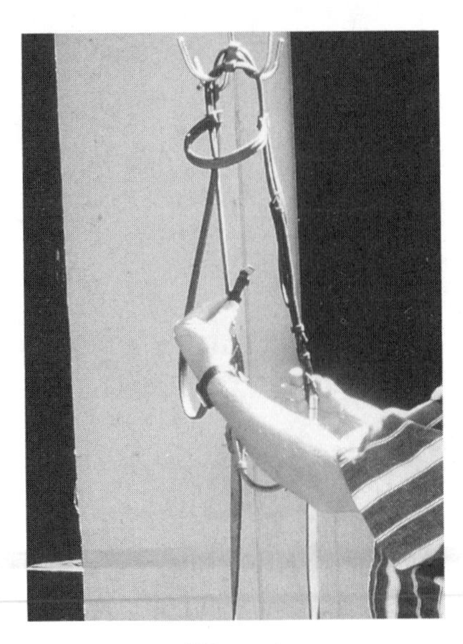

Step 1

Take the buckle end of the throatlatch and pass it around the front of the bridle.

Step 2

Feed the throatlatch through the reins at the point where they are buckled together.

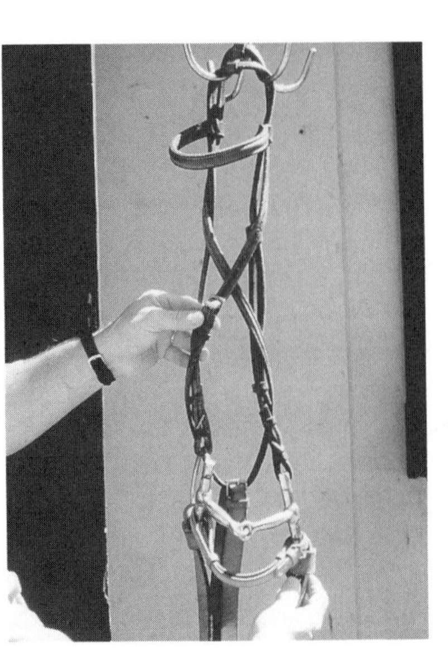

Step 3

Continue passing the throatlatch to the front of the bridle, forming a figure-eight pattern. Buckle the throatlatch at the upper left side of the bridle.

CLEANING OTHER EQUIPMENT

Both running horse and harness horse grooms must clean the racing accessories, such as blinkers, bandages, saddle cloths, etc. *(See Chapter 16 for more information on daily laundry.)* It is especially important to clean the exercise boots thoroughly. (Boots are discussed in Chapter 10.) The insides of exercise boots collect dirt. A dirty boot can chafe the horse if it is not cleaned, inside and out, after each exercise session. (Standardbred grooms should also carefully wipe the boots between training heats, although it is not usually necessary to completely clean them until after the training session is finished for the day.) After cleaning, lay the boots out to dry. When you return to the stable in the afternoon, put them back in the storage trunk.

The cart or sulky and mud apron must also be washed daily with a mild soap and hosed off. They are then dried with a rag, especially the wheel spokes. While cleaning jog carts and sulkies, inspect them for wear and tear. Note problems such as a tire that is low or worn, a crack in a shaft, or a loose bolt. Periodically, spray the wheel axle with WD-40 or a spray silicone. Some carts are "maintenance-free," however, that does not eliminate the need for daily safety inspections. Sulkies are usually hung on a special form designed for that purpose. Jog carts are stored with their shafts pointing upward, near the barn, but out of the way of the shedrow.

SUMMARY

The groom should know the correct procedure for untacking after a race and a workout, as well as the appropriate cooling out measures. Again, these tasks may vary depending on the weather and on the individual horse's needs. Some important reminders: Do not let the horse drink too much water too soon after a strenuous race or workout, and never put a horse back in its stall while it is still hot to the touch, breathing rapidly, or sweating. You cannot do any harm by taking a little more time than necessary to cool the horse out. However, if you do not take enough time, there could be serious consequences—for you and the horse. Moreover, you are much more likely to advance from a hotwalker to a groom if you take care to cool each horse out properly.

Chapter

10

Boots & Bandages

Many people who are unfamiliar with this aspect of horse care immediately think that there must be something physically wrong with a horse that is wearing boots or bandages. On the contrary, there does not have to be anything wrong for a groom to wrap the horse's legs or apply boots. Boots and bandages may be used for several different reasons besides medical treatment, including protection and support. Boots and bandages that are used for therapeutic purposes are discussed in Chapter 12.

The trainer supplies all the boots and bandages needed for the horses in each groom's care, and each groom has a box (which is usually kept

Fig. 10–1. Bandages are stored in the groom's box.

under the shedrow) to store these boots and bandages in. In some stables, every horse may have its own set of boots and bandages. In other stables, however, these items may have to be shared.

BOOTS

There are many types of boots used on racehorses, and they are available in several sizes based on circumference of the leg or size of the foot (usually small, medium, and large). They may be made of leather, rubber, or neoprene. They are attached and adjusted to the limbs by means of buckles and straps, Velcro, and in the case of solid rubber boots, are merely "stretched" over the foot.

Most boots fulfill one of three purposes—protection, balance, or therapy. This section discusses the most common types of boots worn for protection in the stable and protection or balance during exercise.

Stable Boots

Shoe Boil Boot

If the horse flexes its legs excessively when lying down, the bottom of the shoe can bang the elbow and cause it to become infected. When this "shoe boil" condition develops, a shoe boil boot is used to protect the horse's elbow and allow it to heal. This boot can also be used regularly to prevent shoe boils from developing.

The shoe boil boot is shaped like a huge doughnut. It is placed around the front pastern while the horse is in its stall so that when the horse lies down, the shoe cannot touch the elbow. This boot is usually made of leather or canvas with one buckle and strap to fasten it.

Fig. 10–2. A shoe boil boot.

Not all horses are predisposed to shoe boils. A horse with the conformation fault of excessively long, sloping pasterns is a prime candidate for shoe boils, because that kind of horse tends to hit its elbows with its feet while trotting or galloping. However, most shoe boils are the result of how a horse lies down in its stall. Although uncommon, it is possible for a horse to develop shoe boils on both elbows, which would require using two shoe boil boots.

Calking Boot

The calking boot prevents injury to the hoof and coronet of a horse that tends to step on its opposite foot in the stall. This type of boot is usually made of leather with a metal shield. It is felt-lined to prevent chafing. The calking boot is placed over the foot and fastened by a single buckle and strap.

Fig. 10–3. A calking boot.

Easyboot®

These boots, which are available in standard horseshoe sizes, cover the entire hoof (ending just below the coronet) and offer some of the same advantages as regular horseshoes. Easyboots are worn to protect the horse on snow, rough surfaces, and hard pavement. The bottom of the traditional Easyboot has small rubber cleats for traction. For better traction on ice or snow, the Easyboot is also available with metal cleats on the bottom.

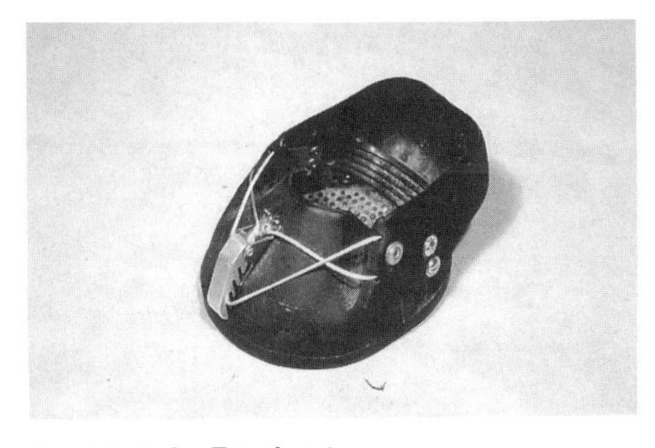

Fig. 10–4. An Easyboot.

Easyboots are recommended for treating a bad case of thrush, a puncture wound, or any other foot ailment. The Easyboot may replace the need for a foot bandage when treating these ailments. Easyboots are not recommended for exercise or racing, as they are too cumbersome and slow the horse down.

The Easyboot is simple to put on. Lift the horse's foot, slip the Easyboot over the hoof, and fasten it with the wire and adjustable metal clip.

Exercise Boots

Because of the nature of their gaits and the interference those gaits cause, harness horses wear boots more often than running horses. In fact, all the boots used on harness horses are too numerous to list in this chapter, so only the most common boots will be discussed. In the following sections, indications are given as to which type of racehorse wears a specific boot. If the boot is common to all racehorses, that is also noted. The boots are listed as they might be applied down the leg, beginning with the elbows, and ending with the feet.

Elbow Boot

This type of boot is generally used on trotters that hit their elbows due to high knee action and excessive flexing of the fetlocks during breakover. High action in front is desirable; however, when the horse folds its front feet too high when breaking over and raising the foot to begin the next stride, it often results in elbow-hitting. In this gait fault, the left forefoot strikes the left elbow, and the right forefoot strikes the right elbow.

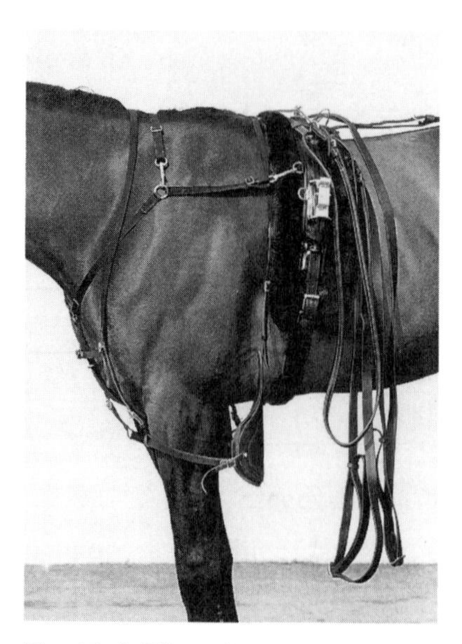

Fig. 10–5. Elbow boots prevent elbow-hitting.

The elbow boot (actually, two boots attached to each other) protects the elbows while the horse is in motion. Two suspender straps are attached to the upper portion of the boots and run up over the withers and neck. The two suspender straps are connected to each other by a single strap running across the chest. Another strap is attached to the lower portion of the boot and buckled over the forearm.

Note: Through selective breeding, corrective shoeing, and the use of advanced training aids, the trotter's gait is more refined today than it was 20 years ago. Thus, elbow-hitting and the use of elbow boots has become less common.

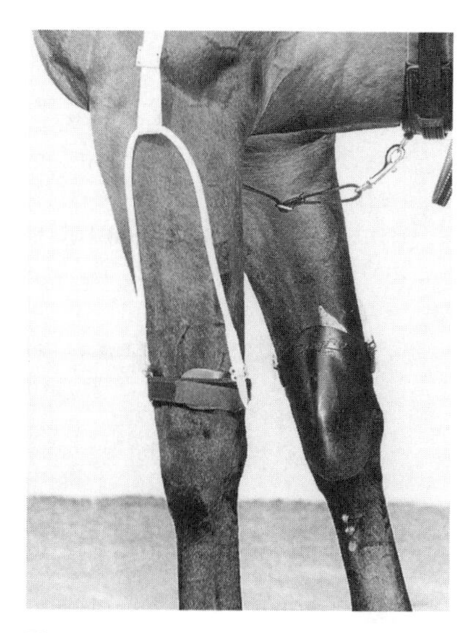

Fig. 10–6. Knee boots prevent knee-hitting.

Knee Boot

The knee boot is used on any horse that hits its knees at speed. The knee boot is held up by two suspender straps on each side connected to a single strap that runs over the front of the withers. Each knee boot is fastened around the leg in one of two ways—with a buckle and strap or with a Velcro strip.

It protects the trotter or pacer's knees from interference with the opposite front foot. The basic types of knee boots are the plain knee boot and the knee and half arm boot. The latter provides protection to the forearm above the knee as well as the knee itself.

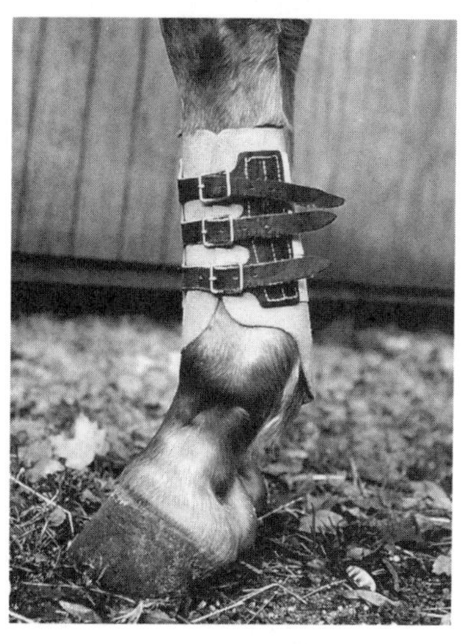

Fig. 10–7. A galloping boot protects the shin and ankle.

Galloping Boot

Galloping boots are sometimes called splint boots or shin and ankle boots. They are applied to running horses. Most are made of leather or rubber and cover the shin and ankle. The galloping boots provide support and protection during a training session. This type of boot is not generally used as racing equipment.

The galloping boot is usually fastened to the horse's leg with three or four buckled straps. When fitting the boot on the leg, be sure that the straps are fastened on the outside of the leg and the ends of the buckled straps are pointing toward the rear of the horse.

Shin Boot

Front Shin Boot

This type of boot is very similar to the galloping boot used on runners. Applied to the front legs of the trotter or pacer, the front shin boot protects the inside of the leg from interference by the opposite front leg. This boot is attached to the leg below the knee and over the ankle by three straps and buckles or by three Velcro strips.

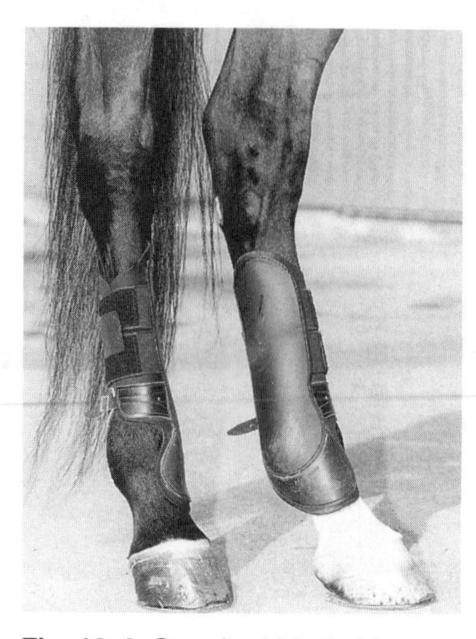

Fig. 10–8. Standard hind shin boots with Velcro closures.

Hind Shin Boot

The hind shin boot is probably the most common boot worn by a trotter. Applied to the hind legs, these boots protect against interference from the front leg on the same side. Even horses that do not generally interfere often wear these boots as a precaution.

The standard hind shin boot covers the leg from just below the hock, down to and including the ankle. The protective padding is on the inside of the leg. For increased protection, trainers may opt for one or both of the following attachments to the hind shin boot:

Speedy cut connection—This attachment extends down over the ankle and pastern to protect the inside of the foot against speedy cutting. A speedy cut connection should never cover the coronet or hoof, as this would limit movement.

Half hock or full hock extension—This attachment extends up to cover the lower part of the hock (half hock) or the entire hock (full hock) on the inside of the hind leg only, for protection against high interference.

Tendon Boot

The tendon boot is similar to the shin boot except that it is constructed to provide extra protection to the tendons at the back of the cannon. These boots are primarily used on pacers that interfere with the opposite hind foot while in motion. They are fastened to the legs in the same manner as the shin boots.

Ankle Boot

Ankle boots applied to runners are usually made of leather and cover the ankles. They support and protect the ankles during exercise. The ankle boot is fastened with two straps: they should be buckled on the outside of the ankle and should point toward the rear of the horse. This boot is generally not considered racing equipment.

A similar type of boot is applied to trotters and pacers. When worn by pacers on the front legs, they protect the insides of the ankles in the event of brushing. They also protect the front ankles of the trotter and pacer from interference with the opposite front leg.

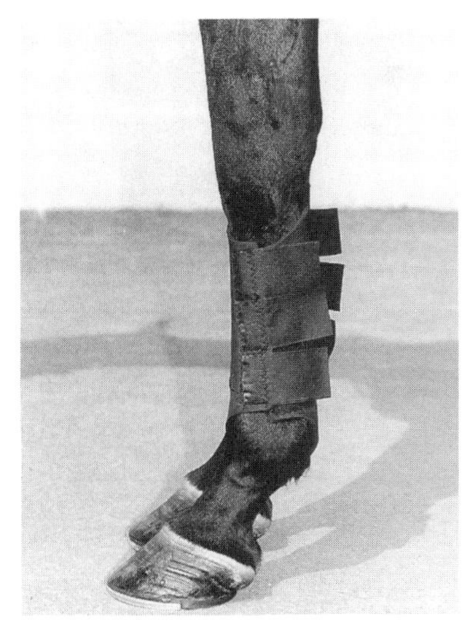

Fig. 10–9. Tendon boots have extra protection at the back of the legs.

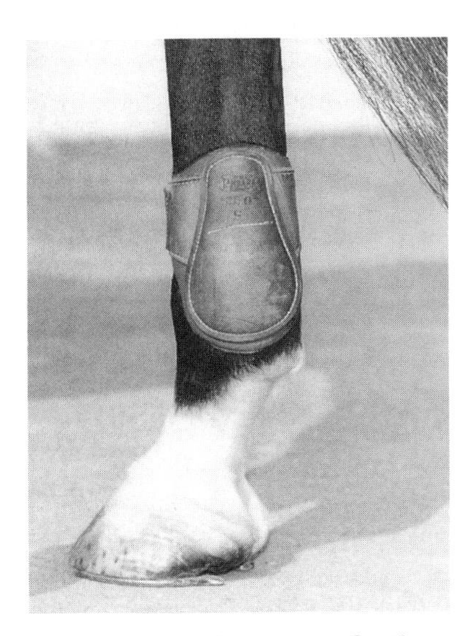

Fig. 10–10. On the left, an ankle boot on the front leg of a runner. On the right, an ankle boot on the hind leg of a trotter or pacer.

Fig. 10–11. On the left, smooth bell boots. On the right, ribbed bell boots.

Bell Boot

Bell boots protect the horse's feet (specifically the coronary band, heel bulbs, and quarters) in cases of over-reaching. *(See Chapter 5.)* They are common racing and training equipment for the harness horse. The trotter in particular may wear bell boots on the front feet to help provide the proper balance to the trotting gait. Often, trainers will cut unnecessary areas off of bell boots to make them less cumbersome for use in races. For running horses, bell boots may be used for exercise, but are never used as racing equipment because they are considered too heavy. In addition to exercising and racing, bell boots are also used to protect the feet (of all breeds) during shipping.

Bell boots are made of rubber and come in various weights and colors. The two basic styles of rubber bell boots are "smooth" and "ribbed." The ribbed bell boot is heavier and affords more protection. Bell boots may be used on all four feet, but are usually applied only in front. They are available with three basic fastening options: closed or pull-on bell boot, open or slotted bell boot, and Velcro open bell boot.

Closed or Pull-On Bell Boot

This type of bell boot is placed on the foot by turning the boot inside out and pulling the wide end of the boot over the toe of the horse's foot *(see Figure 10–12)*. Once the boot is on the foot, the boot is turned "outside in" until properly positioned over the foot. Unlike the other types, this bell boot is used as racing equipment.

Applying a Closed or Pull-On Bell Boot (Fig. 10–12.)

Step 1

Turn the bell boot inside out and pull it over the horse's toe.

Step 2

Slide the boot onto the pastern.

Step 3

Turn the boot "outside in" to cover the coronet and heel.

Fig. 10–13. A slotted bell boot.

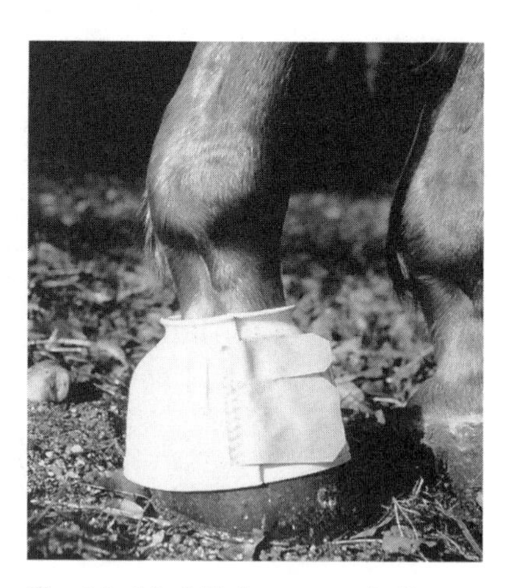

Fig. 10–14. A Velcro open bell boot.

Open or Slotted Bell Boot

This type of bell boot is simply placed around the foot and fastened with a leather strap threaded through two or three metal slots.

Velcro Open Bell Boot

This type of bell boot is open, like the slotted bell boot, but is fastened by a Velcro strap. It should be placed on the foot in the same manner as the slotted bell boot.

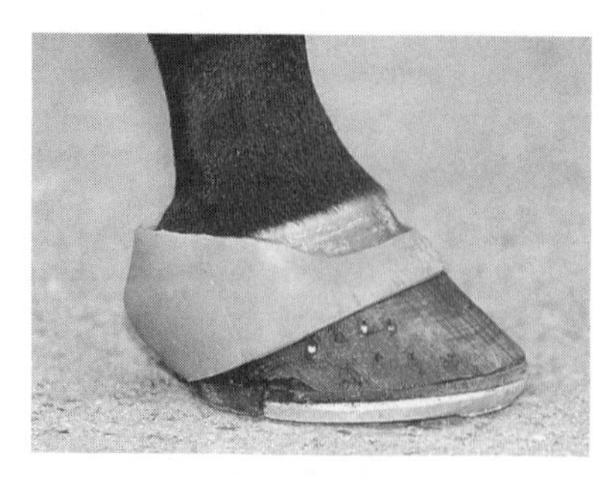

Fig. 10–15. A quarter boot is used on pacers that cross-fire.

Quarter Boot

The quarter boot is made of rubber and is secured to the foot with a single buckle and strap. A hinged quarter boot is largely used on the trotter. It places additional weight on the front heels, providing balance. Occasionally, the quarter boot is applied to the hind heels of the pacer to protect the heels against cross-firing.

Grab Boot

This type of boot is a variation of the quarter boot except that it is placed lower on the horse's heels. It is applied to the trotter on the front heels. The grab boot is made of rubber and is applied by stretching the boot in place over the heels. Its function is to prevent the front shoes from being pulled from the horse's feet in cases of over-reaching.

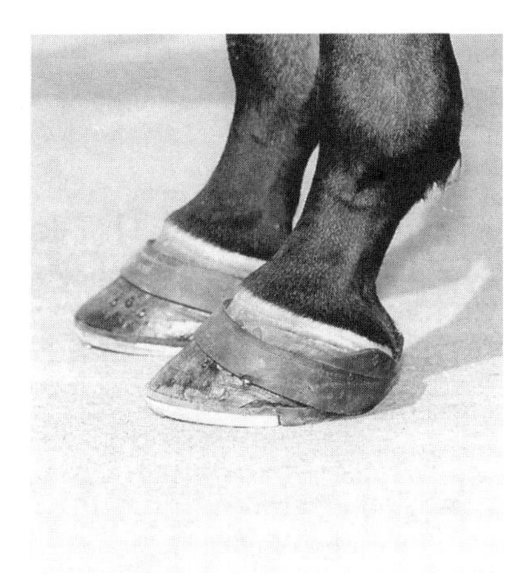

Fig. 10–16. A grab boot.

Fig. 10–17. A scalper.

Scalper

A scalper is made of rubber and is also stretched in place over the foot. When applied to the pacer in front, it protects the inside of the feet in cases of cross-firing. (Only short scalpers should be used in front on the pacer, never high scalpers.) In back, scalpers protect the pacer's coronet and heel. When applied to the hind feet of the trotter, the raised portion of the scalper is turned so that it covers the front of the hind foot. In this way, it protects against scalping from the toe of the front foot. The raised portion of the scalper is available in various heights ranging from four to eight inches high.

BANDAGES

Safety Factors

The following safety factors should be adhered to when wrapping a horse's legs:

- Always tie the horse or have someone hold it.
- Never sit down under the horse. Always stay on your feet.
- When using pins or string ties, always end the bandage on the outside of the leg. This prevents the horse from striking the pin with the opposite leg.
- If you use saddler pins, be sure to insert them parallel to the leg (vertically) to avoid sticking the horse.
- If you are bandaging the horse's legs for medical reasons, always consult the veterinarian on how to apply the required bandage.
- Avoid pulling the wrap against the tendon. The direction of pull should be against the bone, toward the rear of the horse. When you pull toward the front of the horse, you may cause damage to the flexor tendons at the back of the cannon.
- Exercise bandages should be removed immediately after a race or workout. Stable bandages are usually put on after the morning training session and removed the next morning.
- When removing bandages, unwrap them slowly. Always hold the excess bandage in your hand as you unwrap it so it does not get tangled up in the horse's legs if the horse moves around.

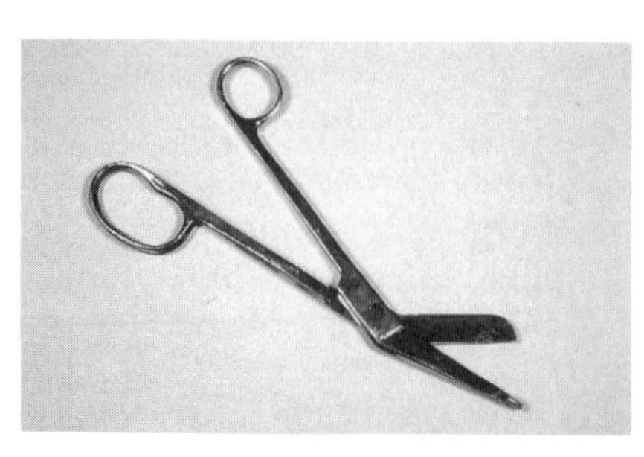

Fig. 10–18. Bandage scissors.

Bandage scissors are used to safely remove some types of bandages from the legs. The tip of one blade is a "spoon bill," which is rounded to avoid pricking the horse while removing a bandage. The "spoon bill" blade is placed against the leg, pointing downward, to prevent injury to the horse.

There is no room for error when wrapping a horse's legs. A wrinkle or tension in the wrong place can result in a tendon injury (appropriately termed a bandage bow). The best way to avoid the dangers of incorrect bandaging is to watch an experienced person apply some, then find a cooperative horse and practice each type of bandage.

Bandaging Hints

Leg bandages are secured to the legs by various means, such as saddler pins, string ties, Velcro, or paper tape. Most professionals apply bandages in pairs because they think that it gives a horse a sense of balance and comfort as opposed to putting only one bandage on the leg. (This feeling would be equivalent to putting on one shoe and walking around the house.) One bandage may be applied to a leg for a specific reason, such as covering stitches or a wound on the leg. However, this is a matter of the trainer's or veterinarian's personal preference.

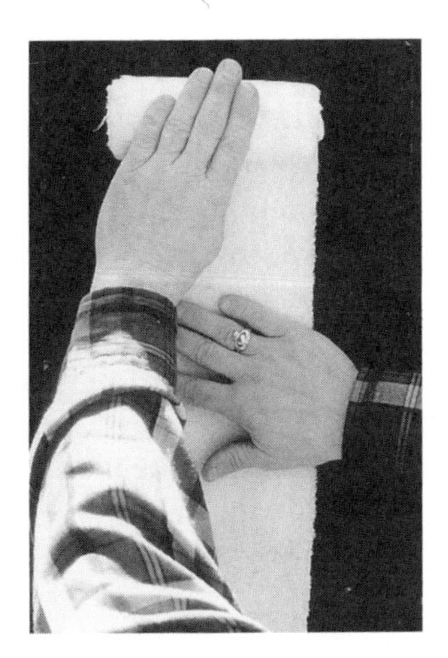

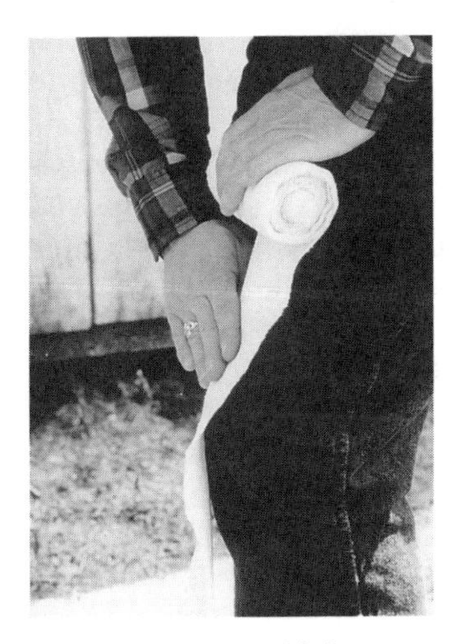

Fig. 10–19. Re-roll bandages on a wall or on your thigh.

When re-rolling a bandage after washing or use, it is important that the bandage be rolled in the correct direction. Otherwise, next time you will wrap the entire leg, only to find that the Velcro does not match up. A good point to remember when re-rolling a bandage is always start with the Velcro or string tie-end first, folding the first roll with the Velcro or string on the inside.

It is helpful to place the bandage on a flat surface such as a door or wall before rolling. Or, lay the bandage on your thigh to roll it. Try to roll the bandage as tightly as possible, as this will aid you in wrapping a smooth bandage on the horse's leg next time.

To better understand the concept of bandaging, the groom should be able to categorize bandages into two basic groups: stable bandages and exercise bandages.

Stable Bandages

Standing Bandage

The standing bandage supports and protects the ankles, ligaments, and tendons while the horse is in the stall. When a liniment is applied to the legs to increase circulation or soothe a particular area, the standing bandage is also used to cover the liniment. This bandage is usually placed over the ankle and cannon. Many running horse trainers apply it to the front legs only, as these legs absorb a greater proportion of stress and concussion on a runner. However, it is not uncommon for a harness horse trainer to use this bandage on all four legs.

The standing bandage is typically placed on the legs after the morning training session and removed the following morning before exercise. The only exception to this practice is if a bandage is changed because it is involved in the treatment of a wound. In this case, the bandage is removed in the afternoon in order to clean the wound or to apply cool water from a hose to the affected area.

It is difficult to make a standing bandage too tight due to the thickness of the cotton roll or quilted pad underneath. Still, it is suggested that you be able to insert a fingertip into the top portion of the bandage. If you are unable to do so, then the bandage is too tight. Also, if a horse begins to stomp its feet immediately after the bandages have been applied, then the bandages may be too tight.

The materials used for the standing bandages are:

- liniment
- cotton with gauze (32 inches x 14 inches) *or* a quilted pad (26 inches x 12 inches)
- flannel bandage (3½ yards long and 5½ – 6 inches wide)
- two saddler pins

Applying the Standing Bandage

This procedure is illustrated in Figure 10–20.

1. Apply the liniment to the leg and gently massage. Be sure to rub only in a downward motion.
2. Roll the cotton or quilted pad neatly on the leg from below the knee or hock down over the ankle.
3. Tuck the beginning of the flannel bandage underneath the cotton or quilted pad just above the ankle and apply the first wrap to anchor the bandage to the leg.
4. Wrap the flannel downward over the ankle. Continue wrapping to just below the ankle, leaving about ½ inch of the cotton or quilt showing. This practice allows proper blood circulation throughout the leg.
5. Wrap the flannel upward, back over the ankle. End just below the knee (front leg) or hock (hind leg). It is also important to leave ½ inch of cotton or quilt showing at the top to allow proper blood circulation.
6. End the flannel bandage on the outside of the horse's leg. When applying the flannel to the leg, be sure to leave one inch of the last wrap showing.
7. Insert the saddler pins vertically, one under the other. When inserting a pin, be sure the point is facing down toward the ground and is parallel to the leg.
8. The finished bandage should look like a tube on the horse's leg.

Applying a Standing Bandage (Fig. 10–20.)

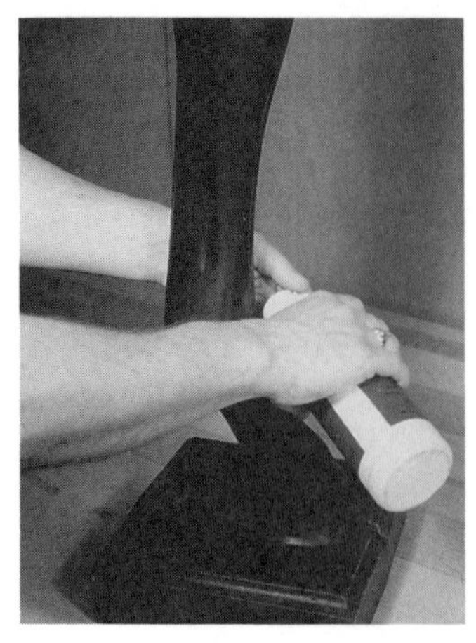

Step 1

Optional: Apply liniment to the leg and gently massage.

Step 2

Roll the cotton or quilted pad neatly on the leg from below the knee or hock down over the ankle.

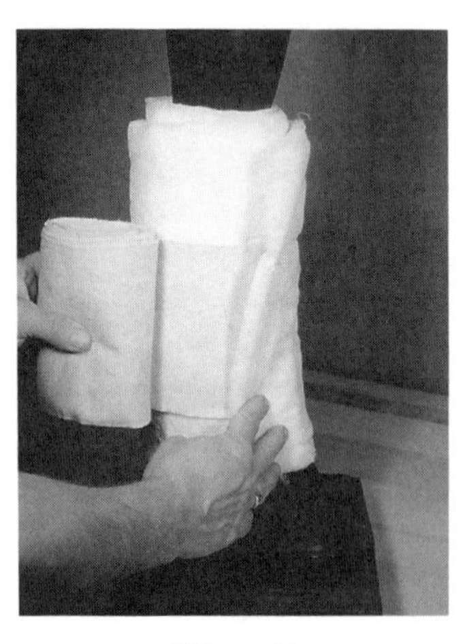

Step 3

Tuck the first wrap underneath the cotton or quilted pad just above the ankle.

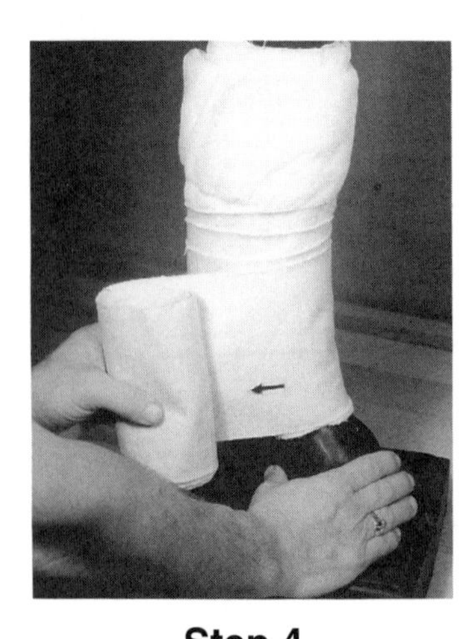

Step 4

Wrap the flannel down to just below the ankle.

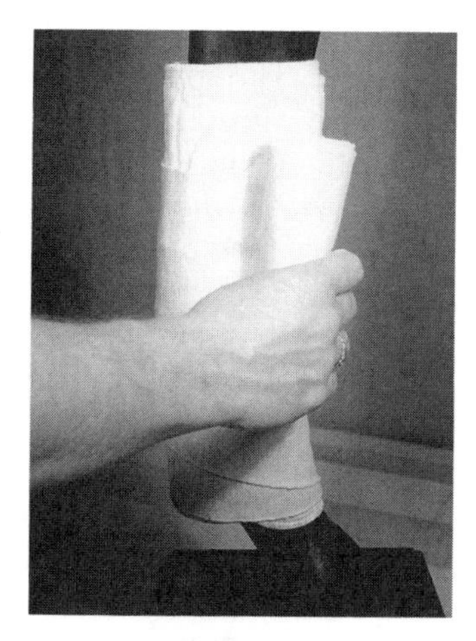

Step 5

Wrap the flannel upward, back over the ankle. End just below the knee or hock.

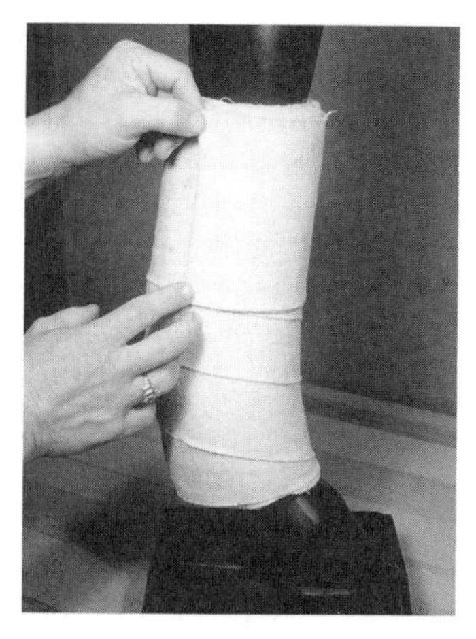

Step 6

End the flannel bandage on the outside of the horse's leg.

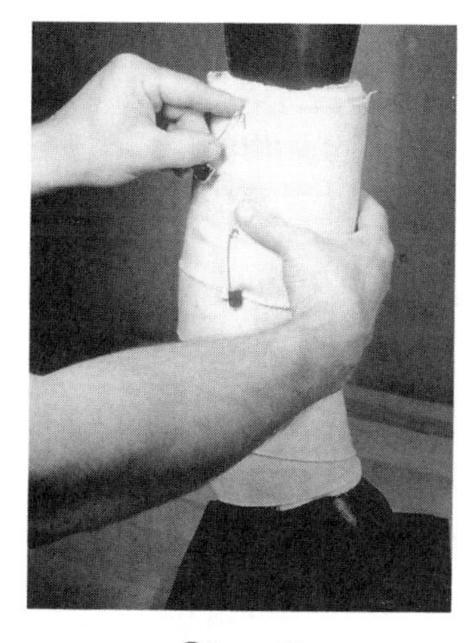

Step 7

Insert the saddler pins vertically, one above the other.

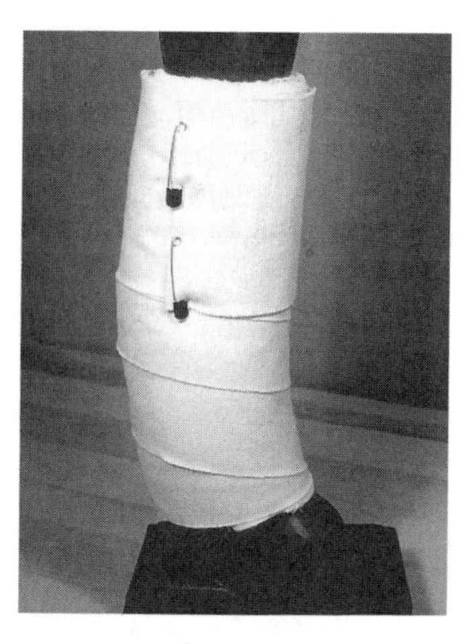

Step 8

The finished bandage should look like a tube on the horse's leg.

Spider Bandage

These bandages are used only on the horse's knees. It is essential to apply them over a standing bandage; without the standing bandage, the spider bandage will slip down off the knee. The spider bandage may be used with liniment on the knee as a precautionary therapeutic measure. It is also commonly used after a bone chip operation involving the knee. *(See Chapter 12 for more information.)*

Making the spider for a spider bandage is very easy. The procedure is outlined in Figure 10–21.

The spider bandage may be removed in one of two ways. If the bandage is soiled from covering a wound, it may be cut off using bandage scissors, and then discarded. If the spider bandage is still fairly clean, it may be untied carefully, washed, and then reused.

Applying the Spider Bandage

Figure 10–22 outlines the procedure for applying a spider bandage to a horse's knee.

1. Position the center of a cotton or quilted pad over the knee. Roll the cotton or quilted pad smoothly over the knee and overlap the top of the standing bandage.
2. Place the spider over the cotton or quilted pad with the strips facing the outside of the horse's leg. Tie the top set of strips together by making a simple knot.
3. After making a simple knot, twist the two ends together and lay them against the leg pointing downward. Tie the next set of strips together over the twisted ends of the previous set.
4. Working down the knee, tie each set of strips together, overlapping the previous sets, until you get to the last two sets. Leaving the next-to-last set of strips open for the moment, tie the very last set of strips together. Twist the ends but this time lay them against the leg pointing upward.
5. Take the next-to-last set of strips (which are the only ones open), and tie a normal shoestring bow on top of the last set you just tied.
6. No strings should be hanging loose on the finished bandage when it is applied correctly.
7. The completed spider bandage allows the horse to flex its leg at the knee. Each set of strips has a certain amount of "give" to allow flexion without coming loose.

Making a Spider (Fig. 10–21.)

Step 1

The spider is made from a sheet of flannel or other strong fabric measuring 16 inches x 26 inches.

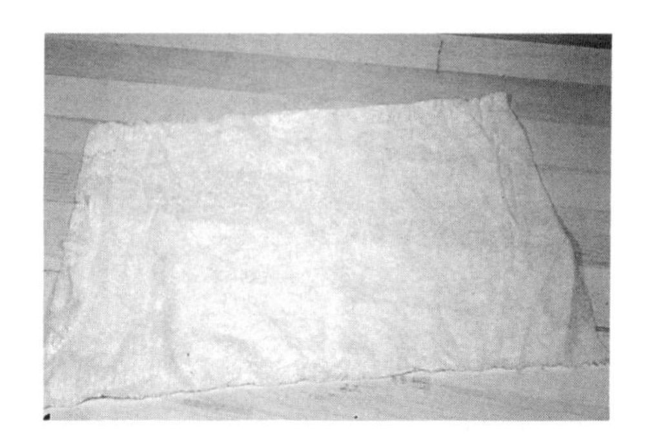

Step 2

Fold the sheet of flannel in half. Cut both layers (top and bottom) into strips toward the center fold, one inch wide and eight inches long. You should be able to cut at least 16 strips.

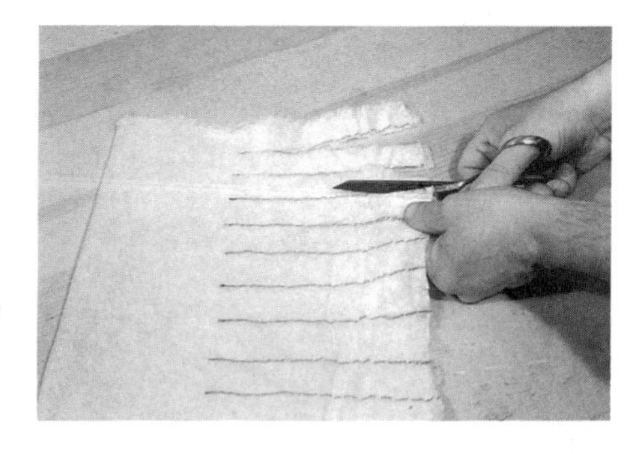

Step 3

Open the sheet of flannel and your spider is ready.

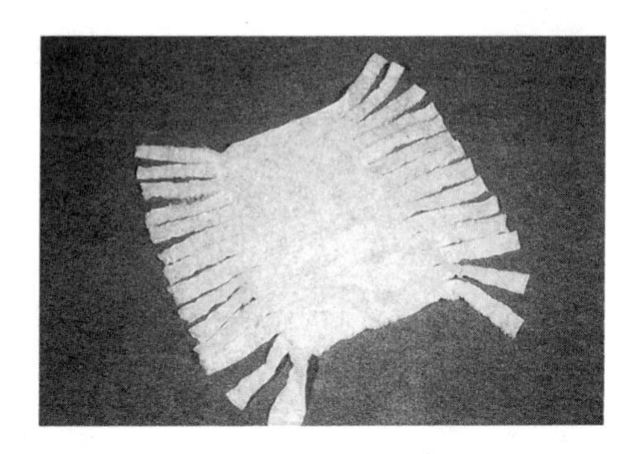

Applying a Spider Bandage (Fig. 10–22.)

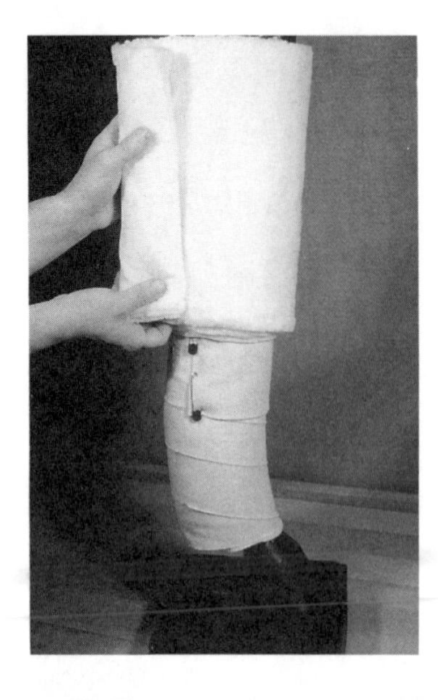

Step 1

Roll the cotton or quilted pad smoothly over the knee and overlap the top of the standing bandage.

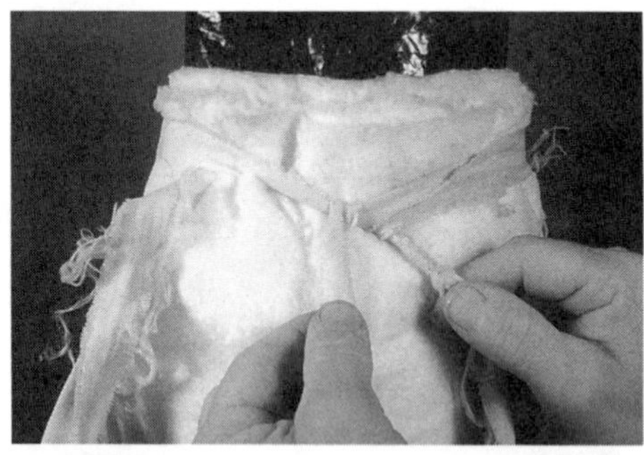

Step 2

Place the spider over the cotton or quilted pad. Tie the top set of strips together by making a simple knot.

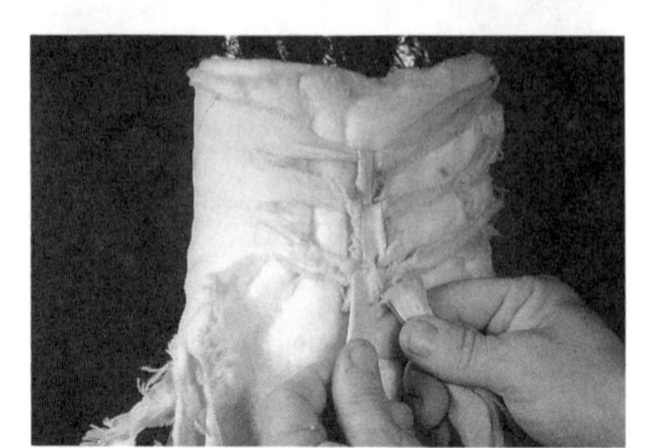

Step 3

Tie each set of strips together over the twisted ends of the previous set.

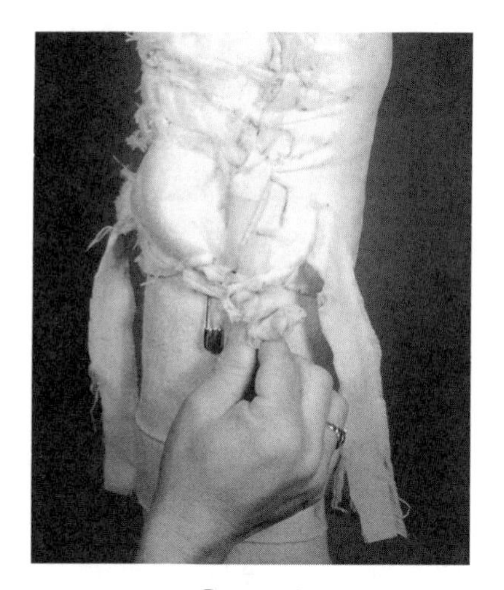

Step 4

Tie the very last set of strips together. Twist the ends but this time lay them against the leg pointing upward.

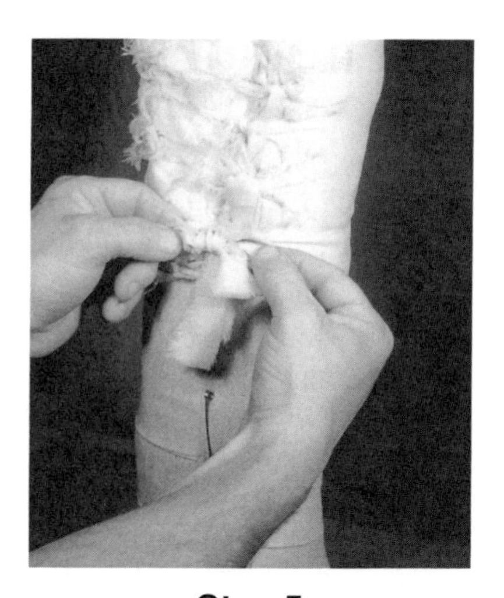

Step 5

Take the next-to-last set of strips (which are the only ones open), and tie a normal shoe string bow on top of the last tied set.

Step 6

No strings should be hanging loose on the finished bandage when it is applied correctly.

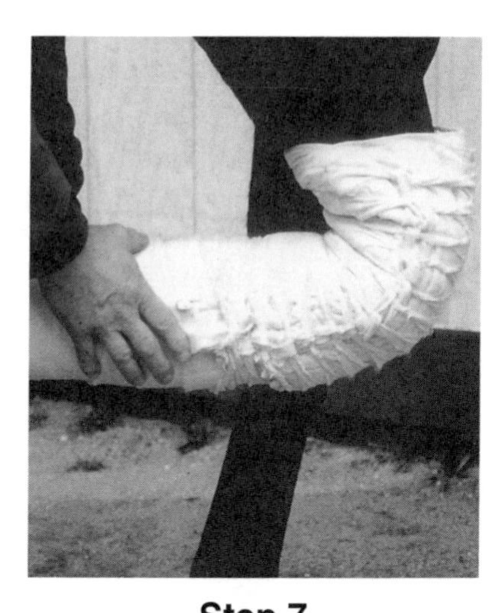

Step 7

The completed spider bandage allows the horse to flex its leg at the knee.

Sweat Bandage

A sweat bandage reduces cold swellings such as windpuffs. *(See Chapter 12 for more information.)* Either a common liniment or a commercially prepared "sweat" may be used with a sweat bandage. It may be applied to all four legs.

The materials needed for the sweat bandage are:

- liniment or a commercial liquid sweat preparation
- Saran Wrap® or a thin plastic sheet (24 inches x 12 inches)
- a cotton or quilted pad
- flannel bandage
- two saddler pins

Note: Many trainers prefer to apply the Saran Wrap around the cotton or quilted pad instead of directly around the leg because there is less danger of any bunched up plastic "cording" the horse's leg. (Cording occurs when something is wrapped so tightly or awkwardly around the horse's lower leg that it restricts the tendon, usually causing it to bow.)

Applying the Sweat Bandage

This procedure is illustrated in Figure 10–23. First apply the liniment or sweat preparation to the leg and massage lightly.

1. Immediately after massaging the leg, wrap a piece of Saran Wrap or a thin plastic sheet around the leg.
2. Place the cotton or quilted pad over the plastic wrap.
3. Roll the flannel on the leg as described with the standing bandage. Insert the saddler pins in the same manner as a standing bandage.
4. The sweat bandage should look exactly like the standing bandage when complete.

Applying a Sweat Bandage (Fig. 10–23.)

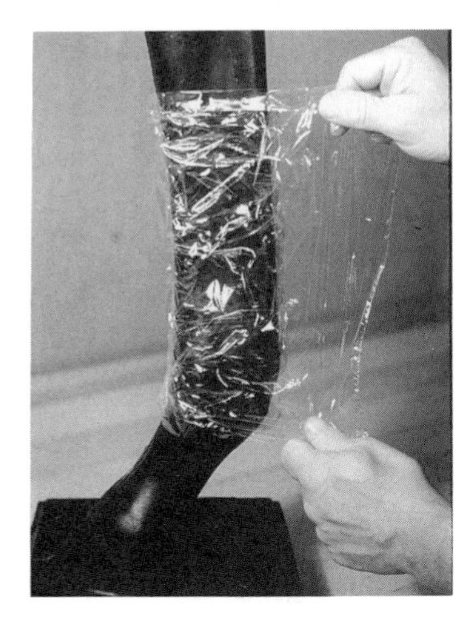

Step 1

After applying the liniment or sweat, wrap a thin plastic sheet around the leg.

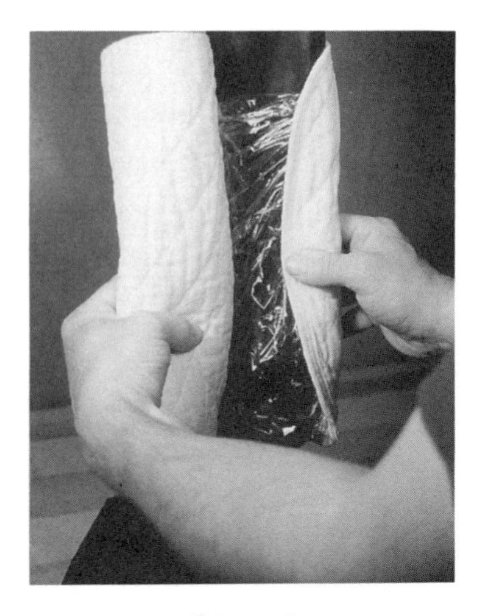

Step 2

Place the cotton or quilted pad over the plastic wrap.

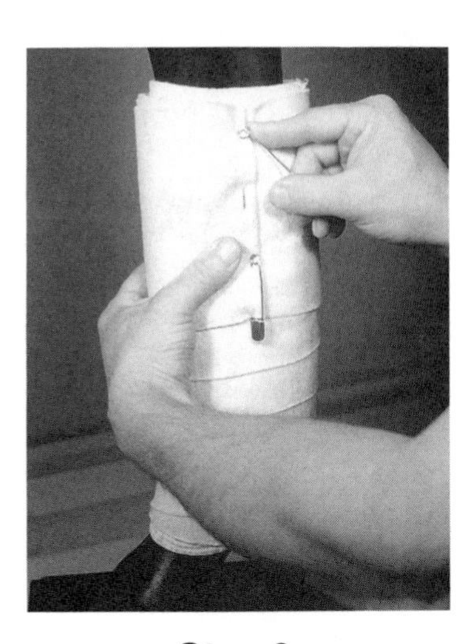

Step 3

Roll the flannel on the leg and insert the pins as with the standing bandage.

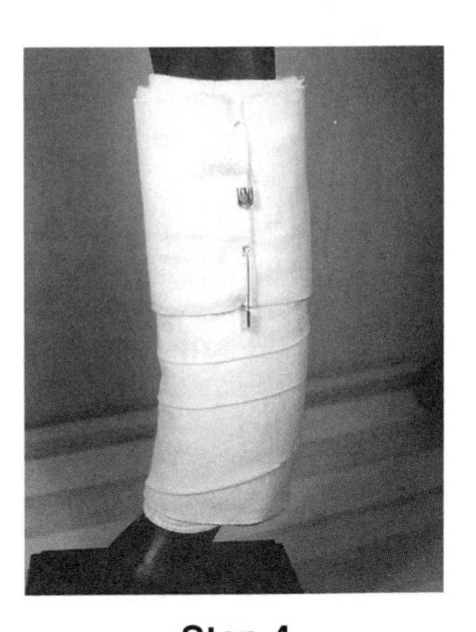

Step 4

The sweat bandage should look just like the standing bandage when complete.

Foot Bandage

The foot bandage protects the foot when treating puncture wounds, thrush, quarter cracks, or any other foot ailment. Normally, this bandage is left on the horse except when exercising or when cleaning or dressing a wound under the bandage.

The materials needed for a foot bandage are:

- burlap or nylon feedbag (12 inches x 12 inches)
- bucket of clean, warm water
- medication
- cotton sheet (15 inches x 18 inches)
- Vetrap® (3½ inches x 2½ yards)

Fig. 10–24. A feedbag keeps the foot clean.

Applying the Foot Bandage

When working outside or inside the stall, place a feedbag on the ground or stall floor and position the horse so that the affected foot rests on the bag. This measure keeps the horse's foot clean while you apply the foot bandage.

This procedure is illustrated in Figure 10–25.

1. Wash the foot thoroughly with clean, warm water and apply any medication recommended by the veterinarian.
2. Flex the affected leg and rest it on your knee so that the foot is off the ground. This position enables you to use both hands to apply the bandage. Place the cotton sheet on the bottom of the foot.
3. Using the self-adhesive (sticks to itself) Vetrap, begin wrapping at the heels, covering the bottom of the foot by crisscrossing in a figure-eight fashion. To do this, wrap from one heel across the bottom of the foot diagonally to the toe, around the hoof wall to the other side of the toe, and back over the bottom of the foot to the other heel. Then wrap the bandage around the back of the foot and start again.
4. Repeat step 3 as many times as necessary. The bandage is complete when the entire bottom of the foot is covered and protected.

Applying a Foot Bandage (Fig. 10–25.)

Step 1

Wash the foot thoroughly and apply any medication.

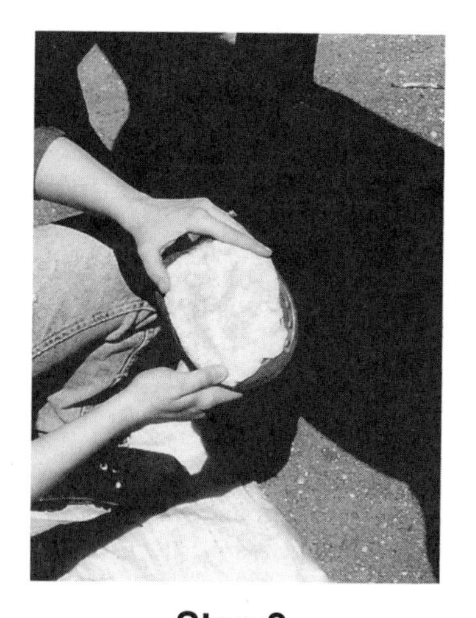

Step 2

Place the cotton sheet on the bottom of the foot.

Step 3

Begin wrapping the Vetrap at the heels, covering the bottom of the foot by crisscrossing figure-eights.

Step 4

When the foot bandage is complete, the entire bottom of the foot is covered and protected.

Exercise Bandages

While harness horses tend to wear more boots during exercise and racing, running horses wear more bandages.

Foam Exercise Bandage (Elastifoam)

The foam exercise bandage supports and protects the ankles, tendons, and shins during a workout. It consists of an elastic bandage with a thin foam backing. When rolled correctly for storage, the foam faces outward, so that when the bandage is applied, the foam is against the leg. This bandage may be used on all four legs.

The materials needed for the foam exercise bandage are just the foam bandage (4 inches x 7 feet) and two saddler pins.

Applying the Foam Exercise Bandage

This procedure is illustrated in Figure 10–26.

1. Lay the foam bandage on the leg just below the knee and anchor it with one complete wrap. Begin wrapping down the leg toward the ankle.
2. Wrap a figure-eight around the ankle. This is accomplished by rolling the bandage *downward* at a slight angle over the front of the ankle. Be sure not to come down onto the pastern. Wrap the bandage around the back of the ankle, and when you come to the front, direct the bandage *upward* at a slight angle. This procedure forms a point in the middle of the ankle. After you have completed the first figure-eight, begin wrapping the bandage over the first wrap in the same manner to form a second figure-eight over the ankle.
3. Slowly and evenly roll the bandage up the cannon until you reach the area just below the knee or hock. Be sure to leave an inch of the last wrap showing. Finish the bandage on the outside of the leg. If necessary, fold the end of the bandage under to end on the outside and insert two saddler pins vertically, one above the other.
4. The finished foam bandage reflects a neat appearance on the leg. The figure-eight wrap around the ankle provides support and yet allows the fetlock to function freely and naturally.

Applying a Foam Exercise Bandage (Fig. 10–26.)

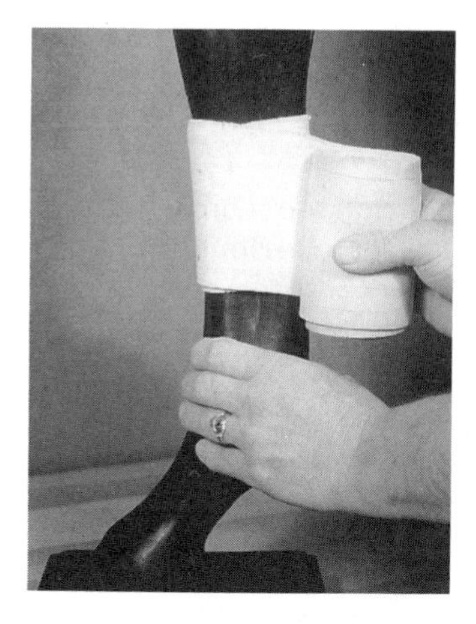

Step 1

Lay the foam bandage on the leg and anchor it with one wrap. Wrap down the leg toward the ankle.

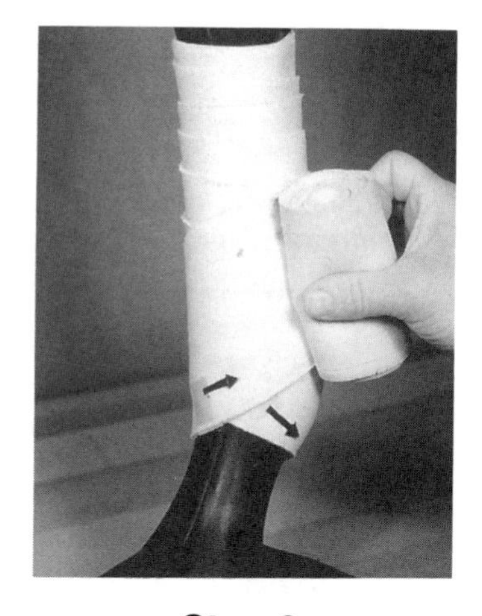

Step 2

Wrap two figure-eight wraps around the ankle.

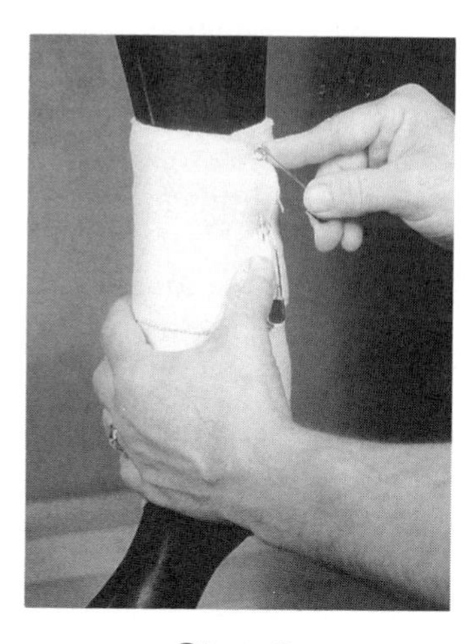

Step 3

Finish wrapping just below the knee (or hock) and insert two pins.

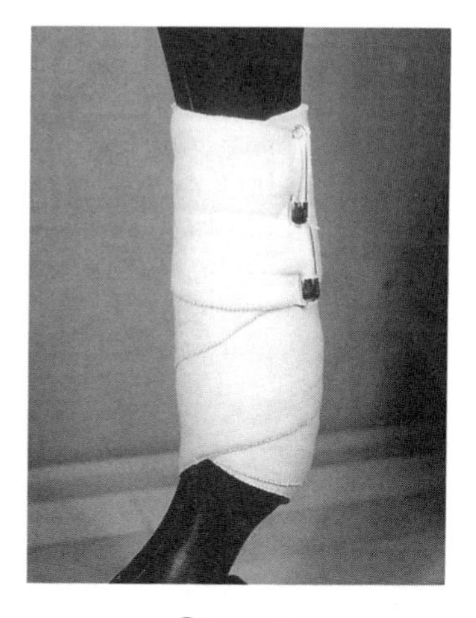

Step 4

The finished foam bandage should reflect a neat appearance on the leg.

Polo Bandage (Legwrap)

The purpose of the polo bandage is exactly the same as that of the foam bandage; when applied properly, it supports and protects the ankles, tendons, and shins on all four legs during an exercise session. It is called a polo bandage because it is commonly used on polo horses.

The materials needed to apply the polo legwrap are merely the polo bandage and one saddler pin. The polo bandage dimensions are 4½ inches x 2½ yards.

Applying the Polo Bandage

This procedure is illustrated in Figure 10–27.

1. Lay the polo bandage on the leg just below the knee and anchor it with one complete wrap. Begin wrapping down the leg toward the ankle.
2. Wrap a figure-eight around the ankle. This is accomplished by rolling the bandage *downward* at a slight angle over the front of the ankle. Be sure not to come down onto the pastern. Wrap the bandage around the back of the ankle, and when you come to the front, direct the bandage *upward* at a slight angle. This procedure forms a point in the middle of the ankle. After you have completed the first figure-eight, begin wrapping the bandage over the first wrap in the same manner to form a second figure-eight over the ankle.
3. Slowly and evenly roll the bandage up the cannon until you reach the area just below the knee or hock. Be sure to leave an inch of the last wrap showing. Finish the polo bandage by fastening the Velcro.
4. To be sure the bandage is held properly, insert a saddler pin vertically on the outside of the leg. The pin is merely a precautionary measure to keep the bandage from unwrapping.

Note: A trainer may use either a foam bandage or a polo bandage for support and protection while exercising a racehorse. The choice is merely a matter of personal preference.

Applying a Polo Bandage (Fig. 10–27.)

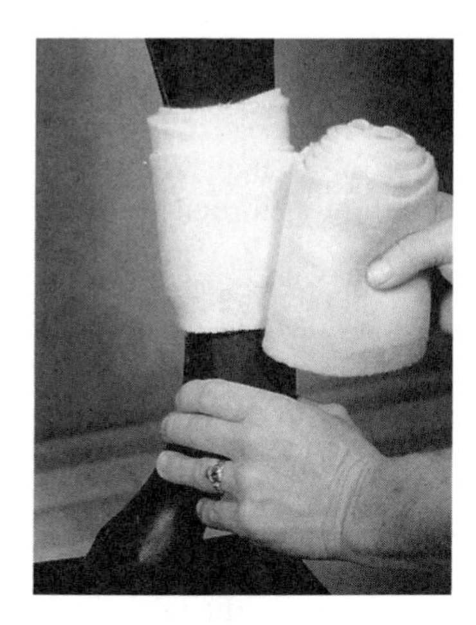

Step 1

Lay the polo bandage on the leg and begin wrapping toward the ankle.

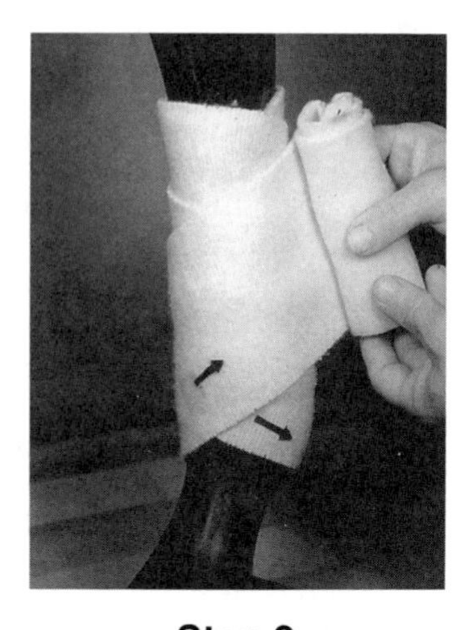

Step 2

Wrap two figure-eights around the ankle.

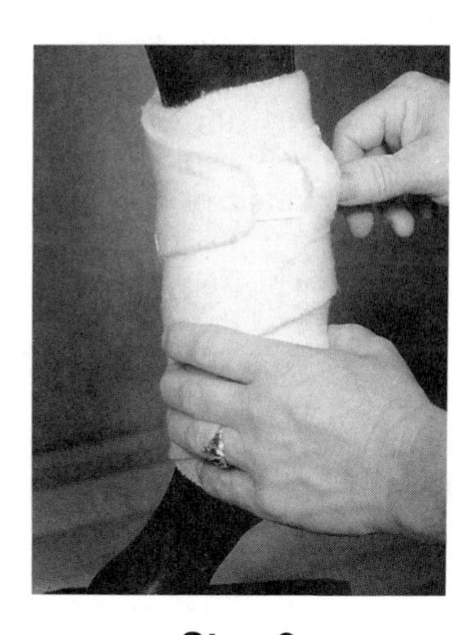

Step 3

Roll the bandage back up the cannon until you reach the area just below the knee or hock.

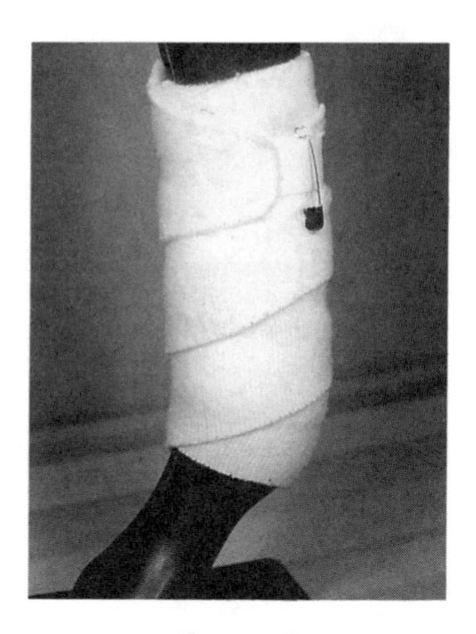

Step 4

Finish the polo bandage by fastening the Velcro. Insert a pin vertically on the outside of the leg.

Brace Bandage

The brace bandage used on harness horses is similar in many ways to the exercise bandages used on running horses; it is applied in the same manner and offers protection and support to the legs. However, the brace bandage is generally used on the hind legs only. As brace bandages weigh less and allow the horse to trot lighter, they are used (instead of shin boots) on trotters that only hit their shins lightly and on occasion.

RUNNING DOWN

Before discussing run down boots, bandages, and patches, perhaps a simple explanation of "running down" is in order. Running down occurs almost exclusively in running horses. The main reason for this is because, during a gallop, all the horse's weight is on one leg at a time. As the horse runs, the back of the pastern "gives" to the concussion of the horse's weight as it pounds the racetrack at considerable speed. (A horse races at about 40 m.p.h.) When the ankle gives too much, or over-extends, its contact with the track is abrasive and the back of the ankle can become raw, open, and very sore.

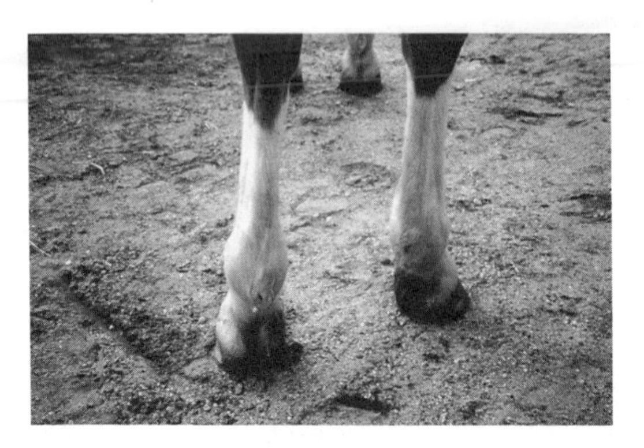

Fig. 10–28. Running down injures the backs of the ankles and makes a racehorse very sore.

Running down is a serious problem with racehorses, as the pain of an open wound on the ankle obviously affects a horse's performance. Conformation (excessively long, sloping pasterns), fatigue, speed, and turf conditions are all contributing factors to running down. Horses that have this problem usually run down on the hind ankles more than the front ankles.

A trainer usually determines whether a horse requires protection from running down based on morning workouts. If so, run down boots or bandages may then become standard exercising equipment for that horse. (Run down boots are used for exercise only; they are not considered racing equipment.) Some trainers prefer to use run down patches rather than bandages for racing as they feel the patches are more comfortable and less action-inhibiting. However, to use run down bandages or patches in a race, the trainer must first obtain the permission of racetrack officials.

Run down bandages require a great deal of skill to apply. Normally, the trainer assumes responsibility for this task. However, if a trainer feels that a groom is capable, then he or she may be asked to apply patches or bandages to the horse before a race.

It is likely that every groom will be required to put run down boots, bandages, or patches on a horse at some point. Therefore, it is important to know how to apply them correctly. If they (run down boots and bandages in particular) are applied incorrectly for a fast workout and the horse runs down, the wounds that result will probably take much time and care to heal properly. Not only will this healing time delay the trainer's schedule in preparing the horse to race; it is also likely to reflect poorly on the horse's caretaker.

The following sections describe the various types of run down boots, bandages, and patches, and give instructions on how to apply them safely and effectively.

Run Down Boot

The run down boot is made of leather or rubber and is lined with soft sponge rubber. It is designed to protect the back of the horse's ankle from running down. It is secured with two straps above the ankle and a third strap below the ankle. The run down boot is used for training only; it is not considered racing equipment.

Fig. 10–29. A leather or rubber run down boot protects the horse's ankles during exercise.

Ace® Elastic Run Down Bandage

The purpose of the Ace elastic run down bandage is to support and protect the fetlock and tendons at high speed.

The materials needed for the Ace elastic run down bandage are:

- water
- sheet of cotton (15 inches x 18 inches)
- Ace elastic bandage (4 inches x 2 yards)
- two saddler pins
- adhesive tape

Applying the Ace Elastic Run Down Bandage

This procedure is illustrated in Figure 10–30.

1. Moisten the leg with water so the cotton sheet will cling to the leg. Wrap the cotton sheet around the leg. Do not worry about the cotton extending below the ankle or above the knee or hock.
2. Lay the Ace elastic bandage on the leg just below the knee or hock and anchor it with one complete wrap. Wrap downward until you get to the ankle.
3. Begin the figure-eight around the ankle. Roll the Ace bandage *downward* at an angle over the front of the ankle. Wrap the bandage around the back of the ankle, and when you come to the front of the ankle direct the bandage *upward* at an angle forming a point in the middle of the ankle. Upon completing the first figure-eight, continue with at least four more figure-eight wraps around the ankle.
4. Slowly wrap the Ace elastic bandage up the cannon until you reach the area just below the knee or hock. Be sure to leave an inch of the last wrap showing. Fold the end of the bandage into a triangle on the outside of the leg. Tuck the end of the triangle so that you create a neat edge.
5. Insert the first saddler pin vertically on the triangular end of the bandage. Insert the second saddler pin next to the first in the same vertical position.
6. Cover both pins with adhesive tape horizontally.
7. Remove the excess cotton by pulling upward at the top of the bandage and downward on the bottom of the bandage. The cotton will tear evenly at the top and bottom of the bandage.
8. The completed Ace elastic bandage supports and protects the ankle during exercise.

Applying an Ace Elastic Run Down Bandage (Fig. 10–30.)

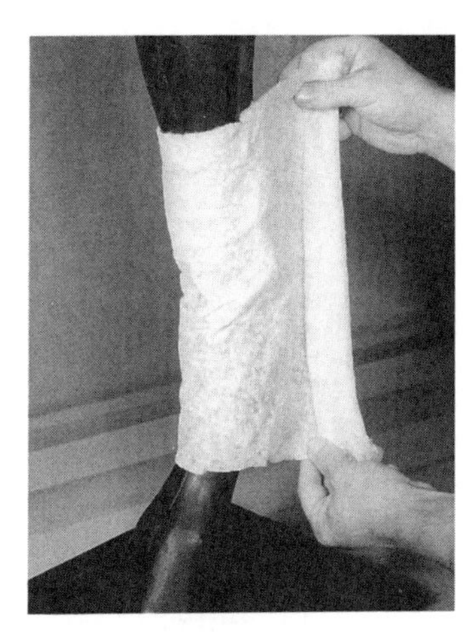

Step 1

Moisten the leg with water. Wrap the cotton sheet around the leg.

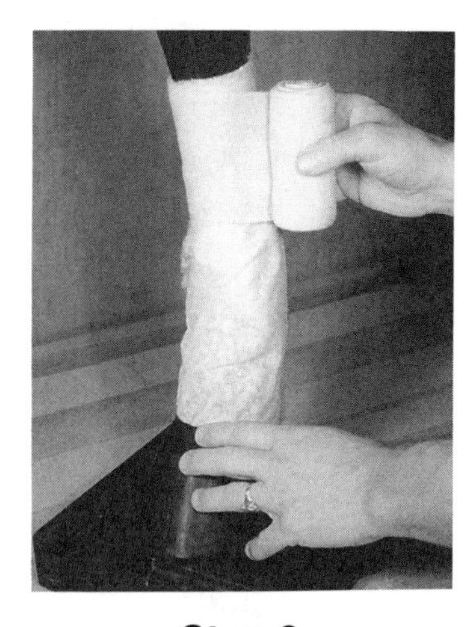

Step 2

Lay the Ace elastic bandage on the leg and anchor it with one wrap.

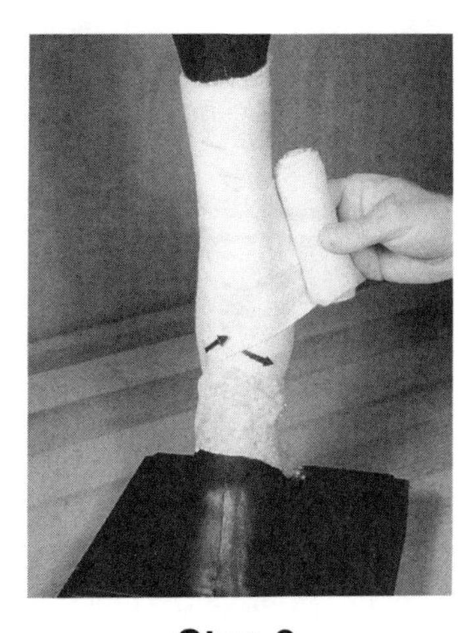

Step 3

Wrap five figure-eight wraps around the ankle. Wrap the bandage back up toward the knee or hock.

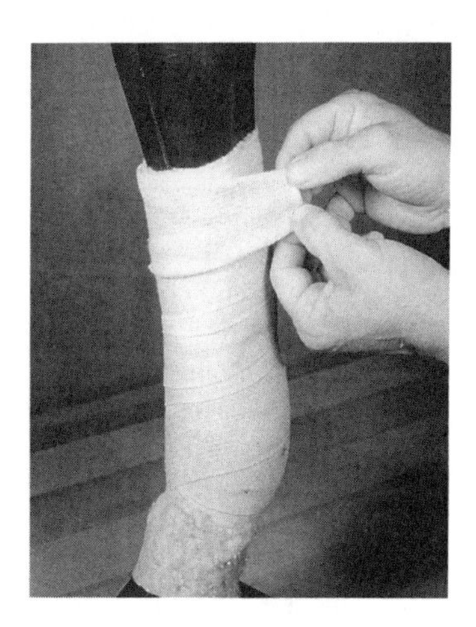

Step 4

Fold the end of the bandage into a triangle and tuck the end of the triangle under.

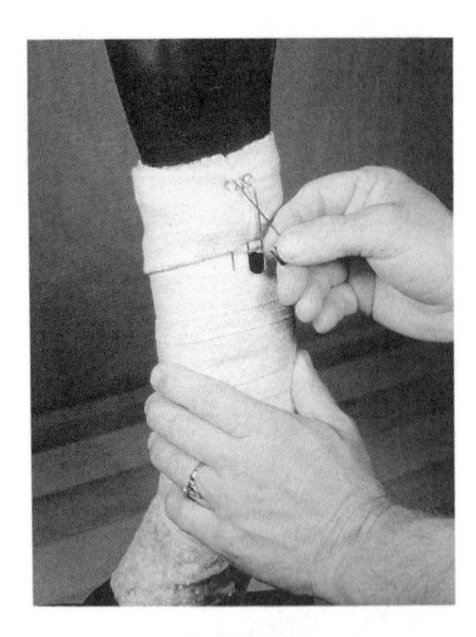

Step 5

Insert the saddler pins vertically on the end of the bandage.

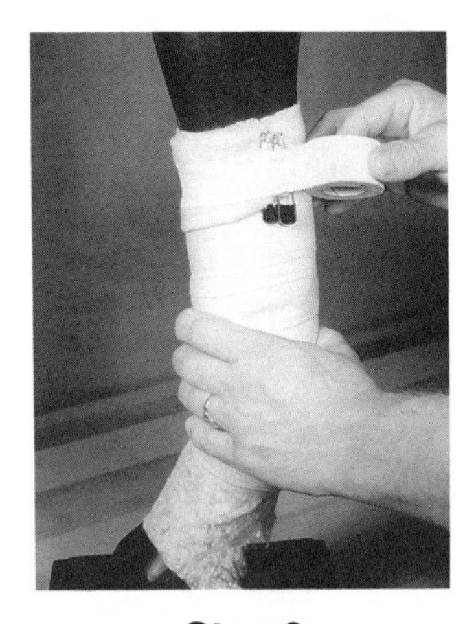

Step 6

Cover both saddler pins with adhesive tape.

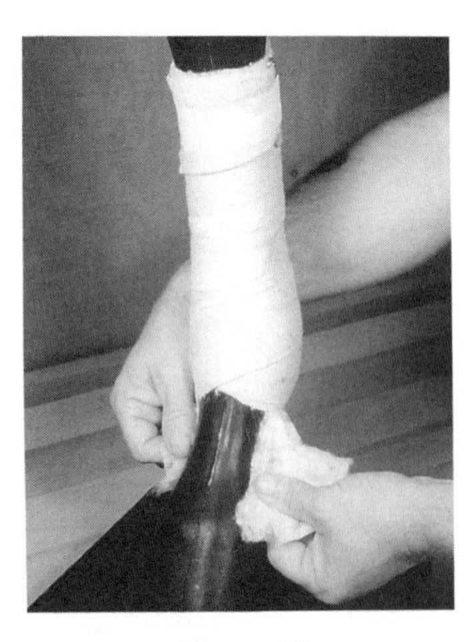

Step 7

Remove the excess cotton by pulling upward at the top of the bandage and downward at the bottom.

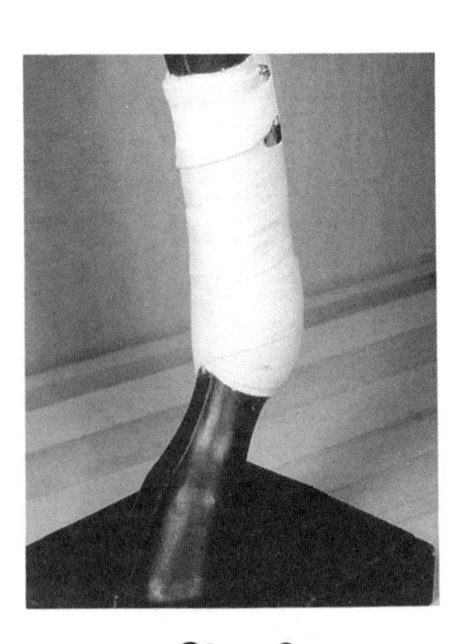

Step 8

The completed Ace elastic bandage supports and protects the ankle during exercise.

Latex Bandage (Sealtex®)

The latex bandage also prevents running down on the back of the ankle, but it is used when the racetrack is wet and sloppy. The latex bandage will not absorb water like the Ace elastic bandage. It is applied to the ankle and cannon. It can be used on all four legs but is generally used on the hind legs only. After the race, the best way to remove the latex bandage is to cut it off the horse's leg using bandage scissors.

The only materials required for the latex bandage are the latex bandage itself and bandage scissors. The latex bandage is 3 inches wide and 36 inches long, and it is made of 100% crepe rubber.

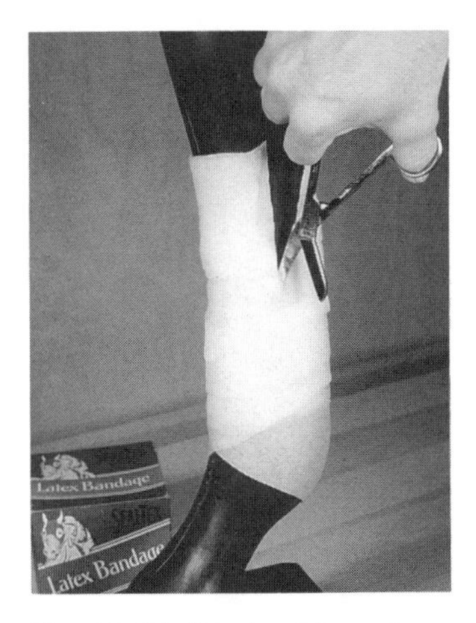

Fig. 10–31. It is best to cut the latex bandage off using bandage scissors.

Applying the Latex Bandage

This procedure is illustrated in Figure 10–32.

1. Separate the paper backing from the latex bandage at the beginning of the bandage. Lay the latex bandage on the leg just above the ankle and anchor the bandage with one complete wrap. Remember to peel off the paper backing as you apply the bandage.
2. Wrap the ankle in a figure-eight as when applying the Ace elastic run down bandage. Slowly stretch and press the latex bandage to the leg, going up the cannon until you reach the area just below the knee or hock. Continue to remove the paper backing as you roll the bandage onto the leg. Be sure to leave an inch of the last wrap showing.
3. To complete the latex bandage, merely stretch and press the end of the bandage to the leg. No pins are required. In cold weather it is necessary to press and warm the bandage with your hands.
4. A properly wrapped latex bandage supports and protects the ankle during exercise.

Applying a Latex Bandage (Fig. 10–32.)

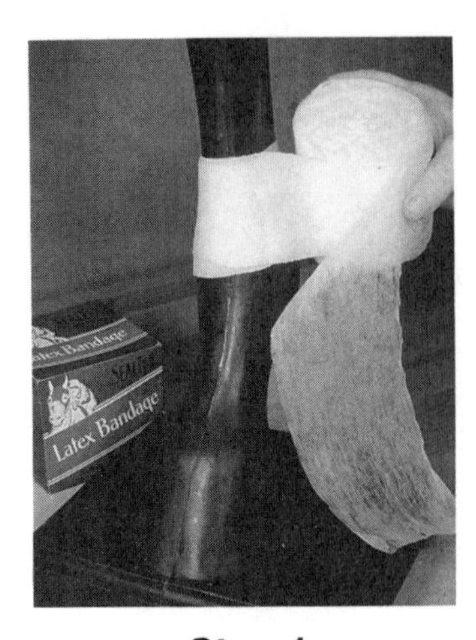

Step 1

Lay the latex bandage on the leg just above the ankle and anchor it with one complete wrap.

Step 2

Wrap the ankle in a figure-eight. Wrap up the cannon, to just below the knee or hock.

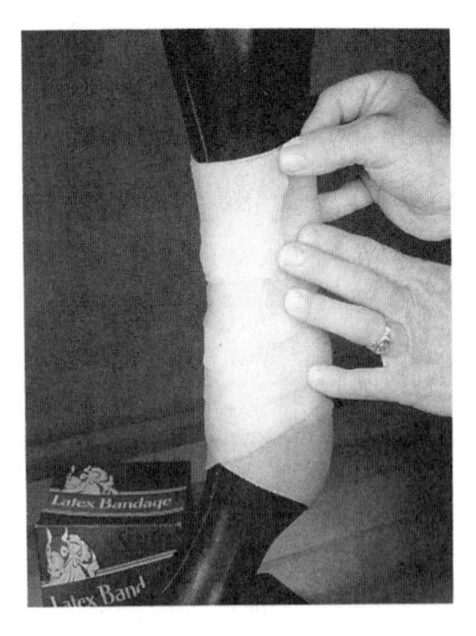

Step 3

Stretch and press the end of the bandage to the leg.

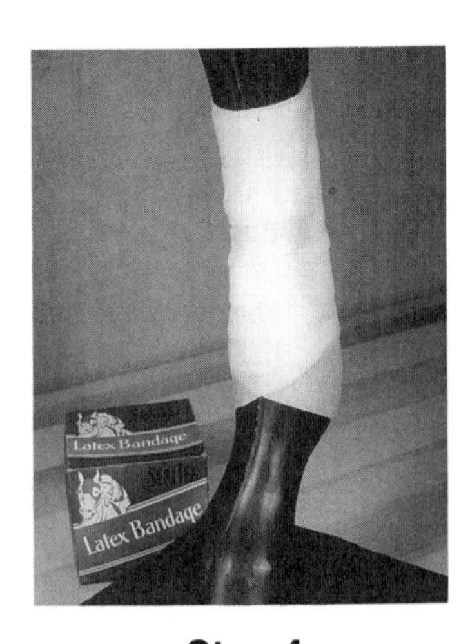

Step 4

A properly wrapped latex bandage protects the ankle during exercise.

Vetrap® Run Down Bandage

Trainers use this bandage to support and protect the ankle and tendons in the same manner as the Ace elastic run down bandage, but using Vetrap instead of the Ace elastic bandage. The Vetrap run down bandage may be used on all four legs.

The materials needed to apply the Vetrap run down bandage are:

- one sheet of cotton (18 inches x 15 inches)
- Vetrap bandage (self-adhesive)
- two saddler pins
- 3M® Rundown Patch (optional)

3M® Patch

The 3M patch is a vinyl disc that prevents injuries due to running down with racehorses. It protects the back side of the ankle during a race or workout. The 3M patch can be used with a Vetrap run down bandage during a race or workout.

Applying the Vetrap Run Down Bandage and 3M Patch

This procedure is illustrated in Figure 10–33.

1. Wrap the cotton sheet around the leg. Be sure to cover the ankle.
2. Begin wrapping the Vetrap just above the ankle. Anchor the bandage with one complete wrap. Make at least two figure-eight wraps around the ankle as described for the Ace elastic run down bandage.
3. At this point, you may apply the 3M run down patch if you wish. The patch will stick to the ankle better if you first make four ¼-inch cuts toward the center of the patch.
4. Peel the vinyl disc from its paper backing.
5. Press it firmly against the back of the ankle, over the Vetrap. Then continue wrapping the Vetrap over the 3M patch with at least three more figure-eight wraps around the ankle. The 3M patch provides extra protection against running down.

6. Slowly wrap the Vetrap bandage upward to the area just below the knee or hock. Be sure to leave at least one inch of the cotton wrap showing. Once you reach the top of the bandage, leave enough Vetrap on the roll for one more wrap. Insert two saddler pins vertically next to each other on the outside of the leg.
7. Cover the two saddler pins with the remaining Vetrap around the upper part of the cannon just below the knee or hock. Press the Vetrap against the leg to end the bandage.
8. Cover the pins with strips of adhesive tape. The adhesive tape is applied to secure the end flap of the Vetrap to the leg, preventing it from coming undone during a race or workout. Tear off the excess cotton sheet by tearing upward at the top and downward at the bottom of the bandage.
9. The completed bandage should look neat and even at both the top and the bottom.

Applying a Vetrap Run Down Bandage and 3M Patch (Fig. 10–33.)

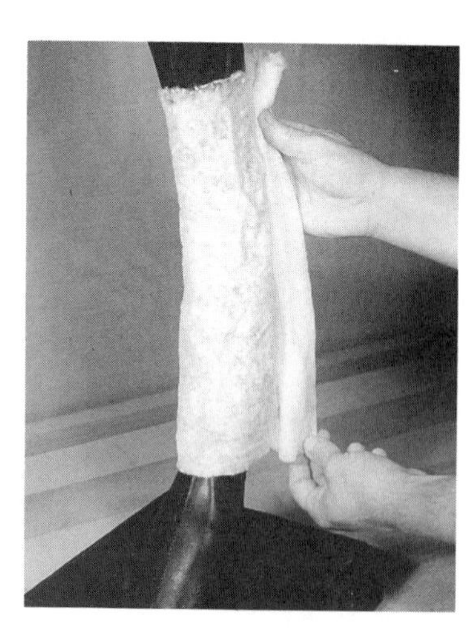

Step 1

Wrap the cotton sheet around the leg. Be sure to cover the ankle.

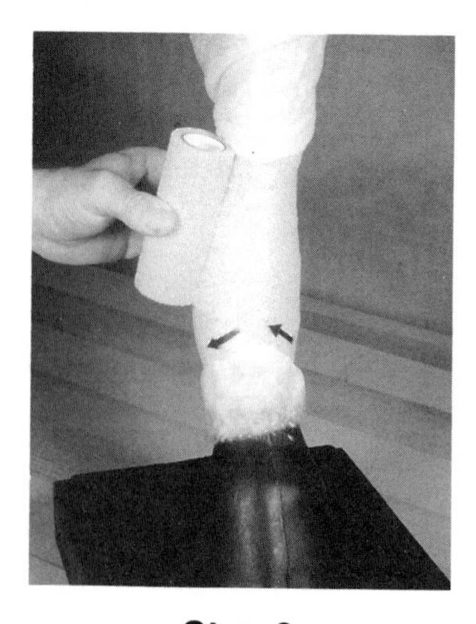

Step 2

Anchor the bandage with one complete wrap. Then make two figure-eight wraps around the ankle.

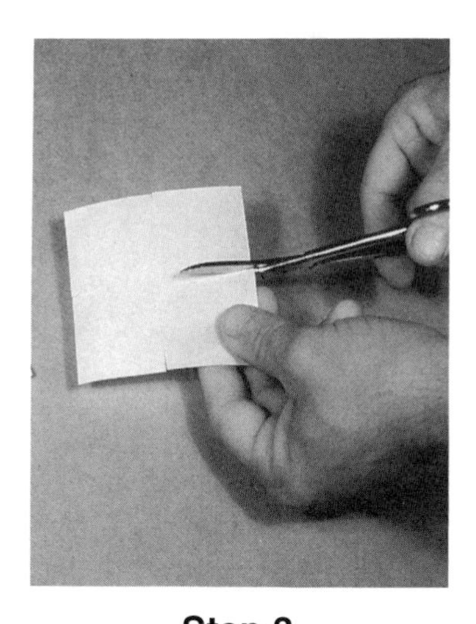

Step 3

At this point, you may apply the 3M run down patch. First, make four ¼-inch cuts toward the center of the patch.

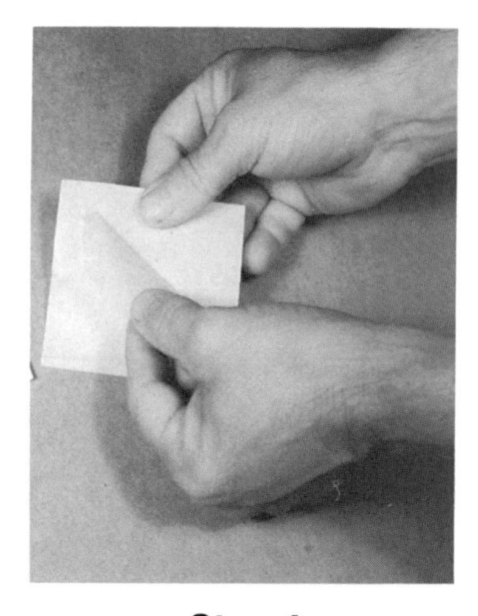

Step 4

Peel the vinyl run down patch from its paper backing.

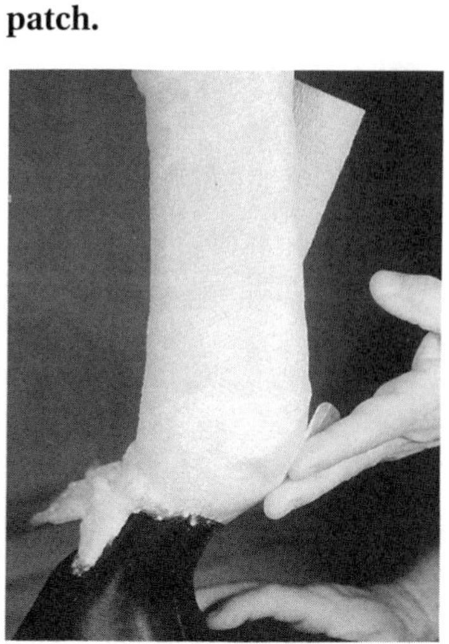

Step 5

Press the 3M patch firmly against the back of the ankle, over the Vetrap.

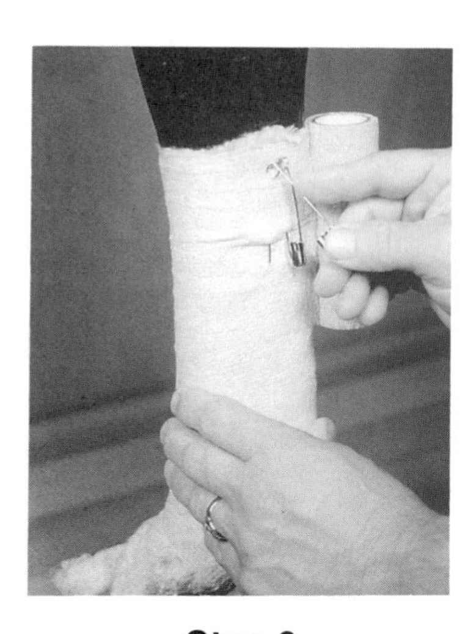

Step 6

Wrap upward to just below the knee or hock. Insert two saddler pins vertically.

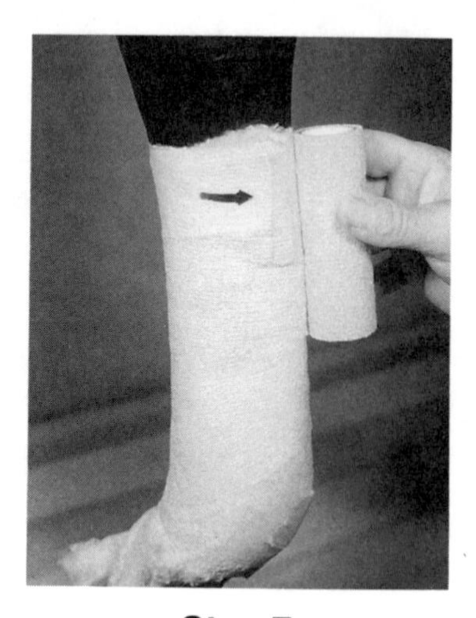

Step 7

Cover the two saddler pins with the remaining Vetrap.

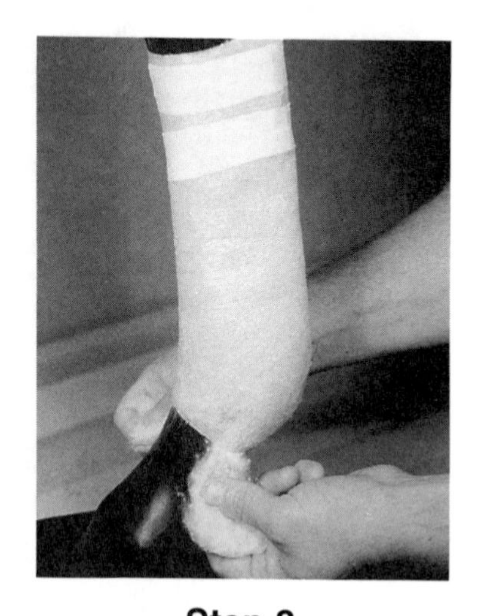

Step 8

Cover the pins with adhesive tape. Tear off the excess cotton.

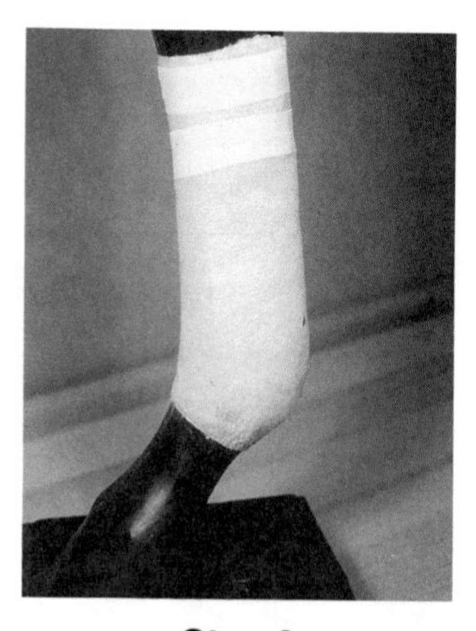

Step 9

The completed bandage should look neat and even at both the top and the bottom.

Rubber Run Down Patch

Rubber run down patches are usually placed on the horse's hind ankles before a race or fast workout. They come in two shapes: disc or flower type. Unlike a 3M patch, a bandage is not left on over the patch while the horse is racing or exercising. The materials needed to apply the rubber run down patch are:

Fig. 10–34. On the left, a flower type run down patch. On the right, a disc-shaped run down patch.

- damp sponge
- spray adhesive
- saddler pin
- rubbing alcohol
- rubber run down patch
- Ace elastic bandage (4 inches x 2 yards)
- adhesive remover

Applying the Rubber Run Down Patch

This procedure is illustrated in Figure 10–35.

1. Remove all dirt and debris from the back of the ankle with a damp sponge. Allow the ankle time to dry.
2. Peel the plastic backing from the rubber run down patch to expose the adhesive material.
3. To be sure the rubber run down patch adheres to the back of the ankle, spray a small amount of adhesive on the patch.
4. Wait until the adhesive becomes sticky and then press the rubber patch firmly against the back of the ankle.
5. Wrap an Ace elastic bandage around the patch and ankle. Secure the Ace elastic bandage with a saddler pin. Leave the bandage on for about ½ hour before removing it. The rubber run down patch should then be firmly affixed to the back of the ankle.
6. To remove the rubber run down patch, merely squirt some adhesive remover between the patch and the ankle. Then clean the back of the ankle with cotton and rubbing alcohol.

Applying a Rubber Run Down Patch (Fig. 10–35.)

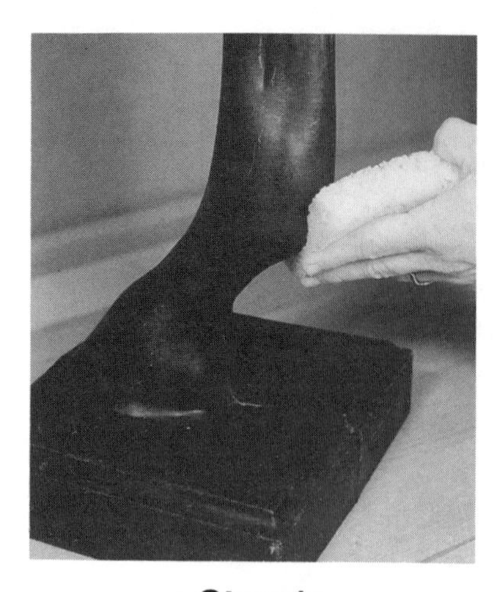

Step 1

Remove all dirt and debris from the back of the ankle with a damp sponge.

Step 2

Peel the plastic backing from the rubber run down patch.

Step 3

Spray a small amount of adhesive on the patch.

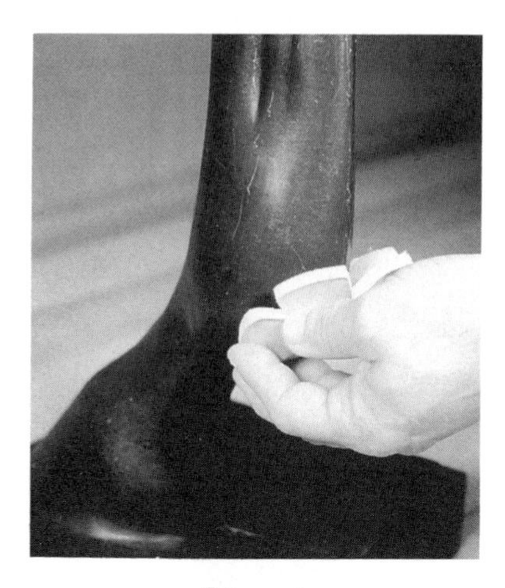

Step 4

Press the patch firmly against the back of the ankle.

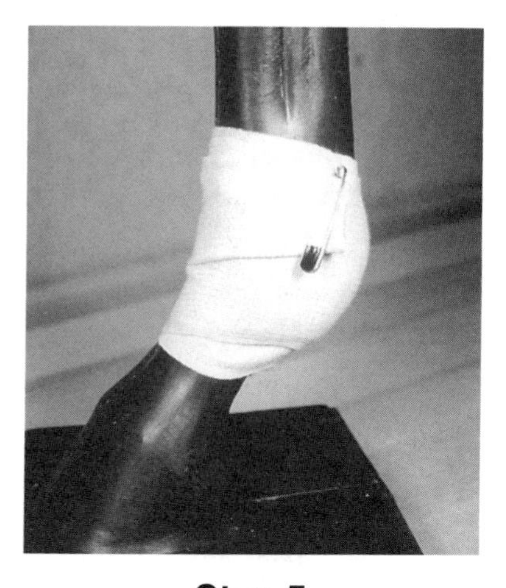

Step 5

Wrap an Ace elastic bandage around the patch and ankle. Secure the bandage with a saddler pin.

Step 6

To remove the patch, squirt adhesive remover between the patch and the ankle. Clean the back of the ankle with cotton and rubbing alcohol.

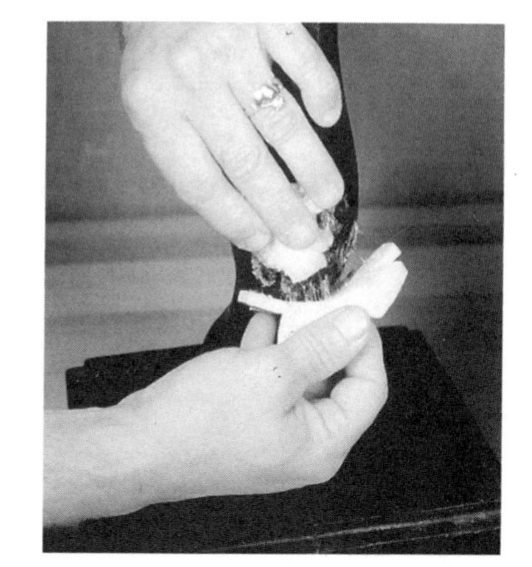

A D Run Down Patch

The A D run down patch protects the horse's ankle from running down and is easier to apply than run down bandages. All that is needed is the A D run down patch itself and a can of spray adhesive. The patch is usually made up of a rubberized felt disc affixed to a large adhesive pad. Four tabs, which are attached to the A D run down patch, wrap around the horse's ankle to secure the patch in place.

Applying the A D Run Down Patch

The correct procedure for applying an A D run down patch is illustrated in Figure 10–36.

1. Remove the protective gauze from the lower part of the patch. Then press the rubber disc section directly against the back of the ankle.
2. Wrap one lower tab of the patch (where the gauze has been removed) around the pastern. Apply some spray adhesive to the tab to secure the patch to the leg, and press the tab against the pastern. Once that tab is in place, affix the lower tab on the other side directly over the first tab in the same manner.
3. Remove the gauze from the upper part of the patch. Spray adhesive on one of the tabs, then press it firmly to the horse's leg just above the fetlock joint. Affix the second upper tab over the first upper tab and the patch will be secure. Gently massage the area around the patch with your hands. Massaging produces heat to make the adhesive stick firmly to the horse's leg.
4. The completed A D Run Down Patch is securely wrapped around the ankle. This patch is not likely to come off during a race or fast workout.

Applying an A D Run Down Patch (Fig. 10–36.)

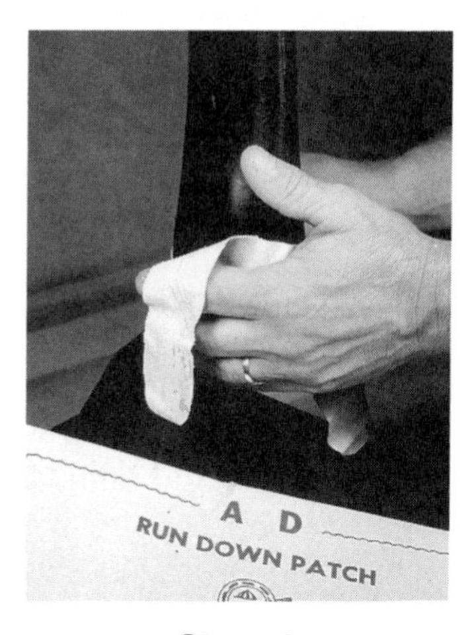

Step 1

Press the rubber disc on the back of the ankle.

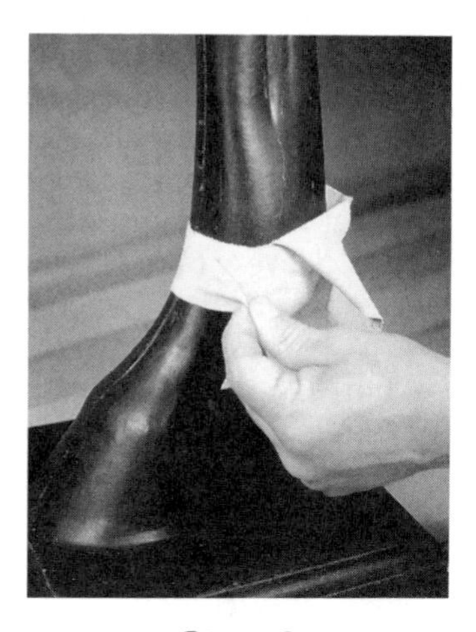

Step 2

Press the lower tab into place, then affix the other lower tab directly over the first tab.

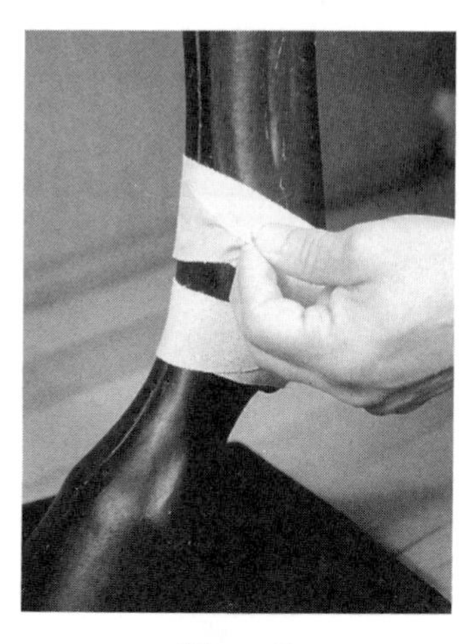

Step 3

Remove the gauze from the upper tabs. Affix them just like the lower tabs.

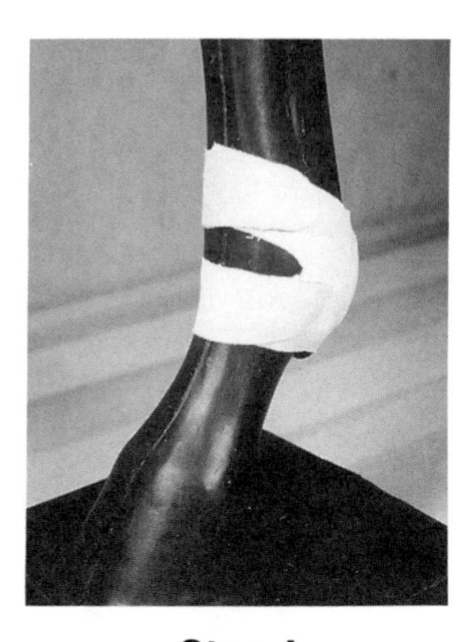

Step 4

The completed A D Run Down patch is secure around the ankle.

SUMMARY

A well-wrapped bandage can prevent a lameness, just as a poorly-wrapped bandage can cause one. Boots and bandages should be in as good a shape when the horse returns from a race or workout as they were when the horse was sent out (except for dirt accumulation). A groom who learns to apply boots and bandages this well is a significant asset to a trainer, who does not have time to double-check every horse's wraps several times a day.

Applying boots and bandages effectively is a valuable skill for other reasons, too. For instance, to pass the practical portion of a running horse trainer's test, you might be required to place a set of run down bandages on a horse. If you have aspirations of someday becoming a trainer, then you should make an earnest effort to master this skill.

Chapter
11

General Health Care

The topic of keeping horses healthy brings to mind an old saying: "An ounce of prevention is worth a pound of cure." As the primary caretaker of the horse, the groom must take every measure to avoid problem situations. Some of these preventive measures may have been discussed in previous chapters:

- adequate exercise
- clean, hazard-free barns and stalls
- consistent feeding
- correct bandaging
- proper foot care
- thorough cooling out after a race or workout

Fig. 11–1.

Besides making every effort to keep the horses safe and sound, a good groom should become familiar with the normal behavior of the horses in his or her care. This practice enables the caretaker to detect abnormal behavior much sooner. In addition, it is good to know what the normal temperature, pulse rate, and respiration rate is for each horse, and check these vital signs daily. The following sections discuss signs of good health as well as warning signs of problems.

(**Note:** These sections are intended as a general overview of equine health. It would take an entire book to completely cover this topic. *The Illustrated Veterinary Encyclopedia for Horsemen,* and *Veterinary Manual for the Performance Horse,* both published by Equine Research, Inc. are two such references.)

SIGNS OF GOOD HEALTH

General characteristics of a healthy horse include the following:

- contentment
- interest in surroundings
- bright eyes
- alert ears
- good appetite
- normal gut sounds
- pliable skin
- shiny and/or dappled coat (dappling is rings of hairs which are darker than the rest of the coat)
- pale pink and moist mucous membranes (eyes and gums)
- normal temperature
- pulse and respiration rates
- normal posture or stance
- normal gait
- absence of wounds, swellings, or pain
- absence of discharge from nose, eyes, ears, or vulva/sheath

Temperature	Pulse	Respiration
99.5°F – 101.5°F	30 – 40 beats per minute	6 – 16 breaths per minute

Fig. 11–2. The vital signs of a resting horse should normally be within the ranges indicated.

Temperature

Each horse's temperature should be taken daily. The normal temperature range for a horse is 99.5°F – 101.5°F (37.5°C – 38.6°C). Like humans, individual horses vary a little in their body temperatures, and the groom will quickly learn what is normal for each horse. The normal temperature of a horse may seem higher on a hot day or after a hot trailer ride, and it will be increased for about an hour after exercise. Also like humans, horses have a daily temperature cycle; it is usually lower in the morning than in the afternoon. Because taking the temperature at different times of the day could cause confusion, it is important to take it at the same times every day. It is also best to take the temperature when the horse is relaxed, such as first thing in the morning and/or in the afternoon before the evening meal. This is because a fever (for example, from a cold), will be hidden when a horse is hot from exercise. It is therefore pointless to take a horse's temperature immediately after a workout or race, as it will be above normal anyway.

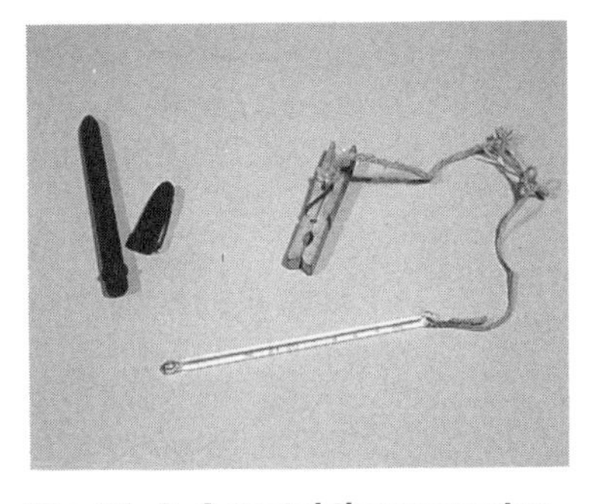

Fig. 11–3. A rectal thermometer with a string and clothespin attached to its glass loop.

To take the horse's temperature, use a rectal thermometer, which is about five inches long with a glass loop in one end. An eight-inch string with a clothespin or alligator clip at one end should be attached to the glass loop. Degrees are marked with large numerals.

Taking the Temperature

Figure 11–4 illustrates the following steps for taking a horse's temperature:

1. Shake the thermometer until the mercury is below normal. Lubricate the bulb end of the thermometer with a small amount of petroleum jelly.
2. Tie the horse in the stall or have someone hold the horse for you. Be sure the horse is standing with its off side against the stall wall. Stand on the near side, facing the rear of the horse.

Cautiously push the tail away from you, exposing the anus. Gently insert the thermometer about three-quarters of the way. Slowly rotating the thermometer as you insert it makes this easier. Do not force the thermometer in—allow the horse time to relax and become accustomed to the thermometer.

3. Release the tail and clip the clothespin to the tail hairs. Leave the thermometer in for at least three minutes to obtain an accurate reading.
4. After three minutes, resume your original position on the near side of the horse. Unclip the clothespin and gently remove the thermometer.
5. Wipe the thermometer with a rag before reading it. Note that the scale is divided into (at least) tenths of a degree. For example, a normal temperature may read 100.4°F. This means that, if your thermometer reads to tenths of a degree, the mercury has risen to the fourth small line after the 100°F mark. With the source of light behind you, slowly rotate the thermometer back and forth between your fingers until you can see the mercury line. Then read the thermometer to the nearest tenth of a degree. *If the horse has a fever (above 101.5°), report it to the trainer immediately.* Clean the thermometer with rubbing alcohol after each use.

Taking the Temperature (Fig. 11–4.)

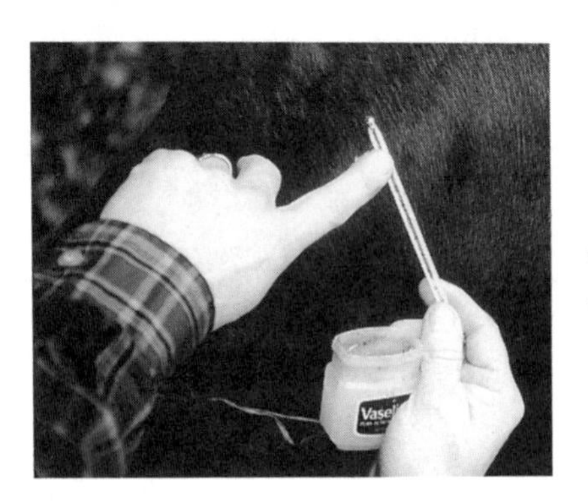

Step 1

Wipe some petroleum jelly on the end of the thermometer.

Step 2

Insert the thermometer in the horse's anus about three-quarters of the way.

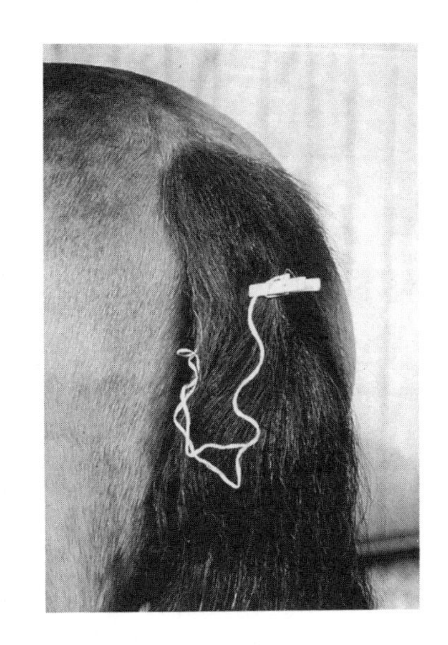

Step 3

Let the horse's tail relax over the thermometer. Then clip the clothespin to the horse's tail.

Pulse

The pulse is the force of blood through the blood vessels. The normal resting pulse rate for a horse is 30 – 40 beats per minute (bpm). The pulse may be taken on several areas of the body—wherever an artery is close to the surface.

Checking the Pulse

The following are eight places on the horse's body where the pulse may be checked, as illustrated in Figure 11–6:

- Maxillary artery—against the edge of the rounded part of the jaw bone
- Facial artery—against the side of the face below the eye or under the straight part of the lower jaw
- Transverse facial artery—against the side of the face just behind the eye
- Heart—behind the elbow against the left side of the chest
- Median artery—inside the forearm
- Digital artery—between the tendons and the cannon bone just below the knee
- Metatarsal artery—outer side of the hind cannon bone
- Coccygeal artery—under the tail close to the body

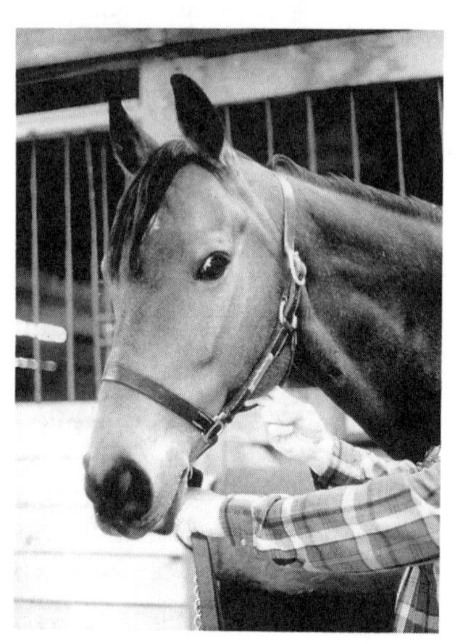

Fig. 11–5. Checking the horse's pulse under the lower jaw.

The most common area where the pulse is taken is under the rounded part of the horse's jaw (1). The artery there is about half the size of a pencil. You should be able to detect a "throb" (which is the result of a heartbeat) by pressing your fingers (do not use your thumb or you will feel your own pulse) against the artery, blocking it momentarily. Then slowly release it until you feel a throb. Once the throb is apparent, tap your foot in cadence with the throb and record the number of beats per minute (or per 30 seconds and multiply that number by 2). If the horse moves a little and you lose the pulse, continue to tap

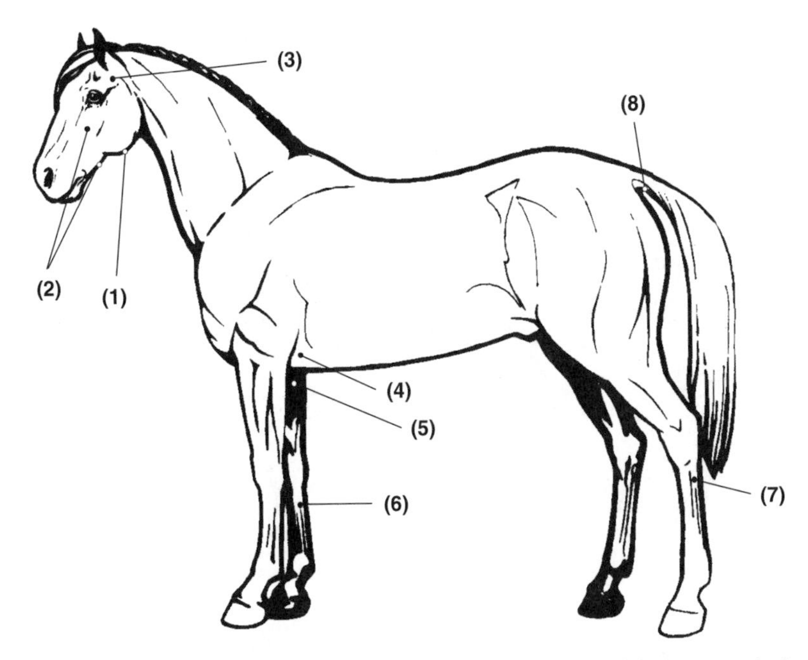

Fig. 11–6. Check the horse's pulse at any of these points: (1) the rounded part of the jaw bone, (2) below the eye or under the straight part of the jaw, (3) behind the eye, (4) behind the elbow on the left side, (5) inside the forearm, (6) just below the knee, (7) outer side of the hind cannon, (8) underside of the tail.

your foot, if possible, in the same rhythm while you try to relocate the artery. If you cannot relocate the pulse within a few seconds, it is probably better to start over.

It is best to check the pulse and respiration in a place where the horse is not distracted. Mild interest in something—another horse passing by—is enough to make the pulse go up noticeably, and outright fear can make the pulse jump by 50 bpm.

Respiration Rate

The normal resting respiration rate for the horse is 6 – 16 breaths per minute, with 12 breaths per minute being average. Several things can make a horse's normal respiration rate increase. Excitement or interest in new surroundings make a horse's respiration rate increase. During exercise, the respiration rate increases dramatically. Hot weather increases a horse's respiration rate, because breathing is also a cooling mechanism. Likewise, transporting a horse in a hot trailer can cause the respiration rate to increase.

Checking the Respiration Rate

Count the respiration rate by observing the movement of the nostrils, flanks, or abdomen. Movement of the nostrils is easy to see after exercise or if the horse is breathing harder than normal due to illness. The nostrils flare out when the horse breathes in. The flanks (located behind the ribcage) show the most movement when the horse is at rest.

Fig. 11–7. A horse's respiration rate can reach 100 breaths per minute during a strenuous race, then return to normal within 15 minutes.

A rise and fall of any of these three areas are considered to be one breath. You merely count the number of breaths taken by the horse for one minute (or for 30 seconds and multiply that number by 2) to determine the respiration rate.

It is best not to touch the horse while checking the respiration rate. The touch of your hand, especially on the flank, may cause normal muscle tremors which make seeing any movement due to breathing difficult.

It is not unusual for a horse to become distracted mid-breath. The observant caretaker may notice that sometimes the horse breathes in half way, stops, then completes the inhalation before exhaling. Do not confuse this for two breaths or rapid breathing.

Mucous Membranes

A mucous membrane is a very thin tissue layer which protects an opening in the horse's body. The mucous membranes which are most easily seen are in the horse's mouth. An experienced horseman can also check the mucous membranes of the eye. However, a horse's eyes can be irritated by a number of things in a stable, so it is a mistake to assume that the horse is ill based on the eyes alone. There are three characteristics of mucous membranes which a groom can easily learn to check: color, capillary refill, and moisture.

The normal horse's mucous membranes are a pale pink color. A horse with a health problem may have dark red mucous membranes (sick or stressed), blue mucous membranes (lack of oxygen), yellow mucous membranes (liver ailment), or white mucous membranes (shock or blood loss). It is normal, however, to see a thin dark red line on the gums around the teeth. This normal feature is not to be confused with a general discoloration of the entire gum.

Condition	Color of Mucous Membranes
normal	pale pink
sick or stressed	dark red
oxygen deprivation	blue
liver ailment	yellow
shock/blood loss	white

Fig. 11–8. Possible explanations for changes in mucous membrane color.

Capillaries are tiny blood vessels. Checking the capillary refill time of the horse's gums indicates whether the circulatory system is properly pumping blood to all parts of the horse's body. To check the capillary refill time, stand on the horse's near side facing the head.

Fig. 11–9. Check the capillary refill time by pressing your thumb against the gum and releasing.

Gently but firmly hold the lower jaw in your right hand. Place your left hand on the horse's nose. While holding the horse's nose with the fingers of your left hand, lift the horse's upper lip on the near side with your left thumb. Press your right thumb against the horse's gum for about one second, then release it and count how long it takes for the spot you pressed to return to a normal pink color. A normal horse's capillary refill time is two seconds or less.

While you are pressing with your thumb, note whether the gums and lips are moist or dry. A normal horse has moist mucous membranes. A horse with dry mucous membranes is dehydrated.

Hydration

A horse should always be allowed free access to clean water, except for immediately after exercise when too much water too soon can cause colic, or in extreme cases, founder. During a one-mile race, a horse may lose half a gallon of water, or more on a hot day. This water loss, due mostly to sweat, naturally produces a mild dehydration which is not serious if the horse is cooled down and watered off properly. A horse can also lose a lot of water through sweat when being transported in a trailer, even though the horse's coat does not feel very wet. This just means that the sweat is evaporating as quickly as it is produced.

Dry mucous membranes and prolonged capillary refill time (three seconds or longer) are two of the first symptoms of dehydration. Another is the skin pinch test. If you gently pinch the skin of a normal horse, the skin immediately snaps back into place when you release it. With a dehydrated horse, the skin stays pinched for a second or more. Of course, pinching different places on a horse's skin gives different results. The most important place to test is over the point of the shoulder. If a horse's skin in this area stays pinched, the horse is approaching a serious level of dehydration. To test for milder dehydration, pinch the skin in the middle of the neck, and on the neck near the throatlatch. Make sure the neck is straight when you do this or you might get a false impression of dehydration. Thin-skinned horses like Thoroughbreds have a slightly slower response time than other breeds. Also, every horse is different, so it is important to test the horse in the various places along the shoulder and neck when you are certain that it is *not* dehydrated to learn what a normal response time for that individual horse is.

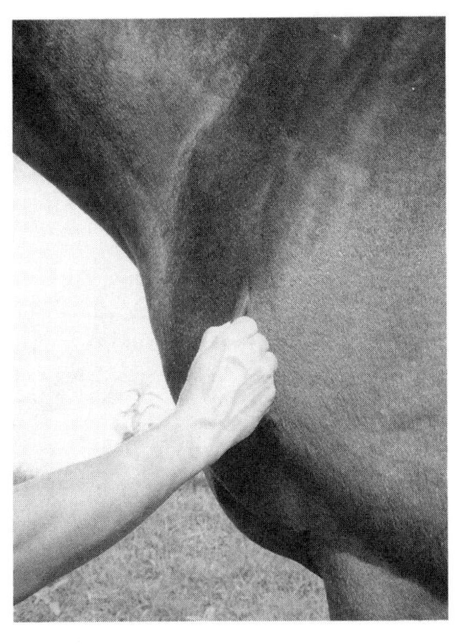

Fig. 11–10. Check the level of hydration by pinching the skin over the point of the shoulder.

Gut Sounds

The horse's body produces gut sounds every 15 – 30 seconds as the feed passes through the intestines. If you press your ear to the horse's side, you should be able to hear them as they are usually quite loud. There are two places to listen for gut sounds on each side of the horse. (It is important to listen to both sides.) The first place is below and in front of the point of the hip. The second place is about six to eight inches lower on the flank.

To listen to gut sounds on the near side, face the rear of the horse and loop your left arm as far as possible over the horse's back. Bend at the waist and press your ear to the horse's side. Keep your eyes open, watching the hind leg to be sure the horse is not going to kick while you are listening. Be prepared to get out of the way if the horse shifts suddenly.

There are several factors which cause changes in a normal horse's gut sounds. The frequency and intensity of gut sounds are, of course, increased for one or two hours after feeding. They may also increase with diarrhea, change of feed, or some types of colic. Normal gut

Fig. 11–11. To listen to gut sounds, face the rear of the horse and loop your arm as far as possible over the horse's back.

sounds are fewer and softer for one or two hours after exercise and at rest between meals. Gut sounds should never be absent altogether.

If you listen for a horse's gut sounds but do not hear anything, wait five minutes and listen again. (If, after five minutes, there are still no gut sounds, refer to Chapter 13 for other symptoms of colic.)

Other Healthy Signs

Most horses have bowel movements at least four times per day. Fresh droppings are slightly moist, free of mucus, and have only a faint odor. The color differs depending on what the horse is fed; most manure is dark green or brown.

An average horse urinates five times or more every day. Horse urine is light yellow, perhaps slightly cloudy, but not milky or strong smelling. The normal exception to this rule is a mare in heat—she urinates small amounts frequently and the urine is thick and very strong smelling.

Of course, all of these indications are very general. Each horse is different, and the observant caretaker soon learns what is normal for the horses in his or her care.

EARLY WARNING SIGNS

It may be difficult to tell that something is wrong unless you know what to look for. Once you know what is "normal" for your horse, you will become alert when something is even slightly abnormal.

Some early warning signs of problems are:

- swellings
- pain or unusual sensitivity to touch or pressure
- lameness or unusual gait variations
- dehydration
- loss of appetite
- coughing or discharge from the nose
- abnormal body temperature, pulse, or respiration
- sweating without obvious cause
- abnormal or absent gut sounds
- change in mucous membrane color
- abnormal behavior: rolling repeatedly, looking at flanks, standing stretched out, shifting weight

In addition to these signs, a horse's general body language often reveals when pain is present. For instance, if a horse is standing in an atypical area of its stall with its neck held lower than the withers, and its eyelids, ears, and lower lip drooping, this can indicate discomfort. Or it could just mean the horse is sleeping. If the horse awakens and perks up at your presence, there should be no cause for alarm. If you call to it and it shows little or no response after several minutes, there could be trouble. When the eyes and nostrils seem tense and have excessive wrinkles around them, and it appears that the horse is grimacing, it is probably experiencing anxiety. Usually, the cause of this anxiety is pain.

More specific body language may indicate the specific site of pain. For instance, if your horse appears hungry at feeding time, but steps back after taking only one or two bites, it probably means chewing or swallowing is painful, and the horse needs its teeth or throat checked by a veterinarian. If your horse is standing over its feed tub without eating, or standing over a

Fig. 11–12. It is important to pay attention if your horse stands over a water bucket without drinking.

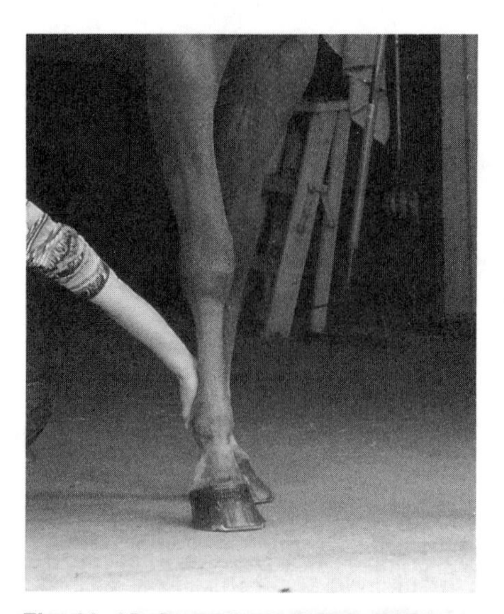

Fig. 11–13. A good caretaker will frequently feel the horse's legs for heat.

water bucket without drinking, it could mean there is pain or stiffness in the horse's neck. If the horse is standing stretched out, with a slightly bowed back and contracted gut, it generally indicates abdominal pain (colic). Stiffness and stubborn reluctance to move can indicate severe muscle cramping (tying-up) or founder. *(See Chapter 13 for information about colic and tying-up.)*

Heat anywhere in a resting horse's legs is a sure sign of trouble brewing. It is the horse's way of telling you that there is some problem developing. A good caretaker frequently feels the horse's legs for heat. As a new groom, you should endeavor to develop a "feel" for heat with your fingers. Lay your hands on as many horses as possible to develop a feel for heat. You will not learn this skill by reading a textbook on the subject; it must be acquired through practical experience.

Check the legs of all of the horses in your care as soon as you arrive at the barn each day. Also check them before they are sent out for a morning workout. When a horse returns from a workout, feel free to ask the exercise rider how it performed. Check the horse's legs for obvious signs of swelling, cuts, etc. After the horse is cooled down, check again for heat.

If you detect heat, tell the trainer immediately. The trainer will probably reduce the horse's work schedule and instruct you to cool the area with hosing, ice packs, etc. until a veterinarian can diagnose the problem.

Another indication that something is wrong with a horse's leg or foot is that it "points" the foot forward, or cocks it slightly backward. The horse dislikes putting any weight on the leg and will probably be lame on that leg when moving around.

PREVENTIVE MEDICINE

Vaccinations

With so many horses coming and going at the racetrack, there is an increased risk of horses being exposed to disease. Consequently, it is important that horses be kept current on all vaccinations. Many tracks require all horses on the grounds to be current on specific vaccinations. These requirements may vary depending on what part of the country the racetrack is located, and state laws regulating incoming and outgoing horses. Figure 11–14 lists the major vaccinations required for horses and a schedule for when the vaccinations may be given. (For more complete information on equine diseases and vaccinations, read *Veterinary Treatments & Medications for Horsemen,* published by Equine Research, Inc.)

A groom is not responsible for giving vaccinations to the horses in his or her care. However, anyone who hopes to become a trainer will want to become familiar with the general vaccination process.

All grooms should know when their horses are scheduled for vaccinations so they can watch for any allergic reactions. A horse that is having an allergic reaction may have some or all of the following symptoms:

- hives—soft, round bumps on the skin, usually beginning at the neck and spreading rapidly down the body
- muscle tremors
- patchy sweating
- anxious or colicky appearance
- respiratory distress—the horse is having difficulty breathing

Respiratory distress of any kind is always an emergency. Do not move the horse. Alert the trainer and/or veterinarian immediately.

Deworming

Internal parasites are bad for the horse's health. All horses have some parasites, but the goal of a good stable manager is to keep the level of infestation to a minimum. A horse with a heavy infestation of

General Vaccination Program	
DISEASE	VACCINATION SCHEDULE
Anthrax	annual, if recommended
Equine Encephalomyelitis (Eastern, Western, and Venezuelan strains)	every 6 – 12 months
Equine Infectious Anemia* (EIA)	Coggins (blood) test every 6 – 12 months
Equine Herpes Virus	every 3 – 6 months for 2-year-olds and under every 6 – 12 months for 3-year-olds and over
Influenza	every 3 – 6 months
Rabies	annual, if recommended
Strangles	rarely recommended
Tetanus	annual
Potomac Horse Fever	at least every 6 – 12 months

Fig. 11–14. All horses should be on a regular vaccination program, as recommended by a veterinarian.

* There is no known vaccination for this infectious disease. A negative blood test every 6 – 12 months confirms the horse was free from this disease at the time the blood sample was taken.

worms has a dull, scraggy coat. Also, the most common reason horses do not gain or hold weight is because of parasite infestation; worms live off the lining of the horse's bowel. The damaged bowel cannot absorb food properly, so the more worms a horse has, the less nutrition it gets from its feed. Worms can also damage other internal organs. The larvae of one kind of worm called *Strongylus vulgaris* can damage the blood vessels in the bowel so badly that circulation to a part of the horse's gut can be blocked off completely. Thus, if horses are not dewormed on a regular basis (at least once every two to three months), parasite infestation can lead to reduced performance, poor appearance, colic, or even death.

Good parasite control consists of daily mucking of the stalls and a regular deworming program. Given a choice, many horses pass manure in a separate area from where they eat: a natural parasite control program. Of course, a horse that is confined to a stall is limited in exercising that option. Therefore, it is up to the groom to prevent heavy parasite loads with regular mucking out.

There are six major types of parasites that plague horses:

- large strongyles (blood worms)
- ascarids (roundworms)
- pinworms
- small strongyles
- stomach worms
- bots

Because different horses may be infected with different types of worms, it is necessary to consult a veterinarian in choosing which deworming method is best. The veterinarian should ultimately use the deworming method that is most effective against the type of parasites with which the horse is infected. Deworming medications are available in five forms: liquids, boluses, feed additives, oral pastes, and injectables. The method of administering the medicine depends on what form is used. The groom may be asked to administer certain kinds of deworming medication, so he or she should be aware of these methods. *(Balling guns and oral dose syringes are discussed and illustrated in Chapter 13.)*

Deworming Methods	
FORM	METHOD OF ADMINISTRATION
Liquid	A veterinarian administers the liquid using a stomach tube. The medication is pumped directly into the stomach. In smaller quantities, it can be mixed with the feed or administered using an oral dose syringe.
Bolus	A veterinarian (or possibly a trainer) administers this large pill by mouth using a balling gun.
Powder or Granules	The trainer or groom mixes the deworming powder or granules with the grain.
Oral Paste	The trainer or groom injects the paste into the back of the horse's mouth with a disposable syringe.
Injectable	A veterinarian administers these about every 60 days through an intramuscular injection.

Fig. 11–15. Parasite infestation is controlled using any of these deworming methods on a regular basis.

Disinfecting Stalls or Equipment

Disinfecting a stall is not a regular responsibility of the groom. However, a stall may be disinfected under the following conditions:

- the stable is moving into a new barn which was previously occupied by other horses
- a horse is being switched from one stall to another within the barn

- the previous occupant died in the stall
- a horse is seriously ill and suffering from severe diarrhea or a pus-like discharge

To disinfect a stall, first remove all bedding and sweep out the stall. Then add disinfectant (closely follow the directions on the bottle) to a bucket of hot water. Dip a rag into the solution, and wipe off any mucus or manure from the walls or stall ledges. Then dip an ordinary broom into the bucket and scrub the solution all over the walls, floor, and ceiling. Do not rinse. Allow the stall to dry thoroughly before rebedding it. Drying may take two days or longer, depending on the climate and weather.

Equipment is disinfected in a similar way, in that it is scrubbed with a brush and disinfectant solution. Do not rinse the solution off. It is best to let the equipment dry thoroughly without rinsing to give the disinfectant time to work. After it is dry, any piece of equipment that will come in contact with the horse's mouth or other mucous membranes should be thoroughly rinsed so that no trace of chemical remains.

The most common circumstance under which all equipment must be disinfected is if the horse using the equipment has been found to have a contagious disease. Otherwise, water buckets, feed tubs, feed scoops, etc. generally only need daily cleaning with a stiff scouring brush and clean, hot water. (Some grooms add household baking soda to the hot water to help clean these items.)

In the past, lye has been used to disinfect stalls, but while it kills viruses and bacteria, it is extremely harmful to the horse (and humans) if swallowed or if it gets in the eyes. Today, it is rarely used.

Lime is used on the stall floor after the stall has been mucked. It acts as a deodorizer, and if sprinkled on wet areas of the stall, it absorbs moisture. However, it is not a disinfectant because it does not kill bacteria.

Commercial Disinfectants

There are several over-the-counter disinfectants, such as pine cleaners, available at any grocery store that effectively disinfect stalls. When selecting such a product, be sure that it says the word "disinfectant" or "antiseptic" somewhere on the label. Otherwise, the product is probably not effective against bacteria, fungi, or viruses. Also, check the label for any indication that it will not work in the presence of organic matter. Organic matter includes small bits of hay, straw, chaff, manure, etc. It is impossible to remove all organic matter from the stall with a broom, so it is important that the disinfectant work in its presence.

Chlorine Bleach

This disinfectant is effective against bacteria, fungi, and viruses. It does not smell as nice as commercial disinfectants, but is effective against more types of microorganisms than many pine cleaners. However, it is only somewhat effective in the presence of organic matter, so it is best used for disinfecting equipment after the equipment has been cleaned, or for disinfecting and whitening laundry. *(See Chapter 16 for information on daily laundry.)*

With this or any other disinfectant, be sure to follow the directions on the bottle for diluting the product with water. Undiluted, these products are toxic to both people and horses.

SUMMARY

A good groom should know each horse well enough to recognize if it is acting abnormally. Careful daily observation of equine behavior is the best way to stay alert to each horse's condition. The caretaker should also know how to check the horse's vital signs and verify that all is normal. It is the groom's job to report any early warning signs of problems to the trainer. The trainer's job is to assess the early warning signs and determine whether or not to call a veterinarian (this is not the groom's job). The trainer is also responsible for maintaining a good preventive vaccination and deworming program.

(**Note:** For complete information on these and other topics on equine health, read *The Illustrated Veterinary Encyclopedia for Horsemen, Veterinary Treatments and Medications for Horsemen,* and *Veterinary Manual for the Performance Horse,* all published by Equine Research, Inc.)

Chapter
12

Detecting & Treating Lameness

Professional grooms should be familiar with the various leg problems that can affect racehorses, and should be able to recognize their different symptoms. Because trainers are often too busy to examine each horse thoroughly themselves, most rely on their employees to detect these problems. A racehorse can only earn money and recognition if it is able to race. If a problem can be detected in its early stages by an alert groom, there is a much greater chance that the horse will recover from it and continue racing.

Fig. 12–1.

Detecting the early warning signs can only be accomplished if you know what a normal horse's leg feels like. Therefore, it is important to feel the legs of as many horses as possible at first. Then, you will know when something is wrong with one of the horses in your care when you conduct routine checks.

Most lamenesses are best detected when a horse is exercising. If you are responsible for exercising the horse, you have the opportunity to notice any of the following problems:

- the horse is stiff when it comes out of the stall in the morning
- the horse carries its head or its hindquarters to one side
- the horse "hunches" its back
- the horse is lame in the turns, or lugs out or in
- the horse swings one leg too far out away from its body in an effort to avoid bending it
- the horse "leans" on one rein or driving line
- the horse refuses to extend itself
- the horse cannot reach or maintain speed

Some of the explanations in this chapter may contain more detail than most trainers expect you to know. Just keep in mind that the object is not to make you into a "textbook veterinarian," but to provide that extra knowledge that can make the difference between an average groom and a top groom. (**Note:** For more detailed information on accurately diagnosing lameness, read *The Illustrated Veterinary Encyclopedia for Horsemen* and *Equine Lameness,* published by Equine Research, Inc.)

WHAT IS LAMENESS?

Lameness is caused by pain. A lame horse may be said to be limping, traveling uneven, or favoring or carrying a leg. Lameness is usually termed mild, moderate, or severe. Pain may arise from sudden injury, or it may be caused by persistent or renewed problems with an old injury. Some lamenesses are caused by congenital defects (problems the foal was born with or developed as it matured, including conformational defects).

A lameness affects the horse's structure or function and interferes with its usefulness. A blemish is a physical defect that **does not** interfere with the usefulness of the horse, but may detract from its appearance or diminish its value. A good groom should be able to distinguish between a lameness and a blemish.

Several factors may contribute to the various lameness problems commonly seen in the racehorse:

- faulty conformation
- accidental injury
- injury due to fatigue or stress
- incorrect shoeing

TYPES OF LAMENESS

The following sections discuss the more common leg problems that plague the racehorse. The first 10 types of lameness discussed may involve either the front or hind legs. The last section is devoted exclusively to hind leg lameness. By learning the specific symptoms of each type of condition, the groom can make quicker and more accurate reports to the trainer. After the sections on lamenesses, we will discuss some common treatment methods.

Bowed Tendons

This is one of the most serious leg problems that can affect a racehorse. In fact, bowed tendons are probably the number one reason why running horses and harness horses are retired from racing. A "bowed tendon" is the term used to describe any damage to one or both of the two flexor tendons (superficial and deep), which are located at the back of the horse's leg below the knee.

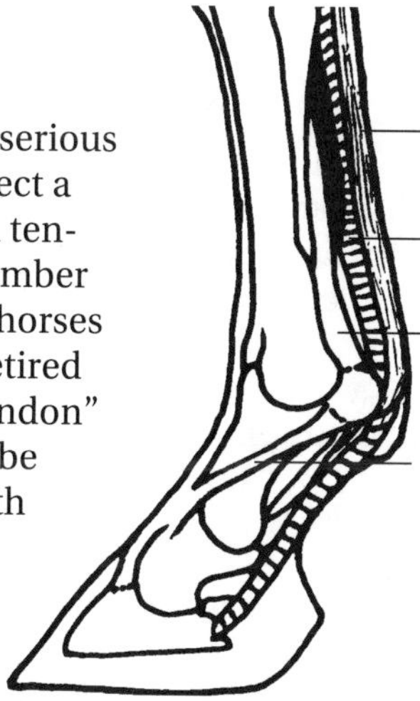

superficial flexor tendon

deep flexor tendon

suspensory ligament

branch of suspensory ligament

Fig. 12–2. Locations of various tendons and ligaments of the leg.

Causes of Bowed Tendons

Bowed tendons are caused by excessive strain and occur primarily on the front legs. The factors which lead to excessive tendon strain include high speed, fatigue, a deep track surface, interference, poor

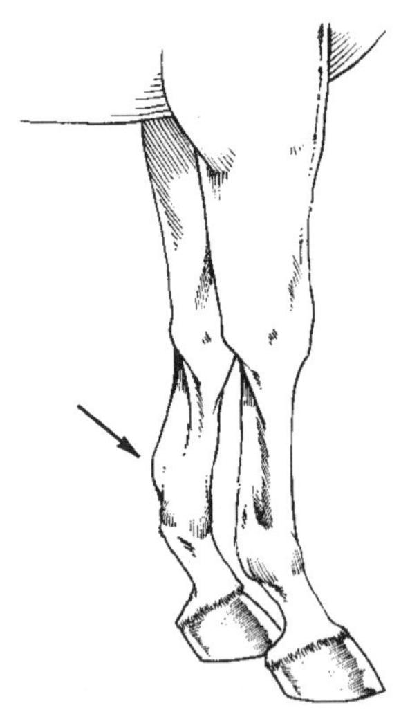

Fig. 12–3. A bowed tendon on the back of the left front cannon.

conformation (long pasterns), and improper shoeing: all contribute to bowed tendons. Occasionally, a poorly applied bandage can "cord" a tendon and cause a "bandage bow."

Symptoms of Bowed Tendons

You should be able to detect a bowed tendon on a horse by noting the swollen or "bowed" appearance of a flexor tendon on the back of the leg. The bow may be high (closer to the knee), in the middle of the cannon, or low (closer to the fetlock). The bowed appearance is accompanied by heat, swelling, and lameness when it first occurs. (Older injuries are not hot, and the horse may or may not be lame.)

Treatment for Bowed Tendons

A veterinarian should supervise the treatment of a bowed tendon. Cold water and ice packs for the first 48 hours on the affected tendon, and anti-inflammatory drugs such as "bute" (phenylbutazone) are usually recommended to reduce inflammation. The horse should be kept as quiet as possible to avoid further damage. After the initial swelling and heat are gone, treatment may entail complete rest, or a very gradual return to light exercise over four to six months. Surgery has also been performed on bowed tendons with some success.

It may take more than a year for a horse to recover from a bowed tendon. Once a tendon bows, it is prone to do so again because of its weakened condition. It is also very possible that the opposite front leg may bow because the horse increases the stress and weight on the good leg to compensate for the injured leg.

It is not uncommon for a horse to resume racing after sustaining such an injury, but as a general rule, a horse is never quite the same after bowing a tendon. Some types of racehorses may be able to race after a tendon injury better than others. For example, a Standardbred may return to racing with better success than a Thoroughbred or Quarter Horse because of the different gaits. A pacer or trotter always has two legs on the ground to support the body, and the racing speed is a little slower than a runner's. Thus, the strain on an individual tendon is less in a Standardbred than in other racehorses. (This is true of many types of lameness when comparing harness horses with running breeds.)

Bucked Shins

This condition frequently occurs in young racehorses (primarily two-year-olds), especially in the last phase of race training. Due to stress trauma on the front legs, the periosteum (membrane covering the bone) on the shins, or front part of the cannon bones, becomes inflamed. The cannon bone may even have suffered some tiny fractures. With untreated or repeat cases of bucked shins, ossification (new bone growth) may occur, giving the front of the shin a thickened appearance.

Causes of Bucked Shins

Bucked shins arise from excessive concussion and stress on the front cannon bones (usually from pushing the horse too fast, too soon). This lameness seems to be most common in running horses.

Symptoms of Bucked Shins

The horse shows obvious signs of pain and heat on the front of the cannon bones. The horse may become lame and shift its weight from one front leg to the other while standing, to alleviate the pain. Be alert to any sensitivity when handling the horse's shins, especially after exercise or a race. If you suspect that a horse may have bucked shins, notify the trainer or stable foreman at once. By the time the horse has a visible swelling on the front of the cannon, the condition has been progressing for some time.

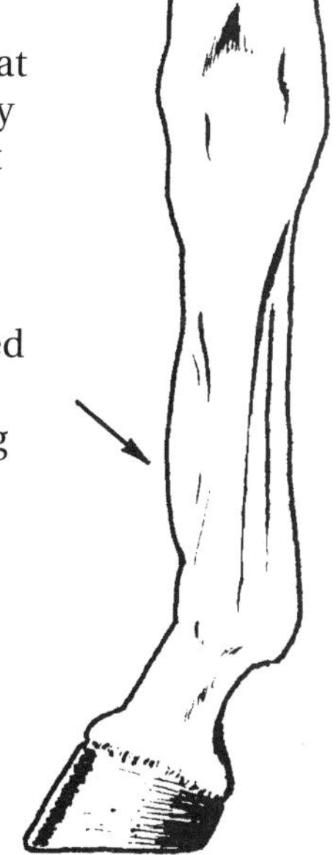

Fig. 12–4. Bucked shin on the left front cannon.

Treatment for Bucked Shins

There are several methods of treating bucked shins, but most professional horsemen agree that complete rest for at least four weeks, followed by a gradual return to exercise, is the key to treating this condition. Resting the horse allows this condition to heal naturally. For the first few days after discovering bucked shins, cold water or cold packs may be applied to the legs, and anti-inflammatory drugs may be given to relieve the pain and swelling.

Splints

This is a common problem which affects racehorses (especially Standardbreds). It is most often found on the front legs, but can be found occasionally on a hind leg. Due to stress and concussion on the legs, ossification occurs at a point of stress somewhere between a splint bone and cannon bone. (There are two splint bones in each leg—one on each side of the cannon bone.) **Note:** A splint can also be classified as a blemish if it does not cause lameness.

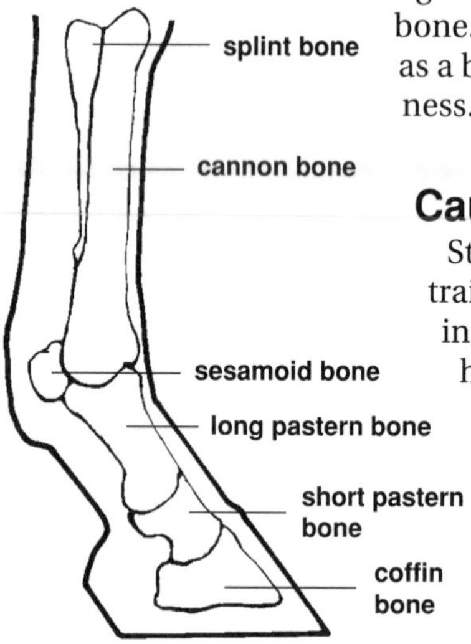

Fig. 12–5. The locations of the foreleg bones.

Causes of Splints

Stress from over-racing or heavy training can cause splints, as can training on a hard surface. Immature horses or horses that are not physically prepared for a certain level of work are prime candidates for a splint, although improper shoeing may also be a cause. Hind leg splints are most often caused by interference.

Symptoms of Splints

Newly "popped" splints, also called green splints, are almost always accompanied by heat, pain, swelling, and perhaps lameness. One can see and feel a single lump or several lumps, and the area is sensitive to the touch. (The lump should not be confused with the natural bony prominence on the end of a splint bone.) Splints are most often found on the inside of the cannon bone a few inches below the knee. In general, the closer to the knee the splint occurs, the more severe the lameness. Professional grooms should learn to run a forefinger and thumb over the horse's leg from just below the knee to just above the fetlock to detect a splint.

During exercise, a Standardbred with a splint often leans on one driving line and/ or carries its head to one side instead of keeping it straight, although other lamenesses may also cause this reaction.

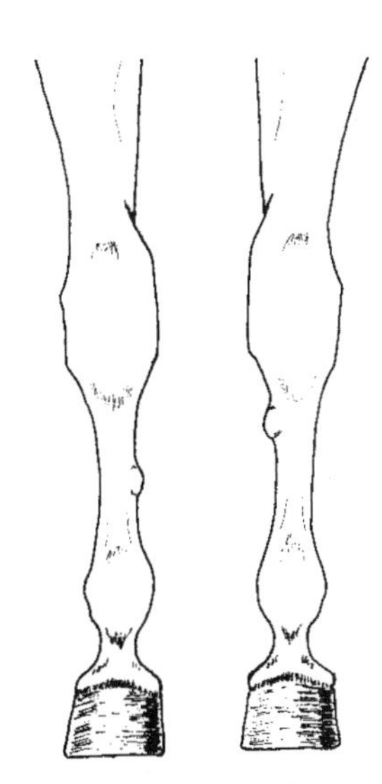

Fig. 12–6. The splint on the right is more likely to cause lameness than the splint on the left.

Treatment for Splints

Cold water and ice packs help to reduce inflammation in the early stages of this condition. Some trainers use counterirritants, such as blisters and leg paints. However, the preferred treatment of splints consists of rest until the inflammation and lameness is resolved.

Suspensory Ligament Strain

The suspensory ligament is one of the main structures that support the fetlock joint. (A joint is where two or more bones meet.) It starts at the back of the cannon bone, just below the knee, and runs down the cannon between the bone and the deep flexor tendon. About two inches above the sesamoids, it divides into two branches. Each branch attaches to the outer edge of a sesamoid bone. Part of the ligament continues across the side of the fetlock and attaches to the short pastern bone and extensor tendon, which runs down the front of the pastern. *(See Figures 12–2 and 12–5 for an illustration of this area.)* Injury to the branches of the suspensory ligament is most common, although injury sometimes occurs near the point of origin.

An injury high on the suspensory ligament (near the point of origin) will cause more lameness and heal more slowly than one lower down. At its point of origin, the suspensory ligament attaches to the cannon bone by tiny fibers. If those fibers tear, or become separated from the bone, they cannot reattach themselves as firmly as before. This area will always be weak in an affected horse.

Causes of Suspensory Ligament Strain

Suspensory ligament injury can be caused by excessive strain as a result of improper shoeing, fatigue (which allows over-extension of the fetlock), tight track turns, deep track surfaces, and pushing a racehorse too fast, too soon. Because of the nature of their gaits, harness horses have a higher rate of injury to this ligament than running horses.

Symptoms of Suspensory Ligament Strain

Heat and swelling over the affected area is most common; thickening of the ligament is usually confined to just one branch, although both may be damaged. Lameness is often present. A veterinarian may ultrasound the ligament and its branches to determine the extent of the damage. He or she may also take x-rays to check for lesions in the sesamoids or splint bones.

Fig. 12–7. Suspensory ligament strain is fairly common among Standardbred racehorses.

Treatment for Suspensory Ligament Strain

Suspensory ligament injuries vary in their treatment because of the other structures of the leg which may also be involved. Cold therapy and anti-inflammatory drugs reduce pain and swelling, and surgery has been tried in some horses. In all cases, rest is very important. A severe ligament tear may require a year to heal properly.

Sesamoiditis

This term refers to inflammation and degeneration of the two sesamoid bones, which are located at the back of the fetlock joint. It is seen in all breeds of racehorses, although it is most common in young Thoroughbreds. As you can see from Figures 12–2 and 12–5, the flexor tendons and the suspensory ligament are closely associated with the sesamoid bones. The flexor tendons run down the back, and the suspensory branches each attach to the outer edge of a sesamoid bone. There are also several smaller ligaments (not illustrated) that attach to the sesamoids.

Causes of Sesamoiditis

Inflammation and eventual degeneration of the sesamoid bones may be caused by tension on the ligaments that attach to those bones, particularly the suspensory branches. Also, sesamoiditis can be caused by pressure from the flexor tendons, particularly at high speed when the fetlock over-extends, also called running down. When the fetlock is over-extended, the sesamoids are pinched, or compressed between the back of the fetlock and the tendons. This pressure is made worse if the back of the fetlock hits the ground. Compression not only stresses the bone structure, but may also compromise the blood supply, which weakens them further. Interference where a horse hits the front sesamoids with the toe of a hind foot (such as over-reaching or cross-firing), can cause sesamoiditis as well.

Sesamoiditis may weaken the bone so much that it is vulnerable to fracture. In some cases, a sesamoid bone may split in two during a race. In other cases, a small piece of bone may be pulled off where a ligament attaches (most commonly a suspensory branch).

Symptoms of Sesamoiditis

The early stages or mild cases of this disease cause only slight lameness and pain when you press directly on the sesamoid bones. In more advanced cases, the affected bone may be a little enlarged. Severe lameness may be accompanied by thickening of the suspensory branch on the affected side. If a fracture has occurred, the entire area at the back of the fetlock will be swollen, hot, and painful, and the horse will be very lame.

Fig. 12–8. Sesamoiditis can be caused by pressure from the flexor tendons, particularly at high speed when the fetlock over-extends. In this case, the left hind fetlock nearly touches the track.

Treatment for Sesamoiditis

Early or mild cases should be treated with cold therapy, anti-inflammatory drugs, and rest. More severe cases should be examined by a veterinarian, who may use x-rays and ultrasound to determine whether bone or ligament damage has occurred. Treatment in these cases depends on the type of damage that is found.

Bone Chips

A bone chip is a small piece of bone that has been fractured and may be dislodged within the joint. Bone chips occur most commonly in the knees, but it is not unusual to find them in the fetlock or hock.

Causes of Bone Chips

Bone chips in the knee are typically caused by over-extension of the knee. Poor conformation, particularly in the knees, also predisposes a horse to bone chips. Chips are more common in young horses that have been pushed to race too soon. Fractures occur because the bones of the knee have not had time to strengthen in response to the stress of training and racing.

Symptoms of Bone Chips

Symptoms of a fracture or bone fragments include swelling, heat, and lameness in the affected joint. The swelling is often confined to one area, and it may appear as a bubble on the front of the knee. During exercise, the horse may swing the affected leg away from its body to avoid bending the knee. While it is important to be aware of the general symptoms, a veterinarian is necessary to properly diagnose this condition.

Treatment for Bone Chips

Treatment for this condition begins with x-rays to determine the exact location of the bone fragment within the joint. Then surgery may be performed to remove the bone fragment or permanently fix the fragment to the bone with surgical screws. An extended period of rest is generally required after the surgery.

Ringbone

Ringbone is a ring of ossified tissue surrounding a bone or joint of the pastern. It is usually found on the front pasterns. Ringbone is a very gradual condition which may go unnoticed for some time. It does not usually affect young horses.

Causes of Ringbone

This problem can be caused by interference, concussion, improper shoeing, or faulty conformation. Horses that are base wide (where the legs are set too far apart) or base narrow (where the legs are set too close together) are predisposed to ringbone. Also, horses that are pigeon toed or splay footed *(see Chapter 5)* are predisposed to this condition. These four conformational problems create uneven pressure on the pastern joints. Horses with upright pasterns are also prone to ringbone because their pasterns are subject to increased concussion.

Symptoms of Ringbone

Ringbone is often characterized by heat, pain, and swelling. However, there is a significant period of time between the cause and the actual new bone growth. Lameness is gradual if the ringbone is a

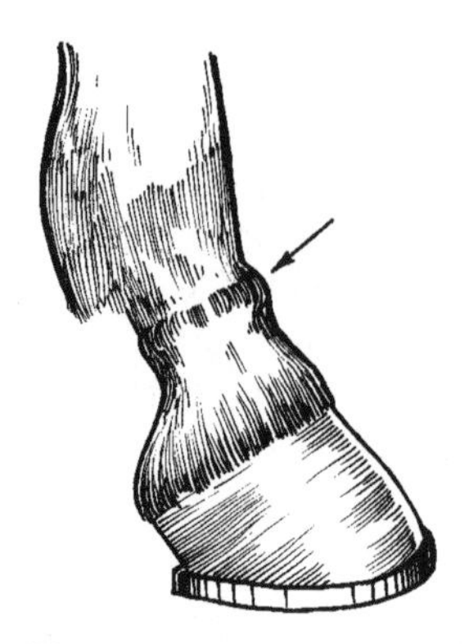

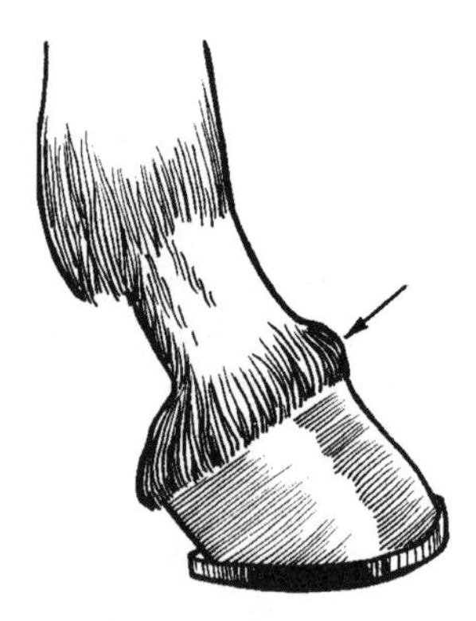

Fig. 12–9. High ringbone.

Fig. 12–10. Low ringbone.

result of poor conformation. If the lameness is sudden, it is usually the result of trauma. For example, if a trotter or pacer already had ringbone, any trauma due to interference could irritate the condition to a point where it causes lameness.

There are two types of ringbone: high and low. High ringbone occurs on the long pastern bone or the upper end of the short pastern bone. To detect high ringbone, run your fingers gently down the pastern from the fetlock to the hoof. You will feel a hard ring of bony growth on the pastern when high ringbone exists. (Do not confuse this with the normal shape of the bone at the first pastern joint.) Low ringbone is found on the lower end of the short pastern bone or the coffin bone. *(See Figure 12–5 for the locations of these bones.)* Low ringbone creates a bony growth just above the coronet.

Treatment for Ringbone

The treatment for ringbone should involve complete rest, and measures to reduce inflammation, such as cold water bandages and anti-inflammatory drugs. In chronic, painful cases, some have a veterinarian "nerve" the affected leg. (The process of "nerving," technically called a neurectomy, is discussed later in this chapter.)

Sidebone

This is an ossification of the lateral cartilage on the wings of the coffin bone. These cartilages are positioned along the sides of the foot, extending above the coronary band and back toward the heel.

Causes of Sidebone

Sidebone occurs in the front feet more often than in the hind feet, and is more likely to occur in heavy horses or horses that are pigeon toed or splay footed. Sidebone may also be caused by concussion (especially with hard surfaces), repeated trauma (as with interference), and poor shoeing. Severe hoof cracks or quittor *(see Chapter 5)* have also been suspected of causing sidebone. It occurs more often in horses over three years old.

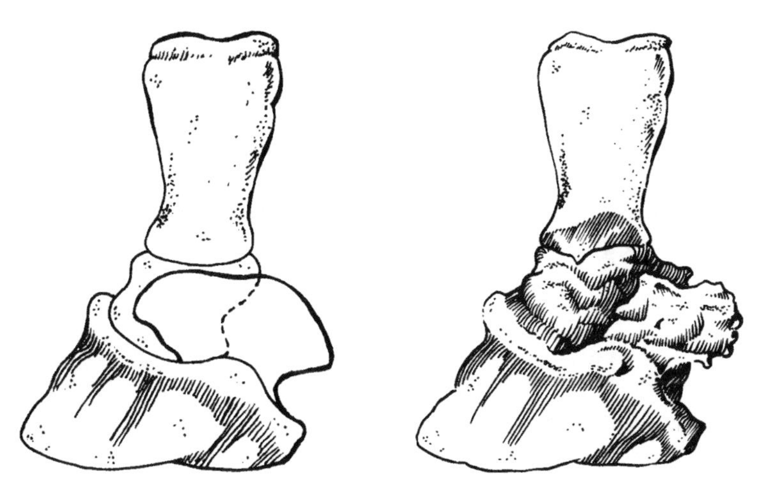

Fig. 12–11. On the left, normal lateral cartilage. On the right, ossified lateral cartilage.

Symptoms of Sidebone

Sidebone appears on the side of the foot (just above the coronet toward the heel). Normal lateral cartilage moves slightly when squeezed. Sidebone can be felt as a hard protrusion that does not "give" when squeezed. The area may also be hot and painful if the condition is active.

Lameness may occur during the period of ossification, particularly when the horse turns. Once ossification is complete, the horse is usually not lame. While there may be no lameness, however, sidebone is a problem because ossified cartilage can inhibit flexibility of the foot.

Treatment for Sidebone

Treatment for sidebone is usually not needed unless the horse is lame. In the case of lameness, treatment may include rest, reducing inflammation with cold water bandages and anti-inflammatory drugs, and corrective shoeing.

Osselets

Osselets begin as an inflammation of the periosteum on the front of the fetlock joint, but chronic cases (cases that persist over time) can result in calcification (thickening due to calcium deposits) and new bone growth in the area.

Note: Another condition that is just as common, looks the same as osselets, and is usually caused by the same factors, is a form of synovitis (inflammation of the lining of the joint). This is a thickening of the small pad of tissue at the front of the fetlock joint. The pad's function is to protect the front edge of the long pastern bone and the end of the cannon bone from trauma during fetlock extension. However, with synovitis, calcification, but no new bone growth, occurs.

Causes of Osselets

Osselets and similar conditions are caused by over-extension of the fetlock joint during exercise. Repeated over-extension results in inflammation and thickening of the tissue pad. In time, the pad may begin to calcify (calcium deposits cause the area to harden) and ossify. Once there is new bone growth, additional trauma may be enough to fracture a small piece of the new bone off the front of the long pastern bone. These bone chips can cause constant irritation to the joint.

Osselets and synovitis are more likely to occur in horses with long, sloping pasterns. (Both conditions are more common in Thoroughbreds than in Standardbreds and Quarter Horses.) Poor shoeing also increases the chances of these conditions occurring and progressing.

If a horse trips during exercise and lands on the front of the fetlock, it may cause tearing of the joint capsule (the fibrous tissue that encloses the ends of bones and provides a barrier for synovial fluid, which lubricates the joint). This injury causes symptoms similar to those seen with osselets and synovitis.

Symptoms of Osselets

These conditions occur almost exclusively in the front legs. The affected area becomes swollen and hot. In fact, when you touch it, the front of the fetlock feels like soft, warm clay. Other symptoms include a short, choppy stride, and pain when pressure is applied to the fetlock or when the fetlock is flexed.

Treatment for Osselets

Some commonly used treatments are anti-inflammatory drugs, ice packs, cold water bandages, poultices, and sweats. Surgery to remove the ossified tissue or bone chips is often very successful. Or, a veterinarian may inject the fetlock with medications which can improve joint mobility and reduce pain and inflammation. If left untreated, flexibility of the fetlock may be permanently impaired.

Windpuffs

Windpuffs (also called wind galls) are soft, puffy enlargements that occur around the fetlocks. They are little pouches of excess fluid in the joint capsule or tendon sheath (fluid-filled protective sleeve surrounding the tendon).

Causes of Windpuffs

Windpuffs are generally caused by overwork. Windpuffs should not be confused with "stocking up," which is a more general swelling of the lower leg caused by poor circulation due to lack of exercise.

Symptoms of Windpuffs

These swellings feel soft to the touch. They may be found on all four fetlocks, usually on the side and toward the back of the joint. Windpuffs usually do not cause lameness, and thus can be considered a blemish. Windpuffs may become permanent blemishes.

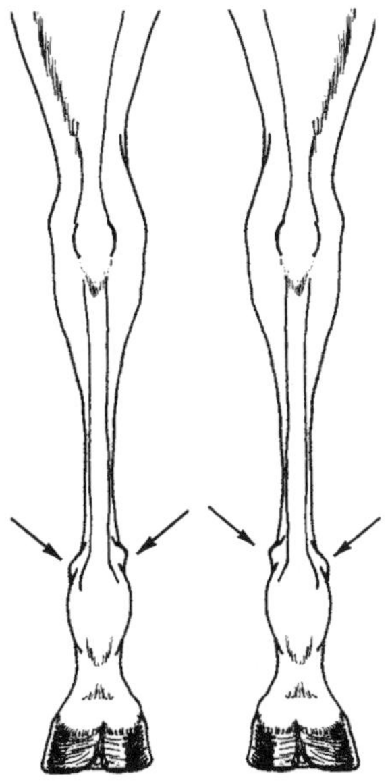

Fig. 12–12. Windpuffs on both hind fetlocks.

Treatment for Windpuffs

Usually, no treatment is necessary. If the horse is lame, there may be some other problem besides windpuffs. However, some trainers treat windpuffs by painting or applying liniment and standing bandages. A sweat bandage using a common liniment or a commercially prepared sweat may also be used. Windpuffs may disappear with a decreased workload, and then suddenly reappear once the workload increases.

Hind Leg Lameness

Most lamenesses plaguing running horses affect the front legs. However, harness horses experience as many hind leg injuries as front leg injuries. Moreover, a hind leg injury may lead to a front leg problem. For example, a pacer sometimes develops a shoulder lameness which has its origin in a hind leg injury. If a hind leg is a little stiff or slow, the front leg must drag the hind leg forward because of the hobbles. In fact, some hind leg lamenesses may go undiagnosed until an associated front leg problem becomes apparent.

Bone Spavin

The hock joint is not just the joining of the two large leg bones: there are many bones which form this joint. Bone spavin involves cartilage destruction and new bone formation within the small, lower joints between the tarsal bones, usually in front and toward the inside where they meet the cannon bone.

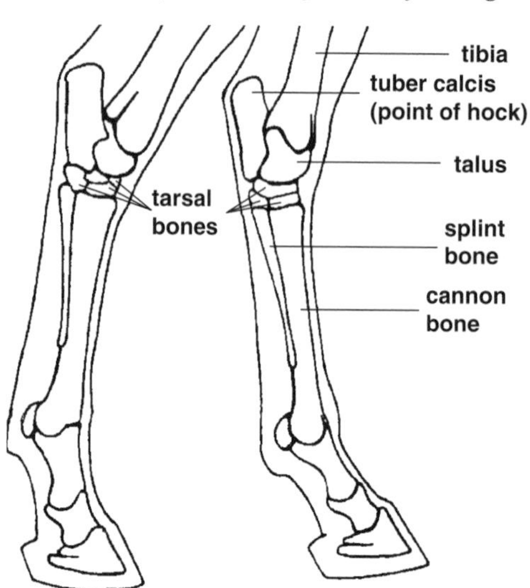

Fig. 12–13. The structures of the hock.

Causes of Bone Spavin

Faulty conformation is the primary cause of bone spavin. Horses with cow hocks (where the fetlocks are farther apart than the hocks) or sickle hocks (where the legs from the hocks down are angled underneath the horse) have increased pressure on the hock joint, which predisposes to bone spavin. It is more common in Standardbreds than in Thoroughbreds or Quarter Horses, although its incidence in all three breeds increases with age.

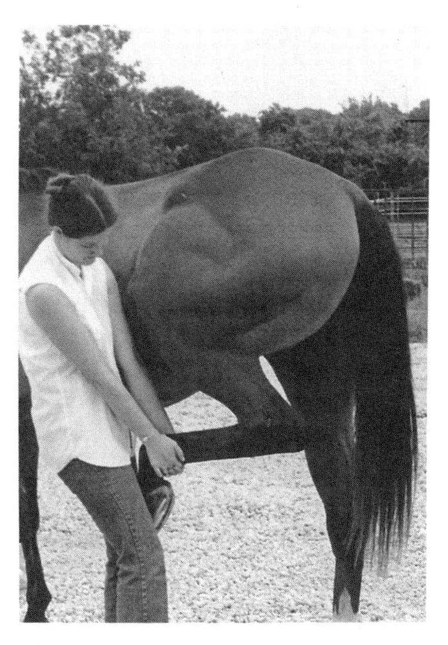

Fig. 12–14. A spavin test is one way to diagnose a bone spavin.

Symptoms of Bone Spavin

Bone spavin is difficult to detect because there is usually no heat present. And rarely does bone spavin produce an obvious bony growth on the hock. But even if there is a bony growth accompanied by lameness, the lameness may be due to another factor. Because this abnormality is mostly internal, it is necessary to have a veterinarian to properly diagnose bone spavin.

One way the veterinarian may diagnose this condition is to do a spavin test. This test involves lifting the hind leg forward and up toward the horse's belly, flexing the hock joint. The leg is held flexed for one or two minutes, then let down, and the horse is immediately trotted away while the veterinarian watches. If the horse shows lameness, this may indicate bone spavin. The veterinarian may take x-rays to confirm the diagnosis.

Treatment for Bone Spavin

Bone spavin may be treated with complete rest and anti-inflammatory drugs (such as bute), although this usually only settles the condition temporarily. Once work resumes, lameness usually returns. Injecting the lower joints of the hock with joint medications can help, and in some cases, surgery to fuse the lower hock joints may be performed. (The joints should fuse on their own in about three years.) However, the condition is irreversible, so treatment is aimed at merely making the horse comfortable.

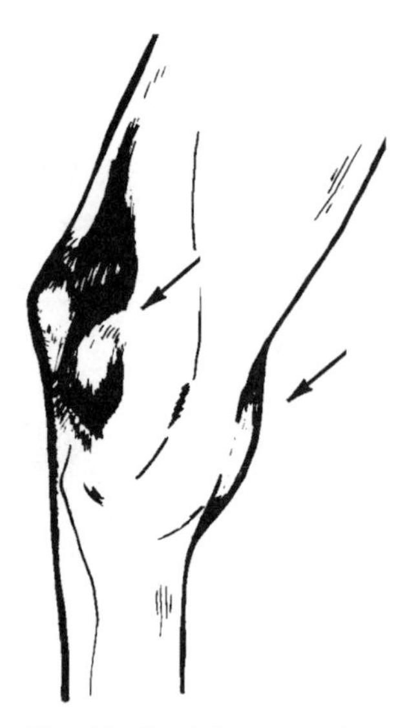

Fig. 12–15. A bog spavin on the hock.

Bog Spavin

This condition is an accumulation of fluid within the joint of the larger hock bones. The fluid pouches out between the tendons that run over the hock, creating several soft swellings.

Causes of Bog Spavin

In young racehorses, the leading cause of bog spavin is a bone disease which can cause small pieces of cartilage and bone to break off within the joint. This, in turn, causes the accumulation of fluid in the joint. In older racehorses, poor hind leg conformation (as described for bone spavin) can cause excess strain on the joint, and this can lead to bog spavin as well.

Symptoms of Bog Spavin

Bog spavin occurs in four locations around the hock:

- a large swelling on the inner part toward the front
- a large swelling on the outer part toward the front
- a smaller swelling on the inner part toward the back
- a smaller swelling on the outer part toward the back

When one swelling is pressed in with the fingers, the other swellings enlarge, although often the smaller swellings are hard to see.

Treatment for Bog Spavin

If there is no lameness with bog spavin, treatment is not usually attempted. Some trainers use cold therapy and/or liniments on bog spavin. However, most attempts to resolve this condition have been unsuccessful. If there are cartilage or bone fragments, surgery is the best treatment. Neither blistering nor firing appear to be successful.

Curb

This problem is an enlargement found just below the point of the hock on the back of the leg. Curbs can initially cause lameness, but full recovery generally leaves only a blemish.

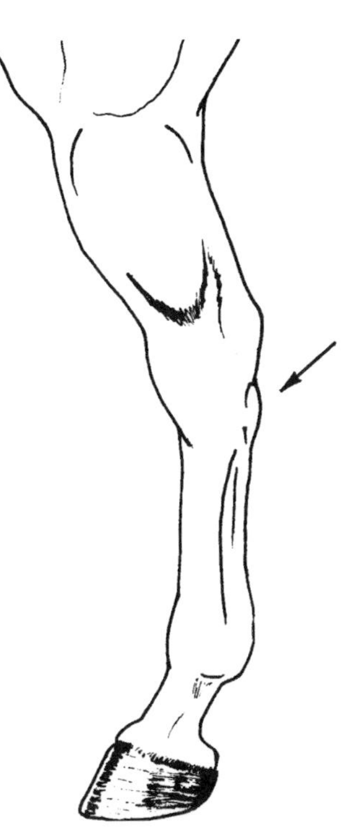

Fig. 12–16. A curb on the left hind leg.

Causes of Curbs

The enlargement is due to the thickening of a ligament. Damage to this ligament can be caused by a horse falling, kicking a hard object, or from overexertion while racing. Poor conformation is generally the cause if the horse has curbs on both hind legs.

Symptoms of Curbs

Most curbs are very distinct and easy to see from the side. Or, when running your hand down the back of the hind leg, a fullness just below the point of the hock that feels like muscle is probably a curb. The horse may be lame at first, and swelling may increase, rather than decrease with exercise.

Treatment for Curbs

Cold packs, sweats, and anti-inflammatory drugs can help reduce the inflammation of a curb. A month's rest is usually the minimum necessary for recovery. In most cases, there will be a permanent blemish, but not permanent lameness.

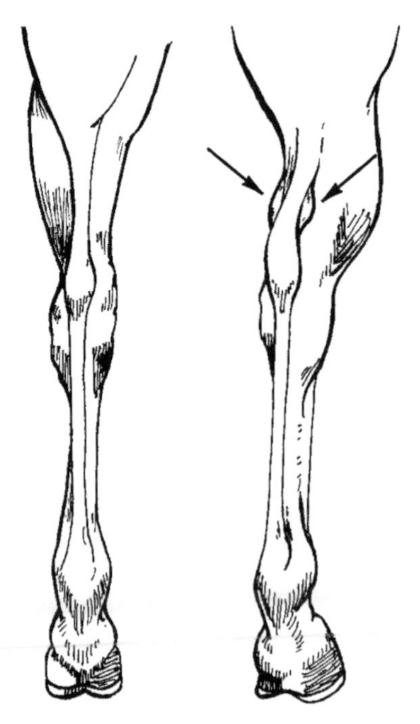

Fig. 12–17. Thoroughpin on the right hind leg.

Thoroughpin

Thoroughpin is a swelling of the sheath of the deep flexor tendon, in the "hollows" above the point of the hock. Because the usefulness of the horse is not affected after the initial trauma, thoroughpin is generally categorized a blemish.

Causes of Thoroughpins

Thoroughpin is caused by strain of the deep flexor tendon sheath, leading to fluid accumulation within the sheath. It is usually found on only one leg.

Symptoms of Thoroughpins

Thoroughpin is located just above the hock in the groove at the back of the leg. This swelling can vary from less than an inch to up to four inches long. While the swelling is present on both sides of the leg, a thoroughpin can be distinguished by movement of the swelling, when pressed, to the opposite side of the hock. Initial lameness may be present. Although the fluid swelling usually remains, any lameness resolves quickly.

Treatment for Thoroughpins

Rest is the best way to allow the condition to heal naturally. Massaging sometimes helps, although the swelling is likely to return. Bandaging is of no use because you cannot apply enough pressure to the grooves of the hock without causing excess pressure over the tendons at the back of the hock.

TREATMENT METHODS

Any of the previously mentioned symptoms detected by the groom should be relayed to the trainer immediately. Never administer any treatment or medication to a horse without a trainer's consent, except in an emergency situation. *(Chapter 13 discusses emergency situations and first aid for the racehorse.)* The following sections discuss some common treatment methods which a groom may encounter and be expected to apply to the racehorse.

Rest

Rest plays a very important role in the healing process of most leg problems. Letting injuries heal naturally, in their own time, is the most effective cure. When racehorses are put back to work too soon after an injury, there is a much greater chance that the injury will recur, doing even more damage to the horse the second time.

Racehorses should rest anytime there is visible lameness, or diagnosis of damage to tendons, ligaments, or bone (whether lameness is present or not). Injuries that involve damage to important tendons or ligaments can take many months to heal. Occasionally, an injured horse is sent to a farm to rest in a more natural environment. Sometimes the horse is retired from racing altogether.

Just because a horse is placed on a rest program at the track does not mean that it requires less care. Daily grooming, bandaging, and walking become even more important when the horse is not in full training. The trainer relies on the groom to keep the horse in its best possible physical and mental condition during this time, because when this rest period is over, it will be the trainer's job to bring the horse back to racing fitness as quickly as possible. Gentle exercise (hand walking for 5 – 20 minutes) twice a day can speed up the healing of many conditions. Or, the veterinarian may show the groom several physical therapy exercises to perform on the horse each day for a few minutes. However, always leave it up to the veterinarian to decide whether walking or physical therapy will help or harm the horse.

Poultices

A poultice is a paste that is applied directly to the horse's skin (poultices are for external use only). Poultices contain ingredients which "pull" swelling from the horse's legs. They are excellent for bruises, abrasions, stiff muscles, sprains, tendinitis (inflammation of a tendon), various foot problems, or minor skin irritations. Poultices contain some or all of the following ingredients:

- methyl salicylate
- oil of peppermint
- kaolin
- glycerine
- thymol
- boric acid
- eucalyptus oil

Some common brands of poultice medications are:

- Antiphilogistine®
- Numotizine®
- Phologo®
- Uptite®

Poultices are available either in a powder form where water must be added, or in an already prepared form. All poultices may be applied either warm or cold to the affected area.

Applying Poultices

To apply a poultice, you need the following materials:

- prepared or mixed poultice
- wet piece of paper
- materials for a standing bandage (a sheet of cotton or a quilted pad, a flannel wrap, and two saddler pins)

The following steps for applying a poultice are illustrated in Figure 12–18.

1. Rub a small amount of poultice on the affected area.
2. Apply another layer of poultice about ¼ inch thick on the affected area.
3. Cover the poultice with a sheet of wet paper to protect the cotton or quilted pad of the bandage.
4. Wrap a basic standing bandage over the area. *(See Chapter 10 for a description of a standing bandage.)*
5. For the poultice to be effective, it is best to leave it on the affected area for 8 – 12 hours. Then wash the poultice from the leg using a hose and sponge.

Applying a Poultice (Fig. 12–18.)

Step 1

Rub a small amount of poultice on the leg.

Step 2

Apply another layer about ¼ inch thick.

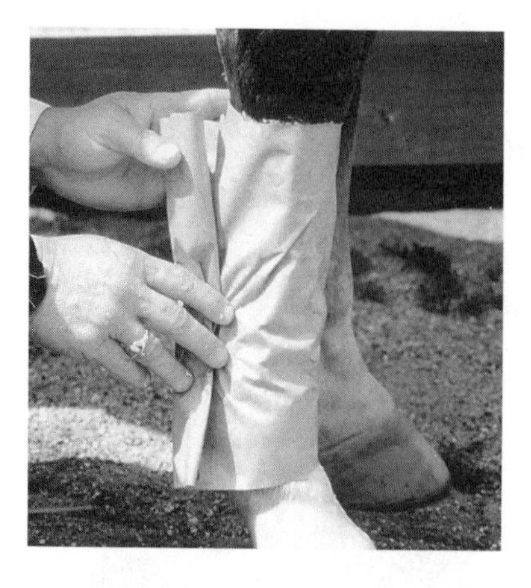

Step 3

Cover the poultice with a sheet of wet paper.

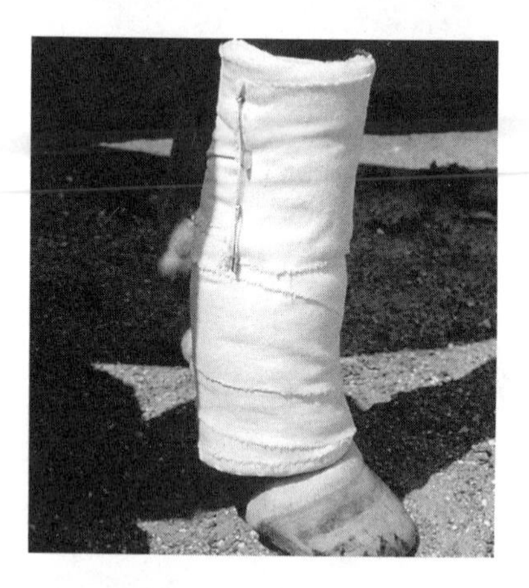

Step 4

Wrap a standing bandage over the leg.

Step 5

Leave the poultice on for 8 – 12 hours. Then wash the poultice from the leg using a hose and sponge.

Poultice Boot

This boot is used primarily when a poultice is applied to treat an injury of the foot or fetlock. The rubber poultice boot keeps the area clean, and prevents dirt or bedding material from adhering to it. The rubber poultice boot can also be used to soak the foot and fetlock in water.

Fig. 12–19. Poultice boots hold the poultice against the foot.

Dimethyl Sulfoxide (DMSO)

DMSO is a topical application which reduces swelling associated with bruises, sprains, and bucked shins as well as arthritic conditions. DMSO is available in liquids or gels. Some veterinarians use DMSO as a vehicle to carry other medications to the site of an injury. However, it is possible to poison a horse by using DMSO with some topical preparations (which are safe when used alone). *Never mix DMSO with anything unless instructed to do so by a veterinarian.*

Always wear disposable latex surgical gloves whenever applying DMSO, as this chemical can be absorbed into the human body through the skin. If absorbed into the body, it will cause an unpleasant "garlic" taste in the mouth, bad breath, and has been known to cause drowsiness. The skin may also be irritated by DMSO. These side effects are temporary and not serious.

Using DMSO more frequently than once per day can cause skin irritation and "burning," particularly on horses with white leg markings. To prevent this, only apply DMSO once per day. Discontinue treatment if the horse becomes sore where the DMSO was applied, and do not wrap or cover the leg while DMSO is on the skin.

Liniments

Many racehorse trainers use liniments before and after heavy training. Liniments increase the circulation to the skin, and are recommended for relieving soreness and/or swelling caused by sprains, windpuffs, sore tendons, osselets, and bucked shins.

Some trainers rub liniment on the shoulder muscles and apply a paddock shoulder blanket to sweat the shoulder muscles just before

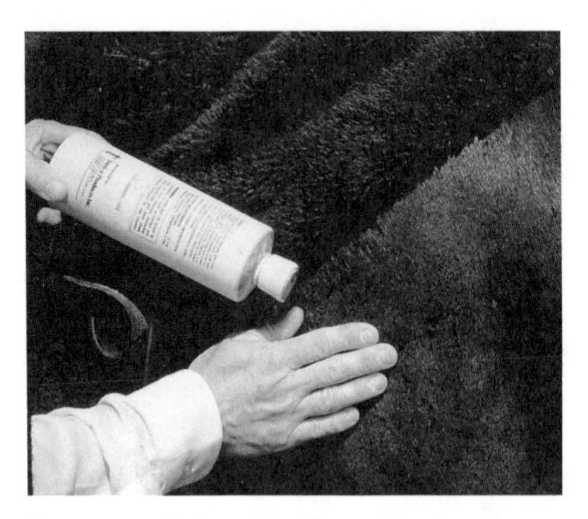

Fig. 12–20. Liniment is often rubbed on the sore muscles of racehorses.

a race. Liniments may also be massaged onto the back muscles to relieve general soreness. Afterward, an infrared lamp hung high on the ceiling of the stall warms the horse's back and increases the blood flow, relieving pain.

Finally, liniments can be added to the wash water as a body wash after a race or workout to cool and stimulate sore, stiff muscles. They also provide relief for minor abrasions. (Do not put liniments on open skin.)

Liniments are applied to the racehorse's legs, shoulders, and back and then are massaged in with the hands. When massaging, it is important to massage in the normal direction of the hair growth as opposed to going against the hair growth. If you rub against the normal direction of the hair growth, not only is it possible for the hair to fall out, but you can unintentionally blister the horse's skin if you rub too vigorously. Massage the liniment until the area is dry. Then apply a blanket or standing bandage.

Most liniments come in the form of a liquid. The most common liniments found in use on racetracks are:

- Absorbine®
- Bigeloil®
- Vetrolin®

Blistering and Firing

Blistering and firing are forms of counterirritation; they inflict pain and inflammation at a site of injury (usually on the horse's legs) in an attempt to increase blood flow to the area and thus promote healing. While the practices of blistering and firing have gone on for many years, their effectiveness is doubtful. In fact, it is now known that the rest period which follows these procedures is the true cause of healing.

Of the two processes, firing is the most painful to the horse. Firing involves inflicting a series of third degree burns over an injury (which may already be inflamed). A hot firing iron is applied to the skin at different depths depending on the location of the injury. There are two basic patterns of firing: pinfiring and linefiring (or barfiring). Pinfiring is the more common method. It involves burning a series of puncture holes on the affected area in a grid spacing pattern. Linefiring is similar to pinfiring, but instead of holes, the firing iron burns straight lines on the affected area.

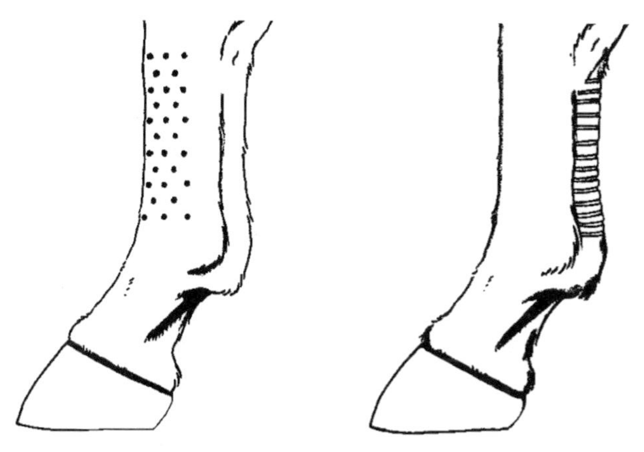

Fig. 12–21. Left: pinfiring. Right: linefiring. A groom can tell if a horse has been fired before because it will have scars in these patterns.

Not only is firing a painful procedure, but there is no proven benefit to the horse. It is the opinion of most veterinarians that firing does not promote healing or strengthen damaged tissue. In fact, firing may even cause additional injury and weakness, including one or more of these serious problems:

- tetanus
- joint infection (a career-halting complication)
- wound infection
- laminitis
- scar tissue that restricts free movement of the tendons

Besides being ineffective (and arguably, inhumane), firing significantly reduces the resale value of a horse. The scars left by this procedure *(see Figure 12–21)* are like an advertisement that the horse has incurred a serious injury.

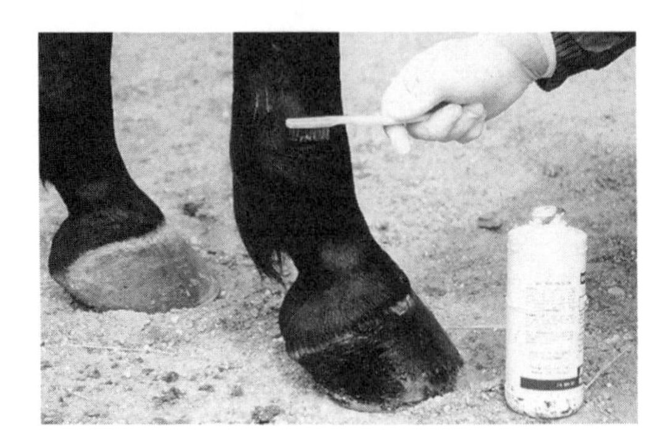

Fig. 12–22. Never use your hands to apply a blistering compound.

While less severe than firing, blistering (and painting) involves brushing a burning liquid onto the horse's leg. Blistering compounds, which consist of red iodine of mercury and cantharides, can cause first or second degree burns and "scurfing" (flaking) of the outer layer of skin. After the "scurfing" has developed on the horse's skin, it can take anywhere from a few weeks to several months to heal, depending on the severity of the burn. Paints, which are made up primarily of iodine, usually cause a less severe irritation and heal sooner. Both paints and blisters are applied with some type of applicator; *they should never be applied with the hands.*

Today, even veterinarians who once used counterirritants are skeptical about their effectiveness. They now see that the enforced rest period, which always accompanies a blistering or firing procedure, is the true cause of healing. The results of many studies indicate that there is no difference in healing between a horse that is blistered or fired and then rested, and a horse that is simply rested (with no blistering or firing) for the same amount of time. In some instances, the horse that was blistered or fired took *longer* to recuperate. Thus, rest is a more humane and more effective treatment than blistering or firing. (It is also less expensive.)

As veterinary schools do not even introduce blistering or firing as a treatment option anymore, more and more veterinarians refuse to perform these procedures. However, because blistering and firing are still practiced in some areas, the groom should be familiar with them and know how to make the horse as comfortable as possible after each procedure.

If the horse's leg has been fired, it is usually bandaged afterward. If a paint or blister was used, do not bandage the leg, as under bandages the chemicals may cause even more severe irritation to the horse's skin. Before firing, the veterinarian should give the horse a local anesthetic. Once the anesthetic wears off, however, the horse will be in extreme pain, for which the veterinarian will probably prescribe anti-inflammatory drugs.

A neck cradle can be used after blistering or firing to prevent the horse from chewing at the site of injury (there is significant pain and itching) or ingesting chemicals that could damage its tongue and lips. (If you do not have a neck cradle, apply petroleum jelly around the horse's lips, muzzle, and eyes, or tie the horse so that it cannot get its head down while the blister is on.) The neck cradle can be removed from a blistered horse after the chemicals are washed from the skin. On a horse that has been fired, the neck cradle should only be used if the horse chews at its bandages.

Neck Cradle

Neck cradles prevent a horse from getting its head down to its front legs to disturb a treatment or bandage. It is not uncommon to place a neck cradle on a horse that has had surgery to its front limbs, to prevent it from chewing at a sensitive area or possibly reopening surgical incisions. If the horse is not restrained by a neck cradle, it may cause further damage to an already sensitive area.

Some trainers feel a neck cradle should be used on a horse that has a habit of chewing and ripping its blanket or bandages during the night. However, there are less extreme methods of curbing this habit which should be tried first. *(See Chapter 14 for more information on stable vices.)* The neck cradle should be reserved for serious situations because it can prevent the horse from resting comfortably.

There are two basic types of neck cradles available: the wooden neck cradle and the metal neck cradle.

Wooden Neck Cradle

This neck cradle is made of 12 wooden rods joined together with rope. At the top of the cradle, each rod is separated by one bead, and at the bottom of the cradle, each rod is separated by two beads. The four center rods are shorter than the other rods at the bottom of the cradle to prevent "picking" in the chest. There are two leather straps attached to one side of the cradle and two buckles on the other side which secure it around the horse's neck.

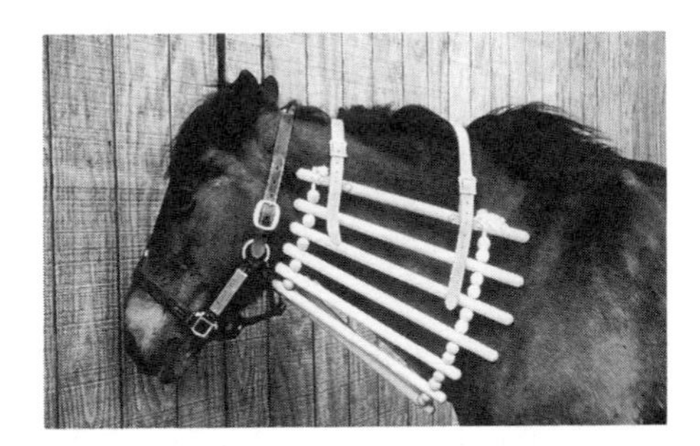

Fig. 12–23. A wooden neck cradle.

Fig. 12–24. A metal neck cradle.

Metal Neck Cradle

The metal neck cradle is made of durable chrome-plated aluminum and is padded for comfort. There are two U-shaped metal rods at each end. The smaller U-shaped rod is placed at the top of the neck and the larger at the base of the neck. These rods are joined by a straight metal rod which runs under the neck. (Some models have an adjustable straight rod to provide more comfort to the horse.) The metal neck cradle is buckled around the neck just like the wooden neck cradle.

Nerving

Nerving is a surgical procedure performed by a veterinarian. The veterinarian severs the nerves leading to the problem area, thus leaving the area without any feeling (including pain). The horse should then move without any physical signs of lameness. This procedure is usually only performed on the nerves at the back of the pastern that supply the back third of the foot with sensation.

Nerving, also called a neurectomy, can be performed in a stall. If so, the veterinarian will have the affected area of the foot under a local anesthetic (so that the area is numbed). This procedure may also be performed in a veterinary hospital or clinic, with the horse under general anesthesia (so that the horse is asleep).

The care of a nerved horse's feet is very important. The horse will not limp when a stone or shoe nail becomes imbedded in the back third of the foot: it is important to check this area more frequently.

Water Therapy

The use of water in treating or preventing ailments is popular at the racetrack. The water's action has a massaging effect and aids blood circulation to the limbs. The cold temperature reduces pain, heat, and inflammation. Water therapy is also called hydrotherapy. There are six basic methods of water therapy used on racehorses:

hosing, ice packs, cold water bandages, ice boots, water tubs, whirlpool units, and swimming.

Hosing

To hose a horse's legs, use a garden hose without the nozzle attachment. If desired, you may hold your thumb on the end of the hose to get two separate streams of water to flow over the knees and down the front of the legs. You can also allow the water to flow over the tendons at the back of the legs.

Fig. 12–25. Hosing the legs with cold water is a common form of water therapy.

Ice Packs

Ice packs or ice water prevents or alleviates some of the pain, swelling, and heat that accompanies lameness. They are especially effective when used to treat bucked shins, and the early stages of splints and bowed tendons. Some trainers believe that soaking a horse's front legs in ice water before a race or workout prevents or alleviates some of the pain, swelling, and injuries which result from associated concussion and stress.

Cold Water Bandages

Cold water bandages cool the horse's legs when they are hot or strained due to a race, workout, or injury. When applied while cooling out after the horse has been washed, they may be used to soothe the legs. Cold water bandages may be applied to all four legs.

The following materials are used for the cold water bandage:

- quilted pad
- ice water
- petroleum jelly
- bandage with either a Velcro ending, such as a polo bandage, or a string ending

Applying the Cold Water Bandage

The following steps are illustrated in Figure 12–26:

1. Soak the quilted pad and bandage in a bucket of ice water.
2. Apply a small amount of petroleum jelly around the coronet and heel to prevent irritation of these areas from water.
3. Roll the quilted pad onto the leg while dripping wet.
4. Begin wrapping the bandage over the wet quilted pad just above the ankle. Continue wrapping downward over the ankle, leaving a ½-inch section of quilted pad showing at the bottom of the bandage. Begin wrapping upward over the ankle, tendons, and shin, leaving a ½-inch section of quilted pad showing at the top of the bandage. Also, be sure to leave 1 inch of the last wrap showing.
5. If a Velcro-type bandage is used, secure the Velcro strip at the end of the wrap onto the leg. If a string-type bandage is used, divide the strings at the end of the bandage and wrap them once around the leg in opposite directions. Tie them in a shoestring bow on the outside of the leg.
6. The completed cold water bandage has a cooling effect on a hot or strained leg.

Note: It is important to periodically soak the cold water bandage while it is on the horse, because this bandage will tighten on the leg as it dries. Also, the horse's body heat causes the bandage to warm up quickly and lose its effectiveness.

Applying a Cold Water Bandage (Fig. 12–26.)

Step 1

Soak the quilted pad and bandage in a bucket of ice water.

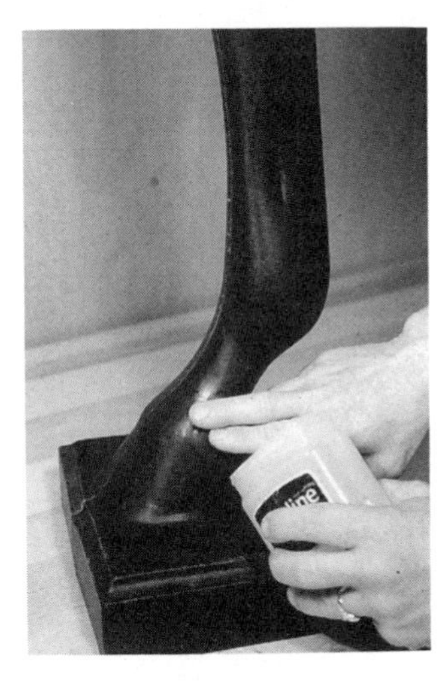

Step 2

Apply a small amount of petroleum jelly around the coronet and heel.

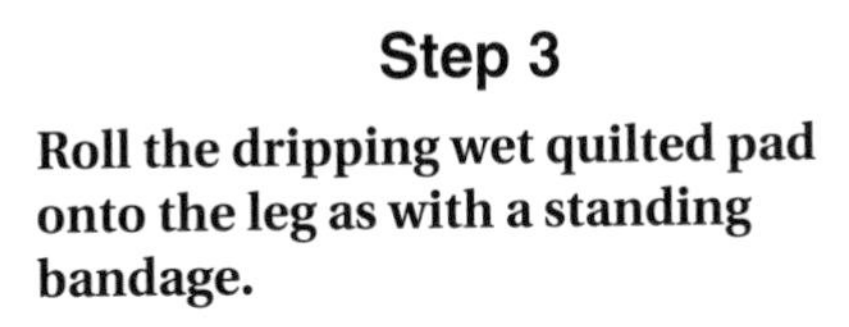

Step 3

Roll the dripping wet quilted pad onto the leg as with a standing bandage.

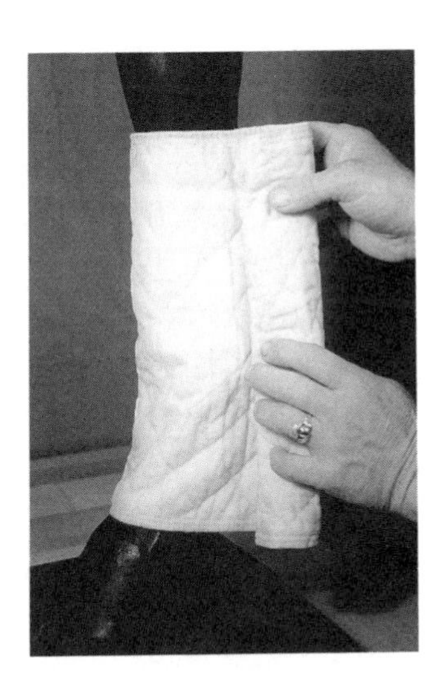

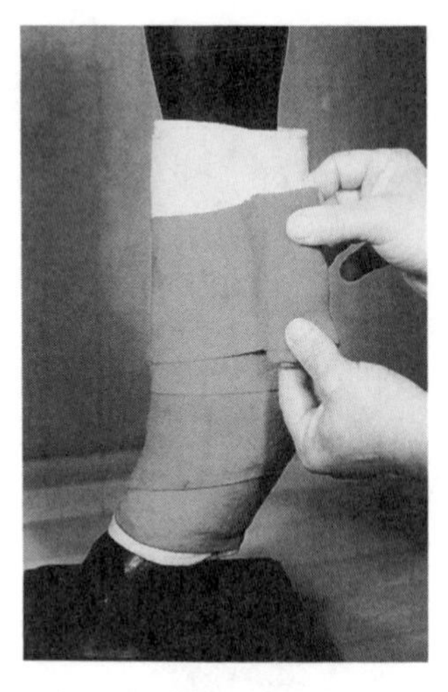

Step 4

Begin wrapping the bandage over the wet quilted pad just above the ankle. Continue down over the ankle, then back up to the knee.

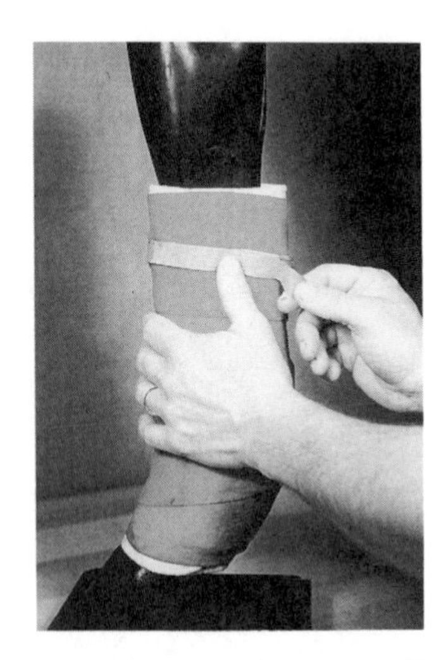

Step 5

If a Velcro-type bandage is used, secure the Velcro strip at the end of the wrap on the leg.

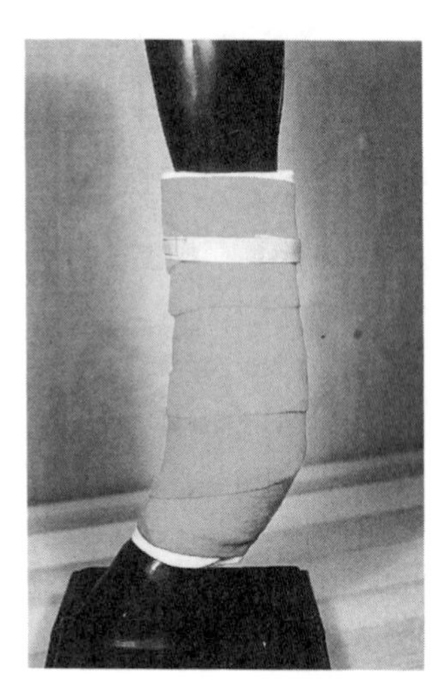

Step 6

The completed cold water bandage has a cooling effect on a hot or strained leg.

Ice Boots

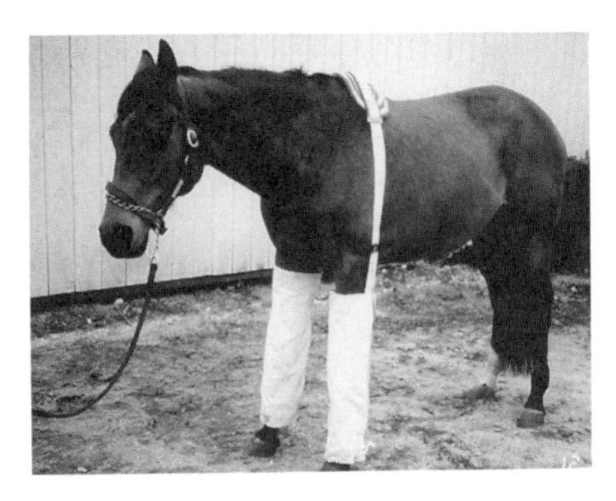

Fig. 12–27. Ice boots are used to reduce swelling on the horse's legs.

Ice boots are made of rubber, plastic, or canvas with heavy-duty metal zippers. The boots cover the front legs and are tied just below the fetlock to allow the melted ice to flow from the bottom of the boot.

Ice boots are secured with an adjustable shoulder strap which passes over the withers. The ice boots are packed with ice cubes or chipped ice. They are used to reduce swelling on the horse's legs.

Some racehorse trainers use ice boots to numb a racehorse's front legs just before a race to make the horse less susceptible to pain. Other trainers feel that this numbing technique is a wasted effort. They contend that the legs completely thaw out between the time the horse leaves the boots and the time it leaves the starting gate.

Water Tubs

There are various preservative or therapeutic reasons why it is necessary to stand a horse's front feet in a tub of water. Essentially, water tubs serve the same purpose as other forms of water therapy—to reduce inflammation and decrease pain in the legs. Water tubs come in three sizes: foot tubs, ankle tubs, and knee tubs.

Be aware of certain safety factors when using water tubs:

- The horse should be tied or cross-tied at the stall entrance with its head facing outward and the safety chain across the stall entrance. The hay net may be hung outside the stall door to occupy the horse while it is standing in the tub.
- The horse should also be held by an attendant with a shank during the entire period of tubbing. Horses have been known to suddenly startle and jump out of the tub, causing further injury to themselves. The attendant's job is to calm the horse and prevent this from happening.
- A thin rubber mat is usually placed on the bottom of the tub. The mat prevents a horse from slipping while standing in the tub.

- Rub some mineral oil or petroleum jelly on the coronet and heel before placing the horse's feet in a tub of water or ice. Too much moisture in these areas of the foot over a long period may result in excessive softness.

Foot Tub

The foot tub is used to soak the horse's feet in water. Foot tubs are usually made of rubber to avoid any injury to the horse. They are usually 9 inches high and 22 – 24 inches in diameter, with a capacity of about eight gallons.

Ankle Tub

The ankle tub is used for soaking the feet, ankles, and tendons of the front legs in water. It is usually made of pliable plastic and is about 28 inches high with a diameter of 22 – 24 inches. The ankle tub may also be used with a whirlpool therapy unit as discussed in the next section.

Knee Tub

Like the ankle tub, the knee tub is also made of pliable plastic. However, it is 30 – 32 inches high and 22 – 24 inches in diameter. The knee tub is used to soak the feet, ankles, tendons, and knees. The knee tub may also be used with a whirlpool therapy unit.

Whirlpool Units

Whirlpool units are very popular with some trainers on the racetrack. They may be used with either hot or cold water depending on the treatment. The constant turbulent action of the water provides an excellent massaging action and aids blood circulation to the horse's limbs.

The basic whirlpool units consist of a whirlpool tub, a blower unit, a flexible hose, and a bubble plate.

Whirlpool Tub—Use either an ankle or knee tub with a whirlpool unit.

Blower Unit—The electric blower unit forces air into the flexible hose at a powerful rate. The blower unit sounds like a vacuum cleaner. It is important that a horse get accustomed to the sound *before* it is led into the whirlpool. The blower unit usually operates at 115 volt AC.

Fig. 12–28. A whirlpool unit consists of a tub, a blower unit, a hose, and a bubble plate.

Flexible Hose—The flexible hose transmits the forced air from the blower unit to the bubble plate. One end of the hose is attached to the blower unit and the other end is attached to the bubble plate. Flexible hoses vary in length, but the average length is 12 feet with a four-inch diameter.

Bubble Plate—This plate is made of heavy-duty metal. It is round (18½ inches in diameter) and is 1 inch thick. The plate is hollow with a series of about 32 holes on top. There is an extended metal hose connection on the outer edge of the plate to connect the flexible hose leading from the blower unit.

1. Place the bubble plate at the bottom of the tub with the holes facing upward.
2. Connect the flexible hose to the bubble plate at one end and the blower unit at the other end.
3. Fill the tub with either hot or cold water above the ankles or knees (depending on the type of tub being used).

Some trainers prefer to place the horse in the tub *before* turning on the machine and others prefer placing the horse in the tub *after* the machine is turned on. This depends on the horse's disposition and the trainer's preference. The length of time a horse stands in a whirlpool tub depends on the ailment being treated or the amount of therapy desired.

To stand a horse in a whirlpool tub, pick up one of the horse's forelegs and place it in the tub. Once the horse is standing straight on that leg, pick up the opposite leg and gently place it into the tub next to the first leg. Have the tub as close to the horse's front legs as possible whenever placing a horse in any type of tub. It is important to center the feet in the middle of the tub so the tendons do not press against the back rim.

There are two other types of whirlpool units. They are whirlpool boots and whirlpool tanks.

Fig. 12–29. Whirlpool boots are excellent for treating sprains.

Whirlpool Boots

These are rubber boots which are filled with water. Air is forced through tubes to each boot from an air compressor unit. The forced air causes the water in the boots to agitate around the legs. It creates strong water turbulence inside the boot. This turbulence relaxes, massages, and soothes the legs. Whirlpool therapy boots are excellent for treating strains and sprains, and reducing swelling in the front legs.

Whirlpool Tanks

One of the latest models of hydrotherapy whirlpool units is the whirlpool tank. The horse must enter and exit through a swinging rear door. Once inside the unit, the horse is secured by two heavy-duty nylon straps with metal snaps on the center ring of the halter. Two larger nylon straps are placed over the withers to prevent the horse from jumping out of the tank. There is a heavy rubber matting on the floor to prevent slips.

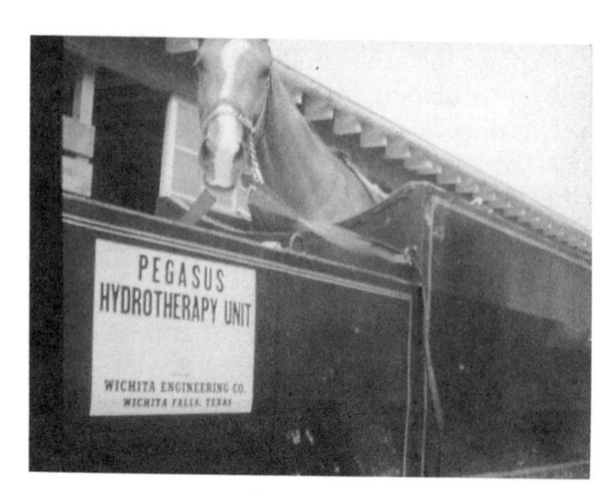

Fig. 12–30. The whirlpool tank is good therapy for sore muscles high on the horse's body.

The water is pumped into the therapy tank *after* the horse is secured in the unit. There are usually four jet nozzles (two on each side) both front and rear; water shoots through these jet nozzles. When the therapy is complete, the water drains from the therapy holding tank through a central drain.

Whirlpool tanks can harbor great quantities of bacteria and fungi, which can be spread rapidly from one horse to another. If the water is not drained after every use, it is important not to put a horse with an open wound, or a sutured wound, in one of these tanks. The tank should be drained periodically (depending on how often it is used, and especially if a horse defecates in the tank), allowed to dry completely, and then disinfected with chlorine bleach or another disinfectant. After the disinfectant has remained on the tank's surface for 15 – 20 minutes, rinse the tank out before filling it again.

Swimming

Swimming is an excellent way to exercise a horse without putting any weight on its back and without causing concussion to the feet and legs. Swimming is hard work for a horse, therefore, neither "bleeders" (discussed in Chapter 13) nor any racehorse with a muscle injury should be required to swim. A few racetracks may have swimming pools for horses, but most are found at farms or training centers.

Fig. 12–31. Swimming does not cause concussion to the feet or legs.

SUMMARY

The number of leg problems that can affect the horses in your care are numerous, as are the treatments that are used to heal them. However, even a general knowledge that something is wrong can be very valuable if reported to the trainer as soon as possible. By checking the horses' legs a couple of times per day, an employee can save the trainer and owner of a valuable racehorse much time, trouble, and worry. (To learn more about the causes and symptoms of lameness, see *The Illustrated Veterinary Encyclopedia for Horsemen* and *Equine Lameness*, published by Equine Research, Inc.)

Chapter

13

Ailments & Remedies

This chapter will familiarize grooms with common ailments, their symptoms, and the remedies for each. Keep in mind, you should never administer treatment or medication to a horse without the trainer's consent, except in an emergency situation. Therefore, it is important that you know how to determine when a problem is a true emergency requiring immediate action.

The next section details a list of situations which are considered emergencies. Following that is a discussion of the various ailments. Deciding whether an ailment is serious enough to

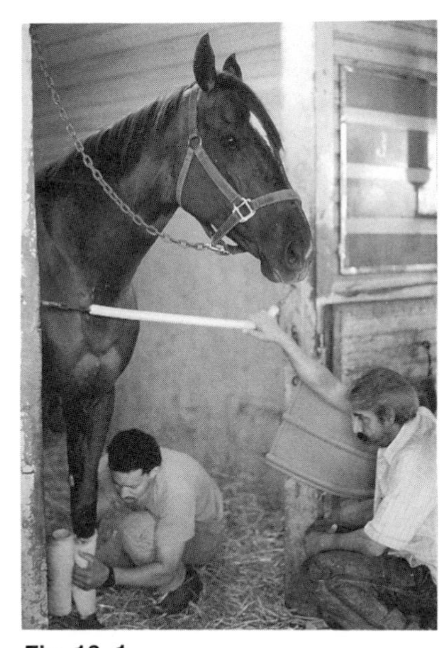

Fig. 13–1.

require veterinary attention is a judgment which must be made by the trainer.

LIFE-THREATENING EMERGENCIES

The following situations are considered emergencies. If the trainer or veterinarian is not at hand, you must ask someone to summon help, and immediately administer first aid in the manner described.

- A horse has a wound that is large or deep and bleeding profusely. Press a thick, clean, cloth pad to the bleeding wound.
- A horse has severe colic, and is thrashing around. Try to get it to stand and walk. If that is not possible or not safe, talk soothingly to the horse until it becomes possible or the veterinarian arrives.
- A horse is struggling in its stall, having difficulty getting up or staying up. The horse may have broken a leg or it may have a neurological problem. Stay with the horse and try to calm it until the veterinarian arrives. Take all safe measures to prevent the horse from struggling to stand up. **Note:** These symptoms should not be confused with those of a cast horse, where the animal cannot stand up because it is lying too close to a wall. *(See Chapter 6 for more information on the cast horse.)*
- A horse *staggers or collapses* due to heat exhaustion during hot, humid weather. Immediately hose the horse's neck, upper torso, and forelegs with cool water.
- A resting horse suddenly begins experiencing heavy or labored breathing. A veterinarian must be summoned immediately. Just like people, horses can have sudden, even fatal allergic reactions. *(See Chapter 11 for more information on allergic reactions.)*

In addition, there are several ailments which can become very serious in just a few hours if not attended to by a veterinarian. These problems can end the horse's racing career or may even be life-threatening:

- severe diarrhea
- injuries to an eye
- founder
- severe tying-up
- injury to a leg where the horse will not put weight on it

Some of these ailments have already been discussed in other chapters, or will be discussed later in this chapter. *(Refer to the Index for more information.)*

WOUNDS

Some wounds are acquired when the horse becomes frightened and struggles to escape from the source of the fear. Many injuries occur in the stable from broken feeders, damaged wall boards, sharp-edged doors, and protruding latches. (Keeping stable equipment in good working order and in its proper place helps to avoid accidents.) A horse can also be wounded in a race or workout. For instance, when horses break from the gate, they sometimes bump into each other, run into the rail, and step on each other's feet.

Checking the horse frequently and carefully for injuries is an important part of a groom's job. Generally, if you discover that a horse in your care has a wound, your top priorities should be:

1. Carefully examine the wound, noting the type and severity (particularly the depth).
2. Notify the trainer, who will determine whether or not to summon a veterinarian.
3. Stop any bleeding by using pressure.
4. Keep the wound clean until the veterinarian arrives.

Note: In cases of severe bleeding, reverse Steps 2 and 3.

Types of Wounds

Wounds are broken down into six categories, based on the cause of the wound and its appearance. Figure 13–2 lists and defines the six categories of wounds. If a horse is wounded, you should be able to recognize and report to the trainer which type of wound the horse has, as each type of wound must be handled differently. For example, an abrasion on the hip may not need anything more than some ointment, whereas a puncture wound on the coronet may call for a thorough cleaning, a poultice boot, and a tetanus shot. (In 1979, a puncture wound from an open safety pin probably cost Spectacular Bid the Triple Crown.)

First Aid for Wounds

When faced with a wounded horse, it is very important to remain calm, because your nervousness is liable to be transmitted to the horse and make the animal even more distressed.

Identifying Wounds	
TYPE	DESCRIPTION
Abrasion	A surface wound (such as a scratch or scrape) that does not penetrate the full thickness of the skin.
Avulsion	A tear with a loose flap of skin, usually caused by a thin object such as wire.
Contusion	A bruise, usually caused by a blow to the skin by a blunt object. The surface of the skin does not break, but there is bleeding beneath the skin. On the surface there may be discoloration and swelling.
Incision	A wound with clean-cut edges, often caused by a sharp object that penetrates the full thickness of the skin. This term is usually limited to surgical wounds.
Laceration	A tear in the skin with jagged or irregular edges, usually caused by a less sharp object that penetrates the full thickness of the skin.
Puncture	A deep hole with a small opening, usually caused by a sharp pointed object such as a nail.

Fig. 13–2. Knowing what type of wound the horse has helps to determine what first aid is necessary.

The following types of wounds will require veterinary attention:

- wounds that are large or deep and bleeding profusely
- wounds on the lower legs that cut the full thickness of the skin and are over 1½ inches long
- wounds near a joint, tendon, or coronet
- wounds that are dirty
- puncture wounds
- wounds that are not healing properly or are infected

As stated earlier, puncture wounds require special consideration, especially those in the bottom of the foot. If a horse has an object in the bottom of its foot, it is best if you do not remove the object. The veterinarian may wish to take x-rays of the object while it is still inside the foot to determine which structures are affected. Also, when you remove an object such as a nail, the hole closes over, making it difficult for the veterinarian to find the hole or to determine the path the object took into the foot.

Although it is always necessary to alert the trainer to any wound on a horse, an inexperienced groom may not realize that, because the horse is such a large animal, it can lose a lot of blood (about two gallons for a 1,000-pound horse) before it goes into shock. However, if there is a deep or large wound and/or profuse bleeding, try to stop the bleeding immediately.

For wounds on the legs, stop the bleeding with a clean cloth. Try not to use cotton or a cloth with loose fibers, as the fibers will cling to the wound and make cleaning difficult. Bandage the cloth firmly in place. Be sure not to bandage too tightly. The goal is to allow the blood to clot, not to cut off circulation entirely.

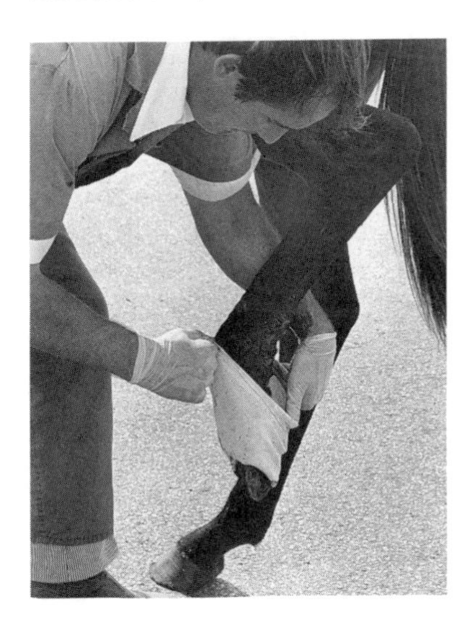

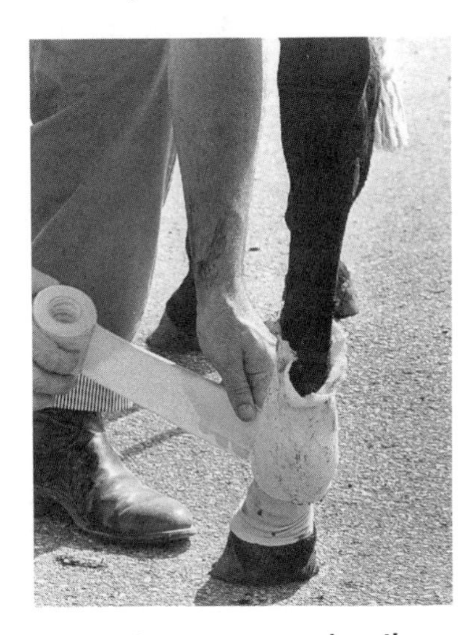

Fig. 13–3. A veterinarian applies a pressure bandage to a wound on the left hind fetlock.

For wounds on the head, neck, or body, press a clean cloth firmly over the wound and hold it there until the bleeding stops, or the vet arrives. *It may take 20 minutes or longer for the blood to clot.*

If the cloth becomes saturated with blood, apply another cloth (and bandage, if necessary) over the first one. Do not remove the saturated cloth from the wound until the veterinarian tells you to.

If a horse bleeds excessively from a wound, especially a vein or artery which spurts blood in rhythm with the heartbeat, it may be suffering from loss of blood (gums are pale). The most serious wounds as far as blood loss is concerned are those on the lower neck and the insides of the forearms. If the wounds are deep enough to cut the major blood vessels located in these places, the blood loss could become significant very quickly.

Treating a Serious (Not Life-Threatening) Wound

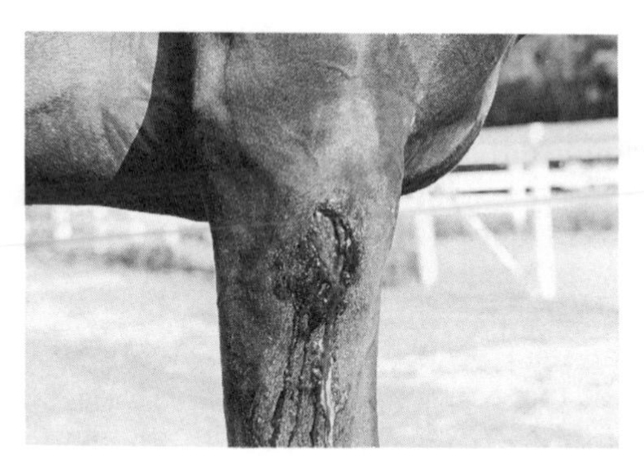

Fig. 13–4. This laceration will require veterinary attention to heal properly, but it is not really an emergency.

For wounds that are not life-threatening, yet will require veterinary attention, take the following measures after notifying the trainer and while waiting for the veterinarian to arrive:

1. If the wound is a few hours old and has stopped bleeding freely, carefully bathe it with clean, warm water to remove any dirt. If possible, add a teaspoon of salt to a quart of water to create a saline solution that is similar to the horse's body tissues—this solution is better than plain tap water. If the wound is still bleeding freely, do not bathe it because water slows the clotting process or washes away the clot. (Do not hose fresh wounds. Excessive water can make wound edges swell, which makes them difficult to suture. A bandage keeps the wound moist enough until a veterinarian arrives.)
2. Apply a sterile, nonstick pad over the wound and bandage or tape it firmly in place. The bandage keeps the flies out of the wound and prevents it from drying out.

Do not apply medication to a wound that may require stitches or has exposed bone. The veterinarian will only have to scrub the medication off to examine and stitch the wound, and most disinfectants and ointments will harm the sensitive outer surface of bone. It is very important to cover any exposed bone with a sterile pad soaked in saline solution and a bandage to prevent serious infection. With deep cuts or cuts with flaps of skin hanging, press the wound edges together and/or cover the underlying tissue with the skin before bandaging.

Treating a Superficial Wound

If the wound is superficial, and the trainer decides that it does not require a veterinarian's attention, use the following procedure to treat the wound:

1. Clip the hair from the area around the wound.
2. Wash the area with Betadine soap and clean, warm water. (Betadine soap is a common antibacterial soap used for gentle cleansing of superficial lacerations, abrasions, and minor burns.) Rinse the wound thoroughly.
3. Apply an antiseptic ointment and bandage if possible. Do not put anything into a fresh wound that you would not put into your own eye.

Wound Dressings

Wound dressings come in many forms and are applied topically (on the skin's surface) to wounds to speed up the healing process. Many wound dressings are antiseptics, which means they either destroy harmful microorganisms (like bacteria or fungi) or prevent them from reproducing. Some wound dressings help keep the skin soft and pliable, while others absorb moisture from the skin. Dressings come in liquids, ointments, sprays, and powders.

Liquids

These wound dressings come in a bottle, sometimes with an applicator, or dauber, in the cap for ease of application. Otherwise, you must apply them to a wound with cotton or gauze.

Hydrogen Peroxide

This antiseptic is commonly used as a topical antiseptic to clean various types of wounds. However, many people are unaware that hydrogen peroxide harms living, healthy cells of horses. There are many other antiseptic scrubs which are more effective and less toxic to equine cells.

Iodine

Liquid iodine is available in various concentrations (1% – 7%). The 1% solution has a low concentration of iodine, while the 7% solution has a higher concentration. *Only the low concentration iodine should be used to treat wounds.* A medium concentration (3% – 4%) can be an effective treatment for thrush, while the high concentration is used on foot infections like abscesses. *(See Chapter 5 for more information on foot abscesses.)* High concentration iodine is sometimes used to disinfect brushes or other grooming tools, however, medium concentrations work just as well and stain less.

Low concentration iodine is an excellent emergency antiseptic treatment of lacerations and abrasions. It can also be used in whirlpool boots to soak a wound on the lower leg—just add enough iodine to the water in the boot so that it looks like weak tea. (On breeding farms, iodine is used as an antiseptic dressing to the navel cord and navel stump of the newborn foal. This practice is a normal part of caring for a newborn foal.)

Methylene Blue

This veterinary antiseptic wound dressing is used to treat ringworm (a fungus), surface wounds, galls (sores caused by chafing of leather against skin), abrasions, chafes, and moist eczema (skin inflammation that oozes serum).

Gentian Violet

This is actually an ingredient found in many commercial wound dressings. It is an antiseptic and excellent protective wound dressing. Gentian violet is a healing aid and is effective against both the bacteria and fungi found in most common skin lesions. This dressing can also be used by itself to treat wire cuts, ringworm, and grease heel (also called scratches or white pastern disease). Any products containing gentian violet should be used with care, as the purple dye tends to stain bandages and other equipment. It also stains skin, so be sure to wear gloves while applying it.

Ointments

There are many ointments available for use on horses. In many cases, the main ingredient is nitrofurazone, which is an effective antibiotic. They are also water based, which means they are easy to wash off when the dressing is being changed. In addition, there are several types of ointments which either prevent or help to heal wounds in horses. Furacin® is an example of an antibacterial agent that prevents infection and promotes wound healing.

Petroleum Jelly

Petroleum jelly (like Vaseline) prevents flies from feeding on a wound and keeps the surrounding skin soft and pliable. It protects surrounding areas of skin from scalding from wound discharges or blistering from a blistering compound. Some caretakers also use petroleum jelly on the heels of a horse when washing to keep them from becoming too moist.

Corona® Ointment

Corona ointment is applied to the skin and contains a lanolin base. It can be used in healing sores, cracked heels, galls, cuts, and burns.

Desitin

This skin ointment contains cod liver oil as one of its main ingredients. Desitin is used for healing cracked heels, skin irritations, cuts, and burns. Like petroleum jelly, it also protects nearby skin from scalding from wound discharges. Trainers often apply this to the heels of a racehorse before it goes out onto a wet, sloppy racetrack to protect the heels from excessive moisture. A disadvantage of this product is that it is very tacky, or difficult to remove from the horse's hair once applied.

Fig. 13–5. An ointment protects the heels of a horse if it is applied before the horse goes out onto a sloppy track.

Sprays

Antibiotic wound sprays are often based on iodine, methylene blue, or gentian violet. Use sprays with caution, as some horses are frightened by the hissing sound and the sudden jet of cold liquid.

Buffered iodine spray is a topical dressing used on wounds, cuts, ringworm, the navel stump of newborn foals, and skin diseases caused by bacteria and fungi. It is also used to disinfect the skin before injections and surgical operations such as castrations. (Betadine Aerosol Spray is a common product.) Methylene blue (Blue Lotion) and gentian violet (Purple Lotion) were discussed in the section on liquids. There is no difference in the product, merely the application.

Powders

Powders for use on wounds may come in different forms. One common form is the puffer: a small plastic bottle that, when squeezed, ejects a small stream of powder onto the wound. Another form is aerosol spray. Although it looks like a spray can, the product is actually in powder form. Some are strictly antibacterial, however, others have special indications.

Caustic Dressing Powder

This type of dressing powder is used for slow-healing surface wounds. It contains copper sulfate sulfathiazole and sulfanilamide in a boric acid talcum base. It is used to control proud flesh, which is excessive granulated and protruding tissue in a healing wound. This powder should not be used on a wound which is healing normally.

Antiseptic Dusting Powder

This powder is an astringent (dries the area) and antiseptic dressing for superficial abrasions, cuts, and wounds.

Wound Infections

Always be alert to signs of infection when changing the dressing on a wound. These signs include swelling, heat, soreness, and pus. If a wound becomes infected, soak it with warm water and Epsom salts for 15 minutes. If this is not possible because the wound is too high on the horse's body, bathe it thoroughly with warm saline solution or a 1% iodine solution. Repeat either of these procedures three times daily until the infection has cleared up. An infection that does not clear up within two or three days must be brought to the attention of a veterinarian, as other treatment may be needed. Also, any infection in a deep wound, puncture wound, or sutured wound must be brought to the veterinarian's attention immediately.

(**Note:** For more information on wound healing and many other topics of interest to the horse person, read *Veterinary Manual for the Performance Horse* and *Veterinary Treatments and Medications for Horsemen,* both published by Equine Research, Inc.)

INTERNAL DISORDERS

EIPH

Many racehorses have what is called Exercise-Induced Pulmonary Hemorrhage (EIPH), which means the horse bleeds from the lungs as a result of intense exercise. Such horses are commonly called "bleeders." This condition may appear as a nosebleed, but the blood often does not show itself until an hour or so after a race or workout. For this reason, it is often the groom who walks into a stall and discovers the horse is bleeding from the nostrils. (Actually the bleeding in the lungs has probably stopped, and the pooled blood is draining out through the nostrils.) While this is not normally an emergency, the groom should notify the trainer as soon as possible.

The following are some measures which will make the horse more comfortable:

- If the bleeding appears while the horse is being walked and cooled down, walk the horse more slowly, allowing it to put its head down (perhaps to graze).
- If the horse has already been cooled out and is in its stall, let the horse rest and encourage it to put its head down (perhaps by placing some hay on the floor of the stall).

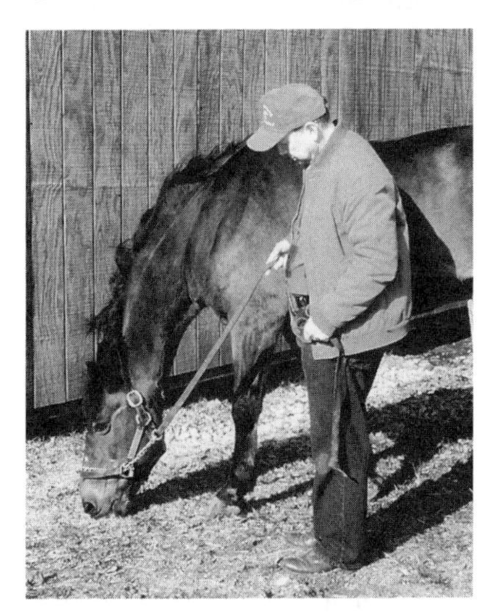

Fig. 13–6. Allowing a bleeder to graze keeps its head down, which facilitates drainage of blood from the lungs.

Note: Do not be overly concerned if the bleeding seems to increase when the horse puts its head down. It is better for the horse to let the blood run out the nose than to let the blood pool in the lungs. It is also easier for the horse to breathe if the blood is allowed to drain.

Rest is especially important for "bleeders." When a horse has exercise-induced bleeding from the lungs, it means the speed and/or intensity of the exercise caused some capillaries (tiny blood vessels) in the lungs to burst. It can take from two to three weeks for these capillaries to repair themselves, and the horse should not undergo any strenuous activity during this time.

Colic

Colic is a general term used to describe any disorder characterized by severe abdominal pain in the horse. There are several different types of colic that affect the horse, but the following are the two most common types:

Impaction—Occurs when materials (such as food) form a complete obstruction within the intestine. This type of colic can be caused by any of the following: lack of exercise, insufficient water intake, or excessively dry, fibrous hay. An obstruction or damage to the intestines may also be caused by parasite infestation. *(See Chapter 11 for more information on internal parasites.)*

Spasmodic—Causes excessive gut activity and mild to moderate cramping pain. You can often hear the bowel gurgle when you are standing nearby. This type of colic is caused by a variety of factors, including unfamiliar or bad feed, too much grain, or parasite infestation.

Causes of Colic

Colic may be caused by numerous factors, including parasite infestation, moldy feed, a drastic change in the weather, insufficient cooling down after a race or workout, nervousness, or a high grain, low hay (fiber) diet.

Symptoms of Colic

The following are some of the many possible symptoms of colic:

- sweating
- loss of appetite
- rapid breathing

- elevated pulse rate (above 60 bpm)
- mucous membranes become bright red or purple
- delayed capillary refill time (two seconds or more)
- elevated temperature
- lack of manure
- pawing
- restlessness
- unmanageable violence
- looking at flanks (or gut) or even attempting to bite flanks
- lying down and rolling violently or repeatedly

Fig. 13–7. Sweating and looking at the flanks are two symptoms of colic.

In addition to these symptoms, the horse's gut sounds change if it is colicking or about to colic. The groom should be alert to the following abnormal characteristics of the gut sounds:

- slower or faster than normal
- softer or louder than normal
- audible only on one side of the horse
- absent

A groom who can relay to the veterinarian or trainer which of these characteristics apply will help the veterinarian to diagnose which kind of colic a horse may have. *(See Chapter 11 for more information on checking vital signs.)* In general, the more severe the symptoms, the more serious the colic, and the more important veterinary attention becomes.

Treatment of Colic

If the horse has some of these symptoms for over 20 minutes, it is probably experiencing a colic attack. Summon a veterinarian immediately. Take away all food and water until the veterinarian says to give it back. Try to keep the horse on its feet and slowly walk the horse until the veterinarian arrives. If a horse just wants to lie down quietly, it should be allowed to do so. However, make every effort to prevent the horse from rolling, as this may further complicate problems in the intestine, or the horse may hurt itself.

All cases of colic should be diagnosed and treated by a veterinarian. Treatment may consist of the veterinarian administering injectable pain killers and sedatives. Mineral oil or fluids may be introduced to the digestive tract by the veterinarian with the use of a stomach tube. Or, the veterinarian may decide upon surgical intervention to correct the cause of the colic.

One of the most common over-the-counter colic medicines is B.E.L.L.® Drops. This medicine may reduce the pain of some types of colic and can be given to a horse at a veterinarian's recommendation. However, without a veterinarian's diagnosis, there is no way to determine what type of colic the horse has. Therefore, using home remedies such as B.E.L.L. Drops, sedatives, or painkillers without the veterinarian's approval can be dangerous; they can mask the symptoms for a time while the underlying cause persists or worsens.

Prevention of Colic

Colic can be prevented to some degree with the following managerial practices:

- Any change in the feed should be done gradually.
- Feed on a regular daily schedule.
- Have the horse's teeth examined on a regular basis.
- Split the daily grain ration into at least three separate feedings.
- Deworm on a regular schedule.
- Provide fresh, clean water at all times.
- Feed only high quality foodstuffs.
- Allow approximately two hours for digestion to take place before exercising.

Choke

Choke is a blockage of the esophagus, which is the tube through which food passes from the mouth to the stomach. During a choke, the esophagus is usually blocked by food or medication, which prevents the horse from swallowing. (The horse can still breathe.)

Causes of Choke

A horse that bolts its feed or does not chew properly because of tooth problems is a prime candidate for choke. Coarse hay may cause choke, as can eating stall bedding. Inadequate water intake has also been known to cause this problem.

Symptoms of Choke

A horse that suddenly stops eating and turns away from the feed tub is probably a victim of choke. The horse may arch and stretch its neck. It appears anxious and may sweat or paw. Saliva or saliva mixed with food may be visible at the nostrils.

Choke can cause pneumonia if the horse inhales the unswallowed food or saliva into its lungs. Since choke causes drooling and prevents drinking, eventually a choked horse suffers from dehydration.

Treatment of Choke

Call a veterinarian. Lead the horse to an incline and face the horse downhill. If you can see the ball of unswallowed food, you may *very gently* massage the neck over the bulge.

The veterinarian will give the horse a sedative and will break up the obstruction by applying a gentle stream of water through a stomach tube. The horse should eat only soft food for a few days to a few weeks after a choke (depending on the severity of the damage to the throat), and the veterinarian may administer antibiotics.

Tying-Up

Tying-up is the damage of the muscles during or after a race or workout. A mild case may go undetected or be dismissed as a little muscle soreness. A severe case may prevent the horse from moving at all.

Causes of Tying-Up

This condition is caused by exercising or racing a horse after a period of inactivity, especially if the horse was fed the same amount of grain while at rest as it was while working. It is usually seen in very fit horses that have been on a regular daily exercise schedule, have missed a few days of work, and have remained on full feed. Or, it can affect less conditioned horses that have been subjected to an unusually difficult workout.

Symptoms of Tying-Up

Signs of tying-up generally appear 15 – 30 minutes into exercise, but may not be seen until exercise is over. The horse usually is reluctant to move and seems stiff all over. Sweating, rapid breathing, and muscle tremors may also be seen. Some horses have tense, painful muscles over the back and hindquarters. Urine may be brownish or red-brownish.

More symptoms include:

- rigidity through the back
- lack of flexibility in the hind legs; reluctance to flex hind legs
- lack of control in the hindquarters
- elevated pulse and respiration rate
- general nervousness and pain
- lying down and unable or unwilling to get up

Treatment of Tying-Up

Treatment in mild cases may consist of slowly walking the horse for 30 – 40 minutes until all signs of tying-up have been relieved. In severe cases, the horse should be blanketed and not moved under any circumstance. Summon the trainer and/or a veterinarian immediately. After a careful diagnosis, the veterinarian may administer tranquilizers, muscle relaxants and analgesics (pain killers), and/or intravenous fluids, depending upon the severity of the case. Muscle damage has definitely occurred. Kidney damage may occur in severe cases. However, this damage may be reversible; it is almost never fatal.

Prevention of Tying-Up

Tying-up can be prevented through the following practices:

- regulating the horse's diet
- properly warming up the horse before exercise or racing

Fig. 13–8. Tying-up is caused by exercising or racing a horse after a period of inactivity, and continuing to give full feed.

- cooling the horse down properly after exercise or racing
- adding sodium bicarbonate (baking soda) to the feed daily
- administering vitamin E and selenium by adding them to the feed or by injection

COMMON AILMENTS & REMEDIES

Fever

In Chapter 11, we discussed taking the horse's temperature. Anytime the horse's temperature rises above 101.5°F without an obvious external reason, the horse is considered to have a fever.

Bute (Phenylbutazone)

Bute reduces pain and fever in horses. However, always consult with the trainer or a veterinarian before administering any medication. (State Racing Commissions test for the presence of bute in racehorses.) If and when a dosage is determined, the bute can be administered by grinding the tablets and mixing them thoroughly with the feed. Bute may also be administered by a trained person using a balling gun and gelatin capsules (discussed later).

Dehydration

Horses are sometimes susceptible to dehydration, especially in hot and humid weather. The most obvious sign of dehydration is that the horse is very thirsty. The horse may also have a dry mouth and prolonged capillary refill time. Another good way to tell if a horse is dehydrated is to conduct a skin pinch test. *(See Chapter 11 for more information on the skin pinch test and capillary refill.)* Practice these tests several times on a healthy horse to know what is normal. **Note:** To avoid dehydration, horses should be allowed free access to clean water at all times except for a period of time before a race (as determined by the trainer), or immediately following a race or workout.

Oral Electrolytes

Oral electrolytes are used to prevent or remedy dehydration resulting from diarrhea, high fever, fatigue, stress of transportation, hot weather, or long periods of illness. Electrolytes come in a powder or liquid form and should be mixed into the horse's drinking water. Some horses do not like the taste of electrolyte-tainted water, and they refuse to drink the water. To ensure adequate water intake for these horses, they should always be given two buckets of water—one normal and one containing electrolytes. Or, disguise the taste by mixing a very small amount of molasses with the water.

Fig. 13–9. A horse's eyes are subject to a possible irritation during a race.

Eye Irritations

Frequently, horses get dirt in their eyes from a race or hay chaff from the stable. Either of these particles in the eye cause a minor irritation. Typically the horse has increased redness in the mucous membranes of the affected eye, blinks more frequently, and has a clear, watery discharge from that eye. The horse may also tend to keep the affected eye shut.

Eye Wash

A common eye wash is applied to each eye by means of an eye dropper or small syringe (with no needle). Eye wash soothes and cleans the horse's eyes. It controls excessive watering, removes dirt and debris, and relieves redness and itching. After a race, it is a common practice to apply an eye wash to a Thoroughbred or Standardbred's eyes since their eyes may be subject to a great deal of irritation during a race. Clear Eyes is one common eye wash that can be used. Or, a homemade saline solution (one teaspoon of salt per quart of warm water) can also be safely used in the eyes.

Fungal Infections

Most fungal infections are not serious, but are unsightly and very contagious. Ringworm is the most common fungal infection affecting horses. Signs of ringworm are round, scaly, or crusty patches that are hairless or have short, broken off hairs. Some kinds of ringworm cause painful, reddened sores. There may or may not be itching. Ringworm is contagious to other horses. Therefore, grooming tools, blankets, and any other tack or equipment used on a horse infected with ringworm should be disinfected with chlorine bleach or iodine and should not be used on any other animal.

Daily grooming, fly control, proper nutrition, and good overall condition are the best measures for preventing fungal infections. Also, nose-to-nose contact between horses should be avoided, as this can spread the infection.

Fig. 13–10. Daily grooming helps to prevent fungal infections.

Fungisan®

Fungisan is a liquid medication used in controlling ringworm, summer itch, girth itch, and other skin fungus problems affecting the horse. It is usually dabbed onto the affected area with sterile cotton. The groom should always wear disposable gloves when treating any fungal infections.

Coughs and Colds

Racehorses are especially susceptible to coughs and colds because they live in a small area with a high concentration of animals coming and going. Also, when there is a drastic change in weather, or if the horse is not blanketed properly, it may become vulnerable to a cough or cold. Frequent coughing, nasal discharge, fever, and lethargy usually indicate that your horse has a cold. Blanket the horse to keep it free from drafts and ask the trainer if you should summon a veterinarian.

If a horse is coughing, but has no other cold symptoms, it may just be especially susceptible to dust in the hay or stable environment, or, in cases of poor mucking, ammonia fumes from the horse's urine. Make sure the stall is thoroughly clean and dry. Lightly wet the horse's hay to keep down the dust. Obtain the trainer's permission to feed the horse from a tub on the floor and to change the horse's bedding to wood shavings. Avoid powdered supplements, or wet them down as well. Adequate stall ventilation is very important for such horses.

Expectorant Cough Mixture

This is a liquid medication used to treat chronic coughing due to colds. It loosens mucus accumulations in the horse's respiratory tract. As with most liquid medications, it may be administered to the horse by means of an oral dose syringe.

Nasal Inhalant

This medication is usually available in the form of an ointment (for example Vicks Vaporub®). It is applied in and about the nostrils to relieve nasal and head congestion as well as difficult breathing due to a severe cold. Some trainers apply nasal inhalant to the nostrils before a race or a workout to help clear the air passages and allow the horse to breathe easier. This type of medication should be used conservatively as overuse can cause nasal irritation or scalding. It may also decrease the appetite of a sick horse because it cannot smell its food.

Nervousness

Many horses exhibit reasonable nervousness when confronted with strange sights and sounds. However, some horses will become increasingly nervous rather than adjusting and settling down. Such anxiety can reach a level where the animal is likely to injure itself or its handler if not quieted in some manner. In this situation, the trainer or a veterinarian may call for some type of tranquilizer, such as Promazine Granules.

Promazine Granules

This is an oral tranquilizer used to quiet excitable, unruly, and hard to handle horses. The calming agent usually takes effect in about one hour and lasts for several hours. It is, of course, not allowed during racing.

Constipation

Constipation is the inability to pass manure. The best way to tell if a horse is constipated is to observe its stall. If there have been no droppings for half a day or more, and the horse has been eating and drinking regularly, it is possible that the horse is constipated. Normally, horses on a proper diet will not become constipated. In most cases, constipation is not an isolated condition, but a symptom of another internal disorder, such as colic.

Milk of Magnesia

This is a liquid laxative used for horses suffering from constipation. For horses that are known to have a problem with constipation, it is a safe and reliable medication for the evacuation of the horse's bowel. Milk of magnesia may be administered by means of an oral dose syringe. However, if a horse has recurrent constipation, the trainer may gradually experiment with the feed until the cause of the problem can be determined.

Mineral Oil

This can be used as a safe oral laxative, or as a coating for the intestines. Mineral oil is usually administered by a veterinarian using a stomach tube. Some trainers may give smaller amounts using an oral dose syringe, but because the oil is odorless and tasteless, the horse has no urge to swallow it when it is placed on the back of the tongue. The mineral oil may go down the wrong way and be inhaled into the lungs.

Diarrhea

Diarrhea is a symptom; it is a means by which the body rids itself of an irritating substance. The most common cause of mild diarrhea is a change in diet, usually switching abruptly from a blander feed to a richer feed. Although oral medications are used frequently, they generally do not cure diarrhea. This condition usually goes away on its own—success which is often attributed to the medication.

Persistent or profuse diarrhea is a much more serious problem, particularly if the horse is depressed, has a fever, and is dehydrated. This situation needs immediate veterinary attention. Diarrhea accelerates dehydration and the horse can die in a matter of hours from toxemia and shock.

Veterinary Kaopectate®

This medication is used as an oral antidiarrheal for treating enteritis (inflammation of the intestinal lining) or mild cases of diarrhea. As a liquid, it may also be administered by the use of an oral dose syringe.

Warts

Warts are caused by viruses. They generally appear on young horses up to three years of age and are usually confined to the nose and lips. They are not infectious to humans but are contagious to other young horses. Contact from horses touching noses or from grooming tools and twitches used on an infected horse seem to be how the virus is spread. Warts tend to go away on their own in about three months.

Wart Remover

There are many wart-removing compounds on the market for human use which are also effective on horses. Castor oil is also a popular remedy which many horsemen use to remove warts.

ADMINISTERING MEDICATION

Oral Dose Syringe

This device is used to administer liquid medication orally (by mouth). It is usually made of nickel- or chrome-plated brass. There are two stationary rings on each side of the barrel of the syringe and a third ring at the end of the plunger stem. The dose pipe is attached to the end of the barrel and is usually six inches long. The oral dose syringe is available with a barrel capacity of either two, four, or six ounces.

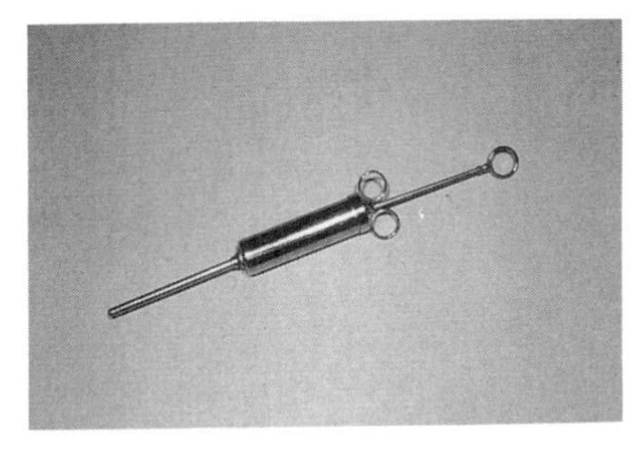

Fig. 13–11. An oral dose syringe.

Liquid medication is simply drawn up into the oral dose syringe by inserting the end of the dose pipe into the liquid and pulling back on the plunger stem until the barrel is filled to the required level.

To administer the medication, stand to the side of the horse, hold the halter, and place the dose pipe into the corner of the horse's mouth. Be very careful not to hit the tip of the syringe against the roof of the mouth. If the tip is rough, attach a short piece of rubber tubing over the end to prevent injury to the mouth. Hold the horse's head up and keep it steady with one hand while you smoothly release the liquid medication into the horse's mouth with the other hand. Remove the syringe as soon as the medication has been released. Wait for a few seconds before allowing the horse's head to drop as some horses do not swallow the medication immediately. Once the head is released, be sure that there is no loss of medication from the horse's mouth.

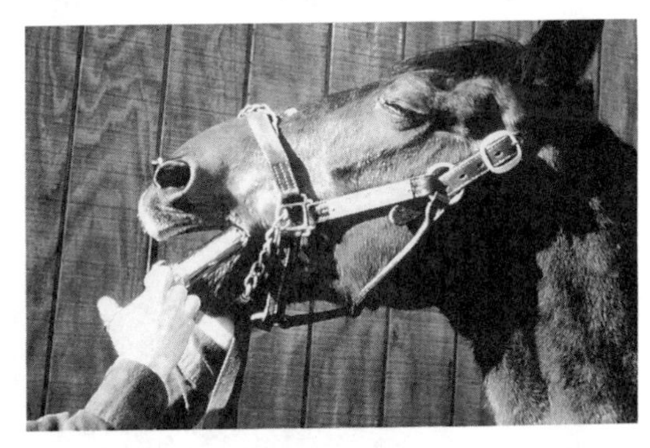

Fig. 13–12. Insert the oral dose syringe into the horse's mouth and press the plunger to release the medication.

It is important to thoroughly clean the syringe with warm water to keep the leather or rubber washers inside the barrel from rotting and cracking. Cleaning is made easy by completely disassembling it. After the syringe has been washed and dried, apply a thin film of Vaseline around the seal to keep it in good condition.

Balling Gun

The balling gun is a veterinary tool used to administer solid medication in the form of a gelatin capsule or a pill (bolus). The use of gelatin capsules allows any form of solid medication to be administered by the use of a balling gun. For example, a round aspirin tablet will not fit properly in the barrel of the balling gun. However, if that same aspirin tablet is placed inside a dissolvable gelatin capsule, it is possible to fit it into the balling gun.

This tool should be used exclusively by a licensed veterinarian or a qualified person who is trained in its use. It is very easy to misuse the balling gun and cause a serious "choke" in the horse or damage to the back of the throat. However, it is possible for a groom to use a balling gun if he or she is properly trained by a veterinarian in its use.

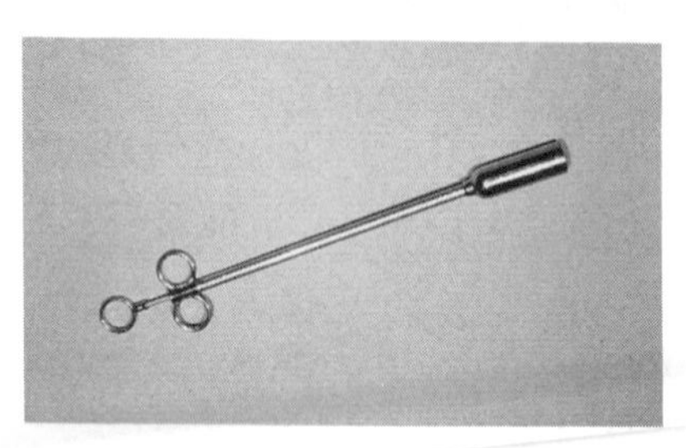

Fig. 13–13. Balling gun.

Like the oral dose syringe, the balling gun is usually made of either nickel- or chrome-plated brass. The holding cup at the end is capable of holding ½- or 1-ounce capsules. There are two rings on the handle for the forefinger and index finger with a third ring on the end of the plunger for the thumb. The total length of the balling gun is about 15 inches.

The holding cup with the medication is inserted into the back of the mouth above the tongue. The thumb then pushes the stem rod forward, which pushes a capsule or a bolus into the horse's throat. The horse is then allowed to drop its head and swallow the medication. The person administering the medication should stay with the horse to ensure that the medicine is swallowed properly and the horse is not choking.

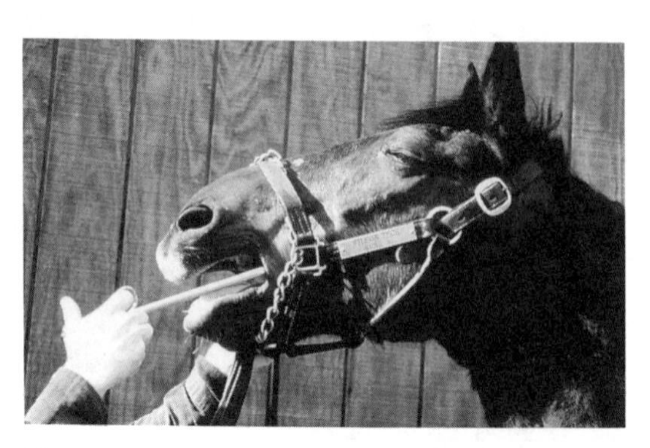

Fig. 13–14. The horse's head should be raised so the medication is more easily swallowed.

When administering a bolus with a balling gun, the horse's head should be raised so that its lower jaw is slightly inclined upward. An assistant standing on the near side would be helpful to prevent injury to both the horse and handler.

DRUG POLICIES

The previous sections discuss many drugs and medications that can be found in a racing stable's medicine cabinet. It is important for all backstretch employees to become aware of the medications which can be administered to the horse in normal, everyday care as well as in emergency situations. On the other hand, there are medications which can only be administered by a veterinarian or under the direct supervision of a veterinarian.

Always follow precise directions and obtain permission from the trainer or veterinarian when administering any medication to the horse. Some medicines may be quite effective in treating a particular condition, but may be considered illegal if found in the horse's system either before or after a race. Each state's Racing Commission issues to trainers and veterinarians a list of those drugs which it considers to be illegal. It determines the presence of these drugs in a horse's system using a blood or urine test. It is the trainer's and veterinarian's responsibility to be aware of those medications and drugs which are deemed illegal by the State Racing Commission. It should be understood that the trainer is considered by most racing jurisdictions as the sole insurer for the well-being of the horses in his or her care. If an illegal drug is found in a test sample of blood or urine after a race, it is the trainer who must face the consequences.

While the trainer must bear the responsibility for the actions of his or her employees, the groom may also face consequences. A groom may be penalized for knowingly administering illegal drugs to a horse and will very likely lose his or her job. Worse, the groom may be fined by the Racing Stewards, and the groom's license may be revoked. Depending upon the severity of the infraction, it is possible for a groom to be barred from entering any racetrack in the country for a specified amount of time designated by the Stewards.

SUMMARY

The caretaker plays an important role in the health of the horse by closely monitoring behavior, checking for injuries, and administering first aid in a competent manner when required. A horse person with these skills and good judgement has acquired them through years of caring for horses—a single chapter cannot hope to substitute for years of practical experience. However, the information in this chapter will serve as a foundation upon which to build your knowledge about which ailments are emergencies and which are not, and the proper actions to take in either case.

(All health issues relating to horses are covered comprehensively in *Veterinary Manual for the Performance Horse, The Illustrated Veterinary Encyclopedia for Horsemen,* and *Veterinary Treatments and Medications for Horsemen,* published by Equine Research, Inc.)

Chapter

14

Common Stable Vices

Racehorses are notorious for developing bad habits, or "stable vices" because of the large amount of time they must spend confined to their stalls. At most racetracks in the United States, racehorses cannot be turned out in a pasture or paddock as they can on the farm. Therefore, horses may become bored and develop vices to alleviate their boredom.

Unfortunately, the horse with a stable vice may injure itself and the people around it. The horse may also cause significant structural damage to stalls. The most detrimental effect of any stable vice is that the horse wastes important energy on performing a particular vice

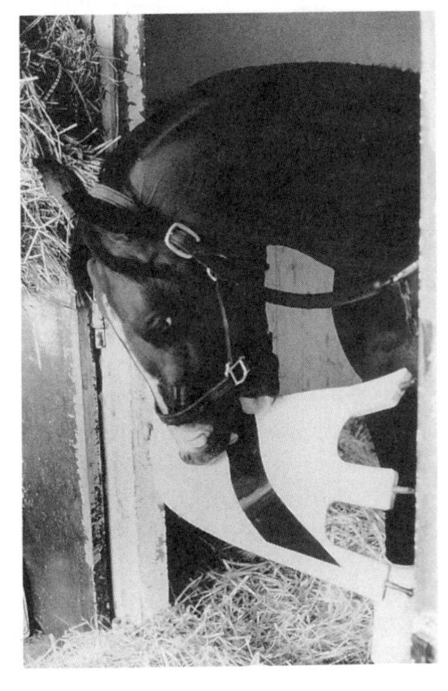

Fig. 14–1.

instead of channeling that energy into winning races. This chapter discusses the most common vices exhibited by the racehorse, and some specific means of discouraging each of them. At the end of the chapter, we will discuss some general preventive measures that horsemen use to curtail the many different stable vices.

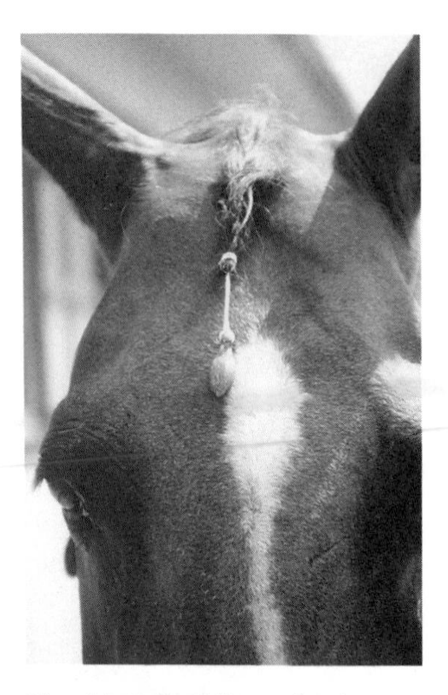

Fig. 14–2. Braiding a lead sinker into the horse's forelock discourages weaving.

WEAVING

This vice is often found in nervous horses. A horse that weaves stands at the stall door with its head out over the webbing and sways its head and neck from side to side. In extreme cases, the horse lifts its legs and places them down with each swaying motion. Weaving causes excessive stress on the legs and has even been known to cause founder.

An old-fashioned but effective method to discourage weaving is to attach a 2-ounce lead fishing sinker to the forelock. Braid the forelock with the sinker at the end of the braid. When the horse attempts to weave, the weight strikes its head and thus discourages the horse from weaving.

STALL WALKING

A horse with this vice walks in a circle around the stall at a fast pace. In an extreme case, this vice can sap a racehorse's energy. There are several methods trainers use to discourage a stall walker.

Some trainers place two bales of straw on each side of the stall in the horse's path. Then the horse becomes discouraged whenever it attempts to stall walk. Other trainers may even tie a horse to the stall wall, although this does not cure stall walking and is not recommended. A better distraction the trainer might try is to hang a hay net by the stall door to encourage the horse to stop to eat. Hopefully, this will break the walking pattern.

PAWING

A horse with this habit constantly digs the floor with its forefeet. There is nothing more frustrating to a groom who has just thoroughly cleaned a stall than to find it a mess after the horse has been pawing. This habit can cause unnecessary pressure on the horse's legs and feet as well as wear down its front shoes.

To prevent a horse from pawing and digging up the stall floor, place rubber mats over the dirt or clay, and spread the bedding on top of the mats. Or instead of rubber mats, floor boards can be used as a cover above the floor, with the bedding spread on top of the boards.

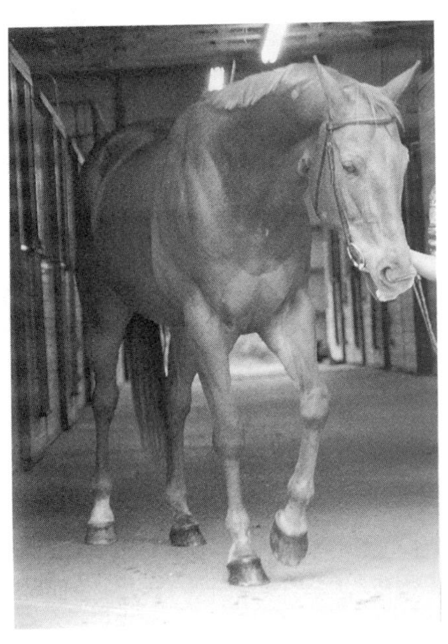

Fig. 14–3. Outside of the stall, the best way to prevent a horse from pawing is to keep it moving.

Outside the stall, the best way to prevent the horse from pawing is to keep it moving. Also, the horse cannot paw if someone is holding up one of its feet.

KICKING

Kicking people is one of the most dangerous vices a horse may have. Most horses develop this habit from kicking a few times and getting away with it (not being punished immediately). Sometimes a horse kicks if someone or something comes up suddenly from behind and surprises it. This horse does not need to be punished because it was frightened. However, mean or aggressive kicking without provocation cannot be tolerated. Whoever is holding the horse or near the horse must punish it *immediately and convincingly.* Give the horse a hard crack (with a whip or lead shank) low on the side of the hindquarters and repeat "No" in a deep, intimidating tone. Generally, after a few such punishments, the horse associates this unpleasant experience with kicking and the vice is cured. *It is important to do this within three seconds of the kick attempt* so that the horse can associate a bad experience with kicking. If you wait longer than three seconds you have missed the window of opportunity—the horse will not know why it is being punished.

If handling a horse that has been known to kick, the groom should approach the horse in its stall with caution, work carefully around the horse's hindquarters, and watch the horse's ears closely. A horse often pins its ears back and swishes its tail right before kicking. If the groom must perform some treatment on the back legs of a kicker, he or she should get another person to hold one of the horse's front legs up so the horse is unable to kick.

MASTURBATION

This habit is confined to stallions. The stallion usually masturbates during the night but may also masturbate during the day. The stallion's penis becomes erect, and he slaps the penis against the abdomen until it ejaculates.

Semen on a stallion's abdomen, then, is a sign that he has been masturbating. Some trainers believe that this habit wastes a great deal of energy and makes the animal sluggish.

Stud Ring

To discourage masturbation, plastic stud rings may be used. These stud rings are available in various sizes. They are placed around the shaft of the penis when it is relaxed. When the penis begins to enlarge, the pressure and pain exerted by the stud ring forces the horse to relax the penis. Since an erect penis is necessary before the stallion is able to masturbate, the stud ring effectively interrupts the natural sequence of events that lead to ejaculation.

Fig. 14–4. A stud ring prevents a colt or stallion from masturbating.

It is important to remove the stud ring periodically (approximately once per month) and wash the penis thoroughly to prevent it from becoming irritated or infected. *(See Chapter 4 for more information on cleaning this area.)*

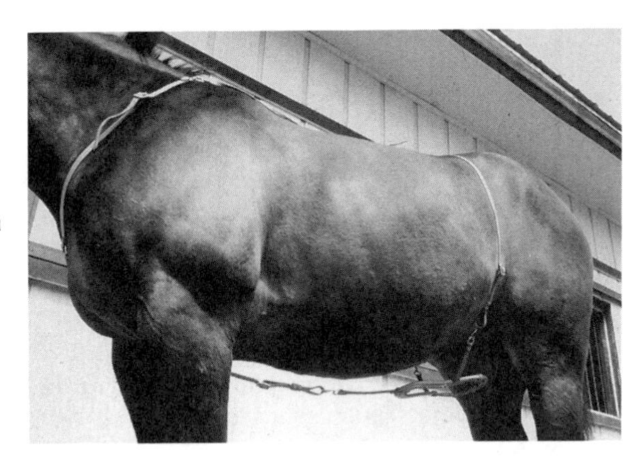

Fig. 14–5. A stallion shield discourages a stallion from masturbating.

Stallion Shield

On the breeding farm, stallions are discouraged from masturbating. Their semen is valuable and good quality semen must be available to impregnate broodmares. If this habit is not curtailed it may lead to impotency.

A stallion shield is commonly used to discourage masturbation. It consists of a spiked metal oval placed under the penis and held in position by a leather harness. When the stallion begins to get an erection, the penis hits the spikes and retracts to the normal relaxed position.

CRIBBING

Many horsemen use the term "windsucking" to describe this habit because the horse literally sucks air into its stomach. The cribbing process can be broken down into three steps. First, the horse grips its upper incisor teeth on an object such as a bucket, ledge, fence, etc. Then the horse arches its neck, expanding the muscles about the throatlatch. Finally, the horse sucks air into its stomach and makes a loud grunting noise.

Although most horses develop this habit after watching other horses crib, the actual cause of cribbing is unknown. And once a horse develops this bad habit, you can only hope to control it; this vice is almost never cured. Cribbing wears down the incisor teeth, may cause a loss of appetite or energy, and can cause colic in a horse that swallows large amounts of air.

Fig. 14–6. First, a cribbing horse grips an object, such as a stall door, with its teeth.

Fig. 14–7. Next, the horse arches its neck. Then, it sucks in air and makes a grunting noise.

To discourage cribbing, horsemen use various types of cribbing straps. Most cribbing straps are buckled over the throatlatch and upper neck, just behind the ears. Cribbing straps are usually kept on horses whenever they are confined to their stalls. However, if a horse has access to a turnout area, it would then be advisable to place a cribbing strap on the horse before it is turned out.

When a horse begins to crib, the groom will most likely be the first one who notices the vice. The groom should then notify the trainer, who will decide whether or not to use a cribbing strap.

The trainer will also specify which type of cribbing strap should be used. The following sections describe five cribbing devices which are designed to interfere in one or more phases of cribbing.

French Cribbing Strap

This cribbing strap is designed to prevent expansion of the wind-pipe which allows the abnormal intake of air, thus breaking up the

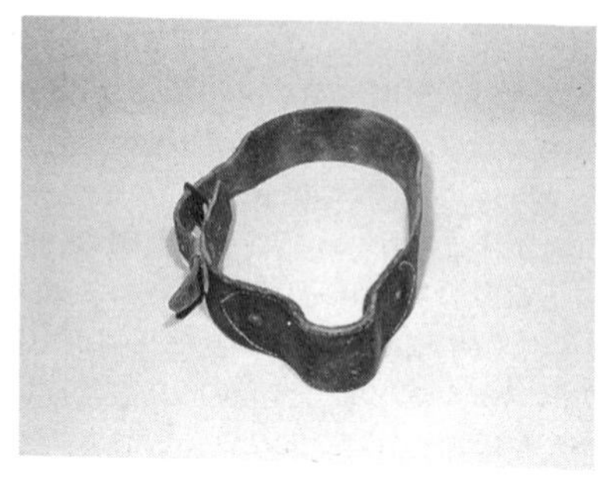

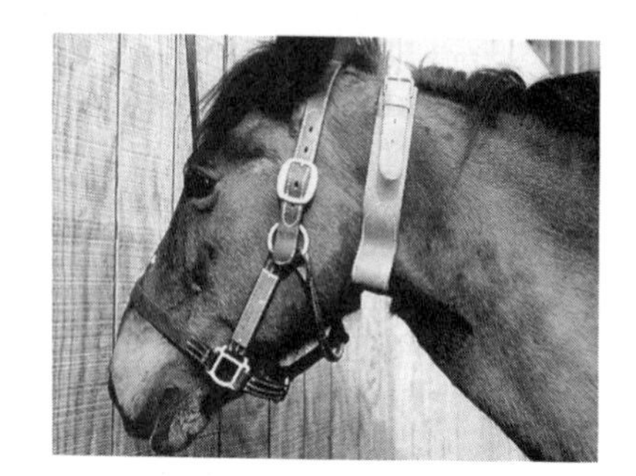

Fig. 14–8. French cribbing strap.

cribbing sequence. The French cribbing strap is buckled around the horse's neck, with the shaped end under the throatlatch. If fitted properly, this device is moderately effective and is not severe.

Metal-Jointed Cribbing Strap

This device is fastened around the throatlatch in the same manner as other cribbing straps. It functions in a similar fashion as the French cribbing strap but is more severe. The metal-jointed, or "nut cracker" cribbing strap exerts more pressure on the windpipe than that of the French cribbing strap. It is most effective in discouraging cribbing.

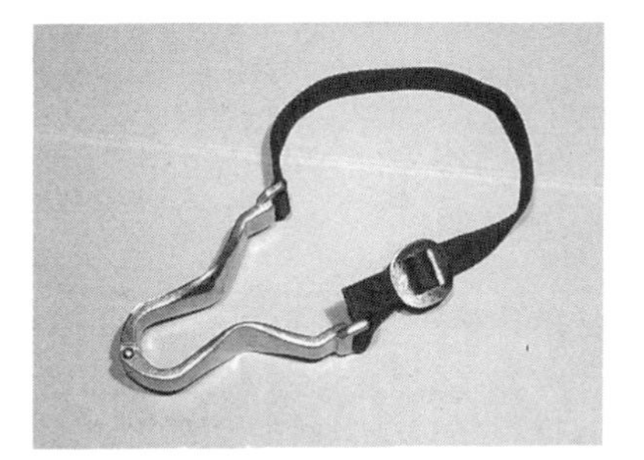
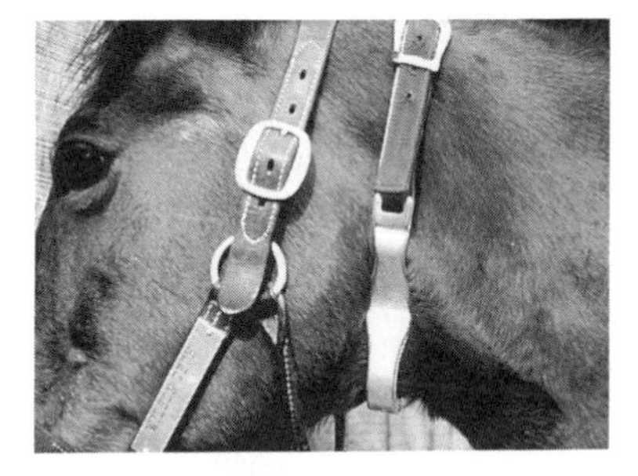

Fig. 14–9. Metal-jointed cribbing strap.

Diamond Cribbing Strap

A diamond cribbing strap prevents the horse from arching its neck during cribbing. The pointed ends of the stiff, leather diamond press into the under jaw and neck when the horse arches its neck.

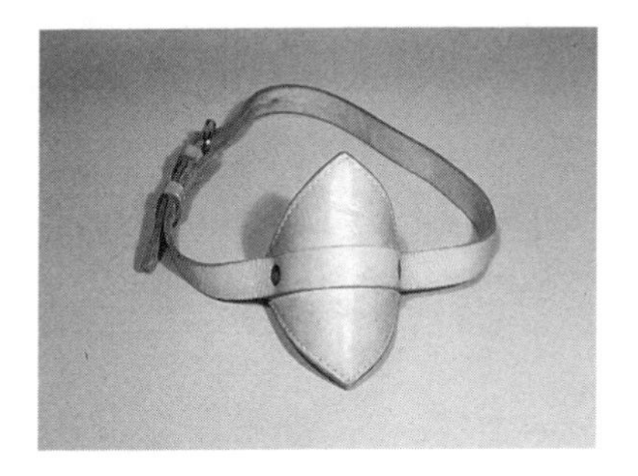
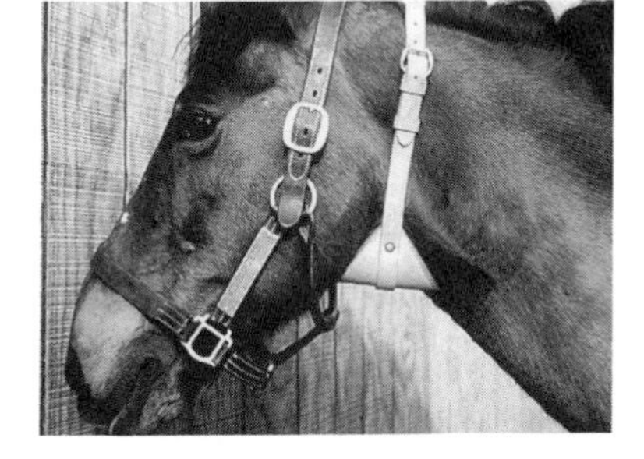

Fig. 14–10. Diamond cribbing strap.

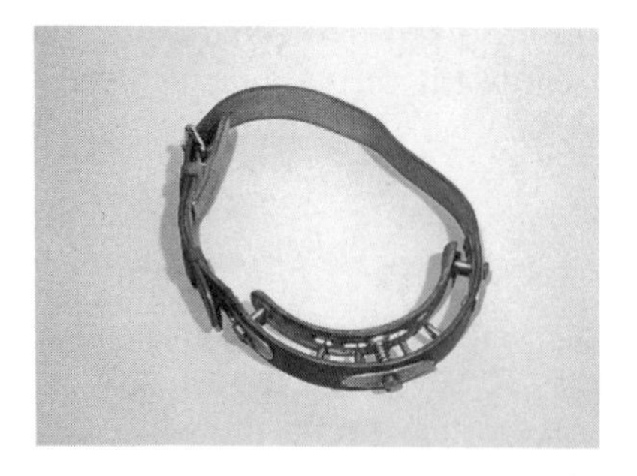

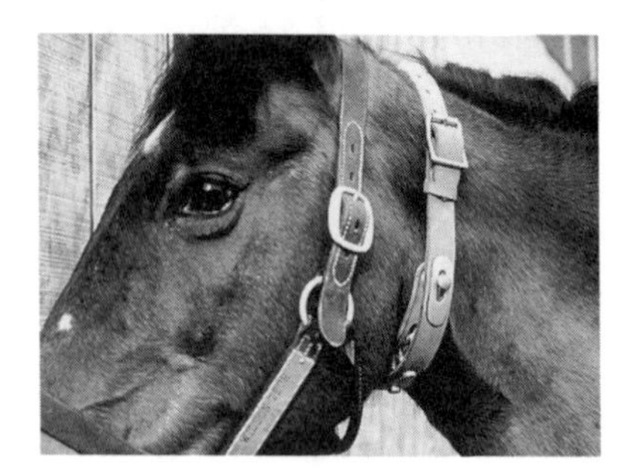

Fig. 14–11. English cribbing strap.

English Cribbing Strap

This type of cribbing strap is probably the most severe. It is placed around the throatlatch and consists of spring-mounted metal spikes. When the horse arches its neck to crib, the spikes are forced upward into the throatlatch. In a normal, "non-cribbing" position, the spikes are retracted below a smooth metal plate and do not irritate the horse in any way.

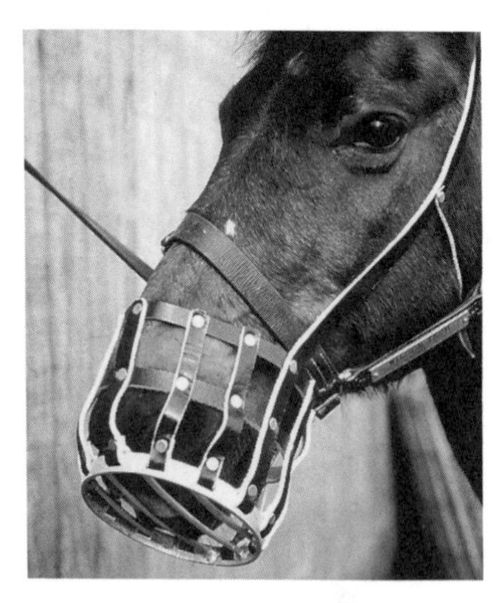

Fig. 14–12. A muzzle prevents cribbing, but does not cure it.

Cribbing Muzzle

Instead of a cribbing strap, a cribbing muzzle may be used to discourage this bad habit. It is placed over the horse's nose (except when the horse is eating) and should prevent the horse from gripping its teeth on an object, thus preventing the first step in the cribbing process.

All of the aforementioned devices are effective in discouraging cribbing, and thus preventing the horse from damaging itself or the stall, but understand that in *no way do they cure the habit.*

GRAIN BOLTING

Fig. 14–13. Placing a large stone in the feed tub forces a horse to eat slowly.

A horse with this bad habit "gulps" down its grain rapidly. At best, the horse fails to obtain the full nutritional value of the grain because the digestive process begins when food is mechanically broken down by the teeth. If the grain is not chewed, the horse cannot digest it. At worst, bolting may cause colic, because the grain is virtually swallowed whole.

There are several ways to discourage grain bolting in racehorses. One way is to place a brick or a large stone in the feed tub to force the horse to reach around the object, thus slowing down its eating. Another method is to use a feed tub with a metal ring in the middle. *(See Chapter 3 for a more complete description of this feed tub.)*

CHEWING

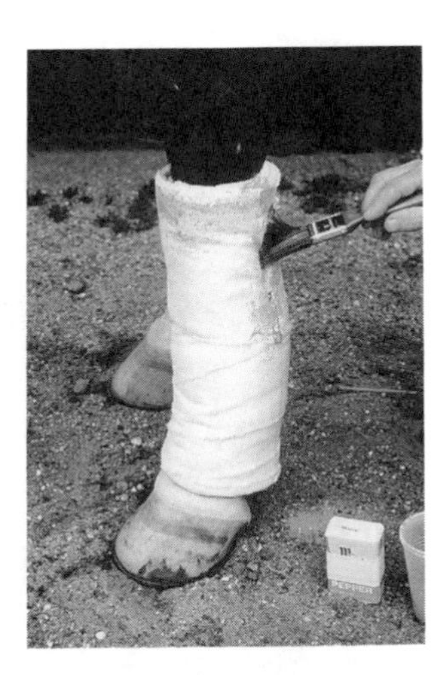

Fig. 14–14. Brushing on a bad-tasting formula discourages chewing.

Another vice is when a horse chews stable blankets, bandages, fences, and wood stall walls. There are several methods to prevent a horse from chewing.

With a horse that chews bandages or blankets, the groom can create a mixture of foul-smelling and bad-tasting items such as hot mustard, red pepper, and liquid soap. (Commercial mixtures can also be purchased.) Brush this mixture on the bandages or on the chest and shoulder areas of the horse's blanket.

The same mixture can be used to stop a horse from chewing wood in the stall or fence rails. Its foul smell and taste discourages most horses from chewing.

Fig. 14–15. Two methods of discouraging wood chewing. On the left, aluminum strips on a stall ledge. On the right, a leather bib.

Applying a soft aluminum covering to the wooden stall ledges also discourages chewing. Or, an electric wire strung along the stall ledge gives a horse a mild shock when it attempts to chew. An electric wire along the top of a fence board is an effective deterrent against a horse chewing wood in a paddock or pasture. The most effective deterrent against chewing is the use of a leather bib. The bib is attached to the halter and covers the lower jaw. It prevents the horse from chewing because the horse is unable to get its teeth on an object.

EATING MANURE

Most horses will not eat manure as long as they have free access to grass hay, as was recommended in Chapter 3. However, if the hay must be removed from the stall (before a race, for instance) a wire muzzle may be used to discourage the horse from eating its manure during this time.

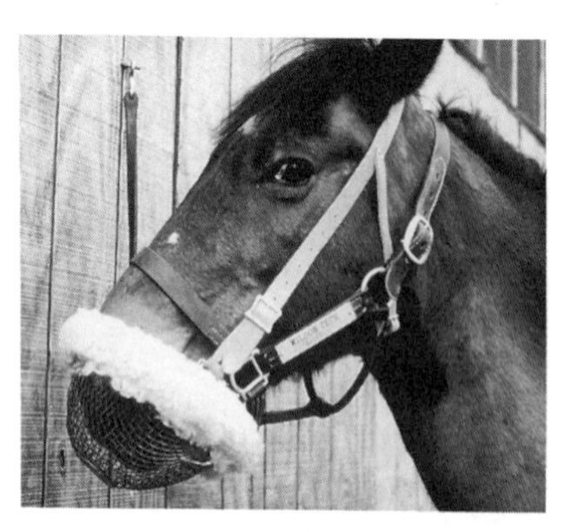

Fig. 14–16. A wire muzzle prevents a horse from eating manure or bedding.

Ideally, of course, the best way to avoid this vice is to make sure the horse has a clean stall at all times, and a properly balanced diet with a mineral salt block. (**Note:** See *Feeding to Win II,* published by Equine Research, Inc. for detailed information about providing the horse with a balanced diet.)

EATING BEDDING

This habit is usually found only among horses with straw bedding. From either hunger or boredom, the horse eats the straw bedding to the point where there is little left in the stall.

To eliminate this problem, the racehorse should have access to grass hay at all times. If necessary, the horse's stall bedding can be switched from straw to wood shavings, which the horse will be less likely to eat. Or, a wire muzzle may be used to discourage this bad habit. See the section entitled, "Preventive Measures" for more information on muzzles.

BITING

Biting is a vice found primarily in young colts. But by no means is biting limited to colts; it may be found in all horses. It is not uncommon to see a horse "lunge" over its stall webbing to bite at anything or anyone passing the stall.

Fig. 14–17. It is not uncommon to see a horse lunge over its stall door to bite another horse or a person.

Some horses get annoyed during grooming and tacking and attempt to bite. Feeding time is also a prime time for many horses to bite. The best method to combat biting is to punish the horse each time it tries to bite. This can be accomplished by lowering your tone of voice and scolding the horse.

A water pistol is also effective in curing a horse of biting. When the horse attempts to bite, squirt water in its face to discourage the act of biting. Again, the reprimand must come within three seconds of the bite attempt, or it will be ineffective.

Never lose your temper with a horse. If you do, you have already lost the battle. *Hitting a horse in the head sometimes causes the horse to become head-shy and very difficult to work with.*

Tying the horse to the wall while grooming keeps the groom from being bitten. A full door screen in place of a stall webbing prevents the horse from lunging and biting someone in the shedrow. In a severe case of biting, a wire muzzle may be placed on the horse.

PREVENTIVE MEASURES

Muzzles

Muzzles are used on horses as a method of restraint—to prevent them from biting, chewing, or eating. There are two basic types of muzzles used on horses: the wire muzzle and the box muzzle.

Wire Muzzle

This muzzle is made of woven wire, its edge lined with fleece for comfort. It is attached to the horse's head by a leather headpiece and throatlatch. Wire muzzles are very popular to use on the day of a race to prevent a horse from eating the straw bedding after its hay is removed. (Some trainers feel that the hay should be taken away on the day of the race so the horse's digestive process will not interfere with its performance.) However, the muzzle does allow the horse to breathe freely and to drink.

Box Muzzle

A box muzzle is usually made of either plastic or leather with large openings for the nostrils. It is secured to the head by three snaps or buckles which attach to the halter. It can also be used to prevent a horse from eating the bedding when its hay is removed. It is very popular with horsemen when shipping horses, because it prevents

Fig. 14–18. On the left, a wire muzzle keeps a horse from eating hay on race day. On the right, a box muzzle is popular when shipping a horse.

the horses from biting each other while they are being transported in close quarters.

Fig. 14–19. Plastic apples or bottles are popular stall toys for racehorses.

Stall Toys

Trainers and grooms have employed numerous methods over the years to discourage stable vices, some of which are still very effective. Horses are curious by nature, so providing them with an alternative form of entertainment is often helpful. For example, some grooms hang a plastic bottle, a rubber ball, or another type of toy inside the stall or outside the stall door. These toys occupy the horse's time and alleviate boredom.

Hay Nets

A hay net hung by the stall door helps to keep the horse's attention. It allows the horse to eat hay without missing any activities going on outside the stall.

Hay nets are also useful in keeping a horse content while shipping. Free access to hay helps avoid boredom and vices in the trailer. *(See Chapter 16 for more information on using hay nets when shipping horses.)*

Fig. 14–20. A hay net hung outside the stall allows a horse to eat while it watches the outdoor activity.

Fig. 14–21. Goats make good companions for horses.

Companion Animals

Some trainers bring in other animals to be companions for the racehorse; to keep it company and alleviate boredom. The presence of a companion animal may also have a calming effect on a nervous racehorse. Among the stable companions commonly found at the racetrack are goats, cats, dogs, ducks, chickens, ponies, etc.

SUMMARY

Racehorses are not born with vices; they acquire them as a result of unnatural or uncomfortable circumstances inflicted by humans. Standing in a stall for over 20 hours a day causes many horses to become bored. As a result, they develop bad habits to alleviate their boredom. The more often grooms can get the horses out of their stalls, the fewer vices these horses are likely to develop. Also, grooms and trainers should work together to devise distractions to keep the horses entertained and at ease in their stalls; the happier and more content racehorses are, the more likely they are to perform well on the racetrack. Finally, a well-balanced diet, frequent meals, and free access to hay helps keep the horse healthy and occupied. (See *Feeding to Win II,* published by Equine Research, Inc. for more detailed information on nutrition and feeding.)

Chapter 15

Restraining Methods

Keeping a 1,000-pound animal under control becomes easy once you know a few simple techniques and a little equine psychology. There are times when the groom may be required to restrain a horse for its own good:

- to control the horse while walking
- to administer medication
- to examine the horse
- to have the horse shod
- to clean the sheath or udder
- to clip the coat with electric clippers

Restraint may be administered by the use of various types of equipment. Some restraining devices are more severe than others, but all can be effective if used properly. The goal is to get the job done safely with the least amount of restraint necessary.

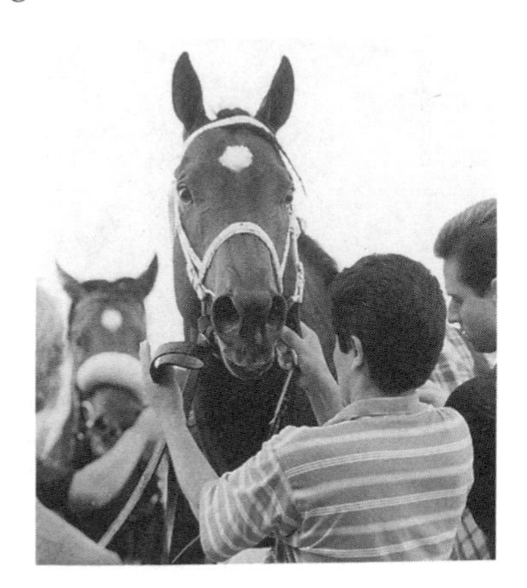

Fig. 15–1.

HALTER & CHAIN SHANK

Restraint begins with the halter. The halter should fit properly, with the noseband lying across the bridge of the nose. Sometimes dropping the halter to a lower position on the nose and allowing the noseband of the halter to rest just above the nostrils can be very effective in restraining the horse.

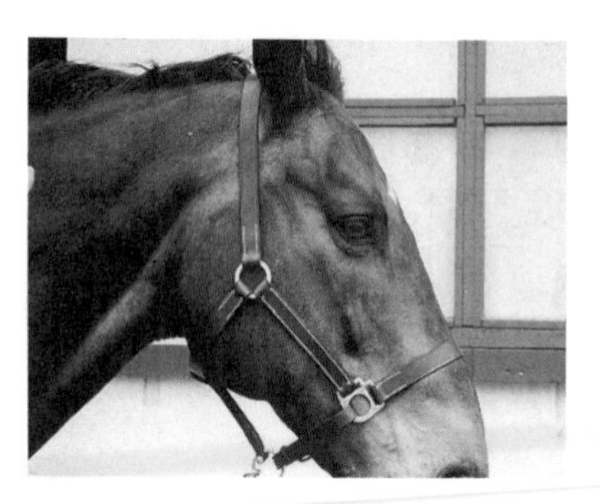

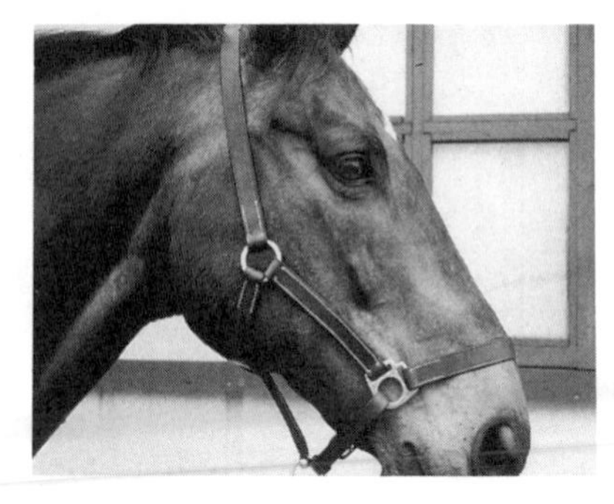

Fig. 15–2. The noseband of the halter on the left is in the normal position, across the bridge of the nose. Dropping the noseband lower on the nose, as in the photo on the right, sometimes makes restraint more effective.

The chain shank is also very effective in restraining and controlling the horse. (Other uses of the chain shank are described in detail in Chapter 2.) To restrain the horse more effectively with the halter and chain shank, slip the chain into the mouth and then pull it forward so the chain lies over the upper gum just above the incisor teeth. This method puts pressure from the chain onto the sensitive upper gum. In this position, you must keep a little pressure on the chain or it will fall from the mouth. This procedure should be reserved for the experienced groom only, as it is very easy to hurt the horse by being too harsh.

Fig. 15–3. A chain shank across the horse's upper gum is a severe method of restraint.

CHIFNEY BIT

The Chifney bit is very effective in controlling the horse while walking. It is used with the halter and chain shank and is placed in the mouth in the same manner as a riding bit. Instead of being attached to the bridle's cheekpieces, however, the Chifney bit is snapped onto the rings of the halter and pressure is applied with a lead shank rather than with reins.

STEEL HALTER

This piece of restraining equipment has become popular recently. The steel halter is made of a large steel oval ring supported by leather straps. It is placed on the horse's head with a small ring on top of the bridge of the nose.

A chain shank is snapped onto the small metal ring, and when pulled by the handler, the steel oval hits the bridge of the nose. At the same time, the bottom of the steel oval hits the bottom of the jaw.

The steel halter can be placed on the head by itself, over a leather halter, or over a bridle. It is very useful over a bridle when leading a horse from the barn to the racetrack and back.

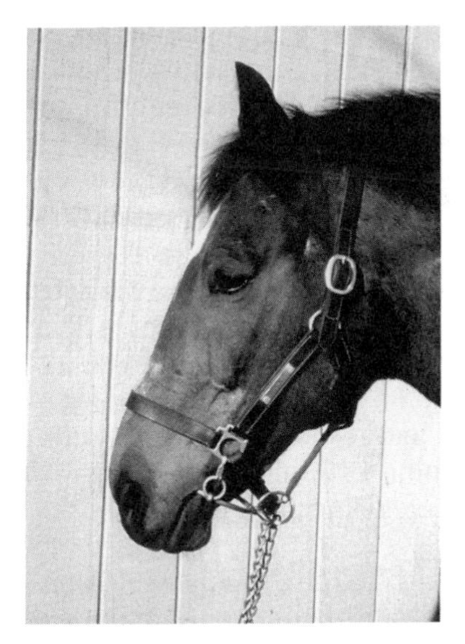

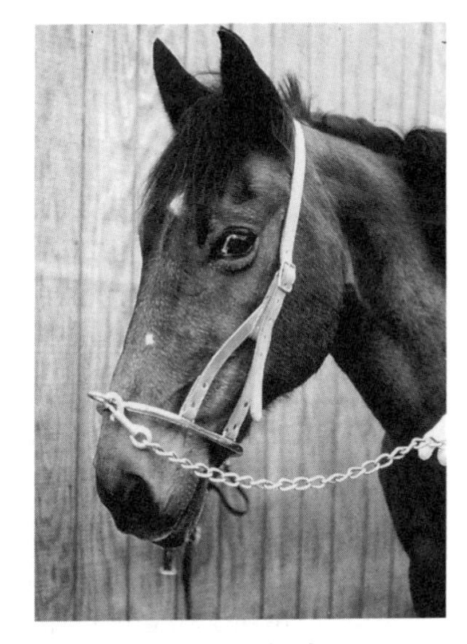

Fig. 15–4. On the left, the Chifney bit attaches to the halter for better control. On the right, the oval of the steel halter encircles the muzzle.

WAR BRIDLE

The war bridle is nothing more than a sturdy nylon rope with a noose in the horse's mouth and a slip knot on the left side of the head. When the groom pulls downward, it puts pressure behind the ears as well as on the mouth. The war bridle may be used by itself or with a halter.

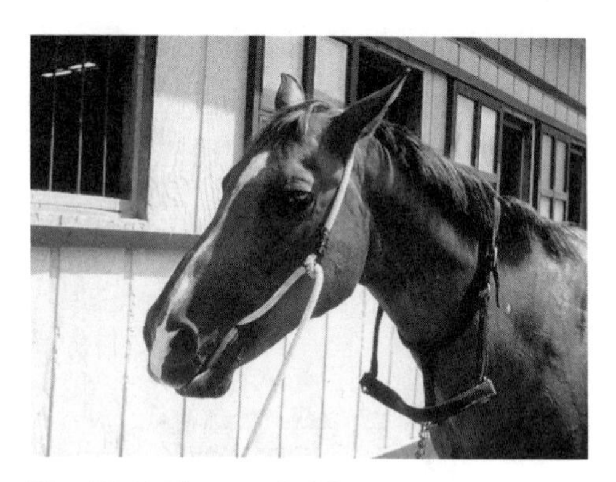

Fig. 15–5. The war bridle puts pressure behind the ears and on the mouth.

TWITCH

The twitch is the most common restraining tool used by horse people. The basic concept of the twitch, which is applied to the upper lip, is to make a horse relax and stand still while it is undergoing an unpleasant procedure. While it was originally believed that the pain of twitching alone diverted a horse's attention from other painful procedures, it has also been found that horses' heart rates decrease during twitching (heart rates increase during other painful experiences). Veterinary experts now think the pressure against the lip causes the body to release chemicals that reduce and suppress perceptions of pain. These natural chemicals cause the horse to relax during the procedure. When the procedure is over and the twitch is removed, the levels of these chemicals return to normal within 30 minutes.

Depending on the procedure the horse is undergoing, the twitch should be loosened about every 3 – 5 minutes. For example, clipping the ears takes approximately 15 minutes. Because clipping is a process that can safely be interrupted, the groom should stop and loosen the tension of the twitch at least two or three times during the 15 minutes. If the horse is undergoing a 10-minute stitching up from a veterinarian, it may not be possible to loosen the twitch at all during this time.

Two of the most common twitches are the chain twitch and the rope twitch. Two other types are the aluminum twitch and the screw end twitch can be used on a horse when you are working alone. It is important to learn how to properly use each kind of twitch; poor technique can cause serious injury to the handler and/or the horse.

Chain Twitch

This twitch is sometimes referred to as the Yorkshire twitch and is sold in most tack shops. It consists of a wooden handle about two feet long and a metal chain looped through the top of the handle. This twitch is usually placed on the upper lip but can also be used on the base of the left ear.

This type of twitch can be very dangerous if the handle escapes the hands of the groom. The horse will usually thrash about and the handle of the twitch will hit the groom and/or the horse and cause serious injury.

Rope Twitch

A rope twitch is usually "homemade" from an ax handle about three feet long and a nylon rope looped through the top of the handle. This twitch can be used on the upper lip or at the base of the left ear. I recommend this type of twitch over the short handled twitch as it provides more leverage and control. Both types of twitches should be used with a halter and chain shank.

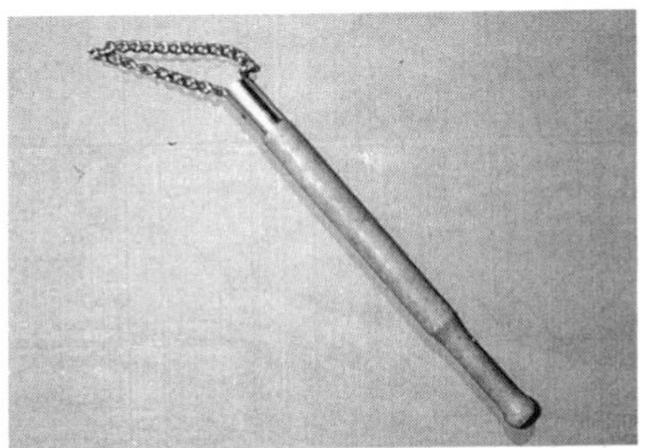

Fig. 15–6. A chain twitch.

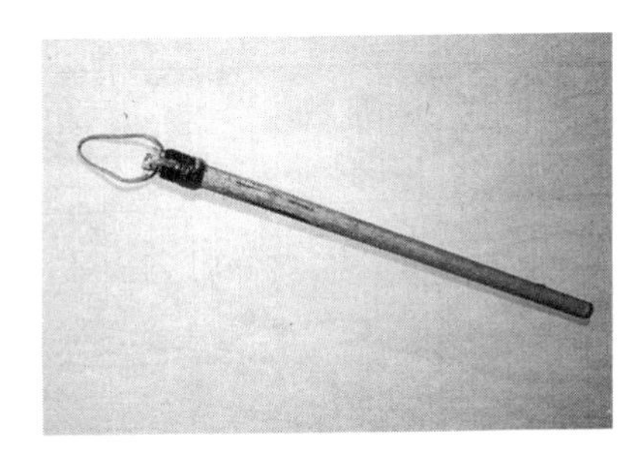

Fig. 15–7. A rope twitch.

Aluminum Twitch

This type of twitch, also called a "humane twitch," "easy twitch," or "tong twitch" is ideal when a groom is working alone. It is designed to clamp the upper lip in one position. After the lip is pulled through the two handles of the twitch, the handles are squeezed together, tied, and then snapped to the upper near ring of the halter. This twitch is helpful when a groom is clipping a horse and the horse will not stand still.

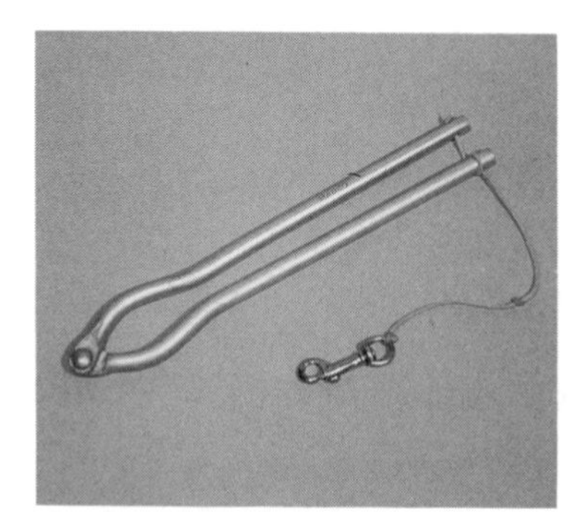

Fig. 15–8. An aluminum twitch.

Fig. 15–9. Snap the twitch onto the upper near ring of the horse's halter.

Screw End Twitch

This type of twitch is also designed to restrain the horse when a groom is working alone. The upper lip is pulled through the opening and the vice-like clamp is tightened until it is properly placed on the horse's lip. This twitch is also called an "English twitch."

Fig. 15–10. A screw end twitch.

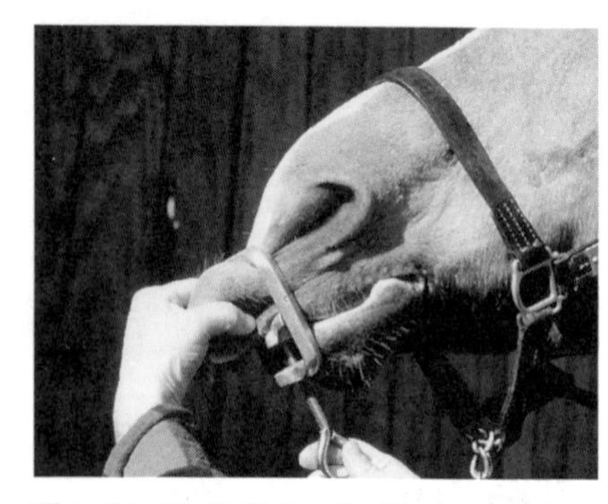

Fig. 15–11. Pull the lip through the opening and tighten the clamp.

Placing a Twitch on a Horse

Instructions for applying the aluminum twitch and the screw end twitch have been discussed briefly above—they are simple to apply. Many backstretch employees, however, may never have been taught how to safely apply the chain or rope twitch. The following steps are illustrated in Figure 15–12:

1. Stand on the horse's near side—never stand directly in front of the horse. Insert just the fingers of your left hand through the loop of the twitch. Be sure that the loop rests on your knuckles.
2. With the fingers of your left hand, gently and quickly squeeze the horse's upper lip upward. This will cause the loop of the twitch to drop over the lip.
3. Once the loop is over the lip, begin to twist the handle of the twitch until the loop is tight around the horse's lip. Stand close to the horse's left shoulder and hold the shank and twitch together.

Once the twitch is applied, it should only be tightened when necessary. If it is constantly tight, the twitch only numbs the lip and has no effect on the horse. **Do not under any circumstances touch that part of the upper lip which is held by the twitch.** It will be very sensitive and the pain could cause the horse to go berserk. Never leave a twitched horse unattended.

Applying a Twitch (Fig. 15–12.)

Step 1

Stand on the horse's near side. Insert the fingers of your left hand through the loop of the twitch.

Step 2

With your left hand, gently squeeze the horse's upper lip upward, causing the chain to drop onto the lip.

Step 3

Twist the handle until it is tight around the lip.

BLINDFOLD

Fig. 15–13. The theory behind the blindfold is that the horse will not fear what it cannot see.

A blindfold is a very popular restraining method, particularly when trying to load a horse onto a van or trailer. The blindfold can be any type of heavy cloth—anything that prevents a horse from seeing normally. An old shirt or light jacket may be used. The theory behind the blindfold is that the horse will not fear what it cannot see.

Caution should be taken when placing the blindfold material over the horse's eyes and securing it to the head. An assistant standing on the off side of the horse is helpful when applying the blindfold over the eyes. It is important that the groom and assistant stand clear of the front feet, as the horse may strike out with its front feet in an attempt to free itself of any means of restraint. The same caution should apply when attempting to remove the blindfold.

OTHER METHODS OF RESTRAINT

In addition to the various types of restraining equipment discussed in this chapter, you may choose to restrain a horse by using your hands and body to apply pressure to the horse's body. The following methods are effective alone or may be used in conjunction with restraining equipment.

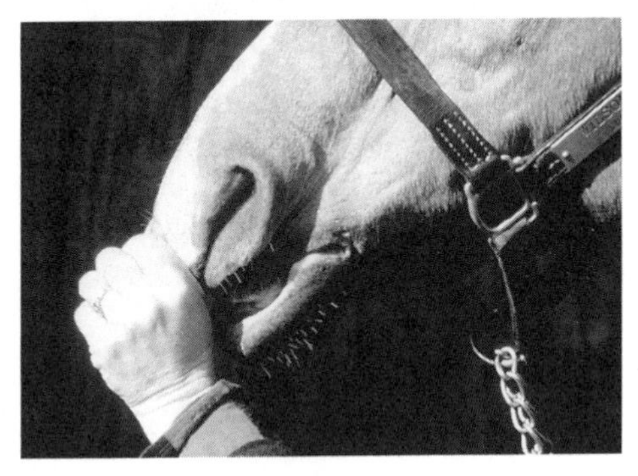

Fig. 15–14. Holding the upper lip with your fist works like a twitch.

Holding the Upper Lip

By simulating a twitch on the upper lip with your

hand, you may effectively restrain a horse for a short period. It is important to use a halter and shank when attempting this method of restraint. Stand on the near side of the horse and hold the upper lip with your left hand and the halter and shank with your right hand. If more restraint is needed, merely twist the lip tighter with your left hand.

Holding the Ear

To restrain the horse in this manner, begin by standing on the near side of the horse and hold the halter (and shank, if necessary) with your left hand. With your right hand, grasp the base of the left ear and squeeze, pulling downward at the same time, for as long as necessary to keep the horse quiet.

Holding the Upper Lip and Ear

This method of restraint is merely a combination of the two previous methods mentioned. With your left hand, grasp the upper lip and with your right hand, hold the base of the left ear. This method is very effective for tilting the head to administer medication to the eye.

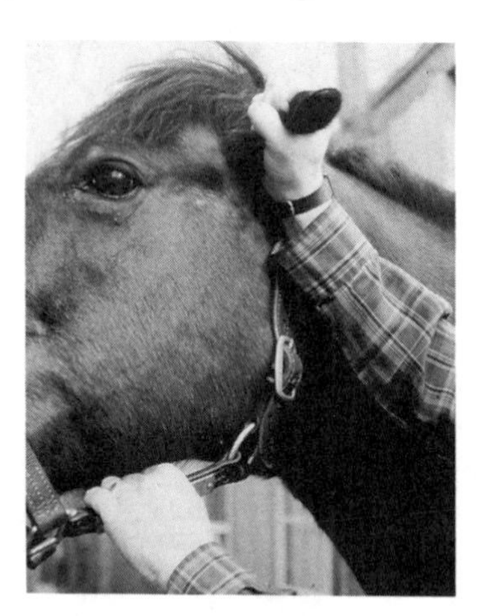

Fig. 15–15. Holding the Ear: Squeeze the base of the left ear and pull down.

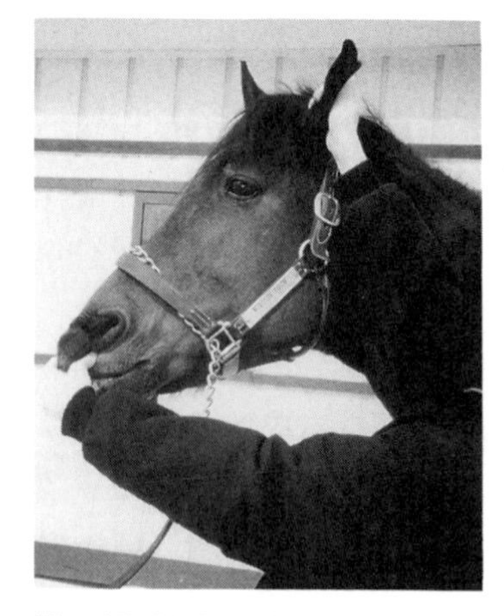

Fig. 15–16. Holding the Lip and Ear: Grasp the lip in your left hand and the ear in your right.

Holding a Front Leg

By lifting and holding one of the front legs, the horse is forced to distribute its weight evenly on the other three legs and is therefore unable to kick. This method is very effective when someone is working on the hind legs. To apply a mild version of this method, stand just behind the horse's shoulder on the near side. Pick up the foot as if you were going to clean it. Holding the fetlock, fold the leg upward as completely as possible. To apply a more stringent version, stand in front of the horse, facing the rear. Lift the front leg off of the ground toward you, and lock your fingers under the fetlock, holding the leg stretched outward. The higher the leg is held, the more weight is placed on the hind legs.

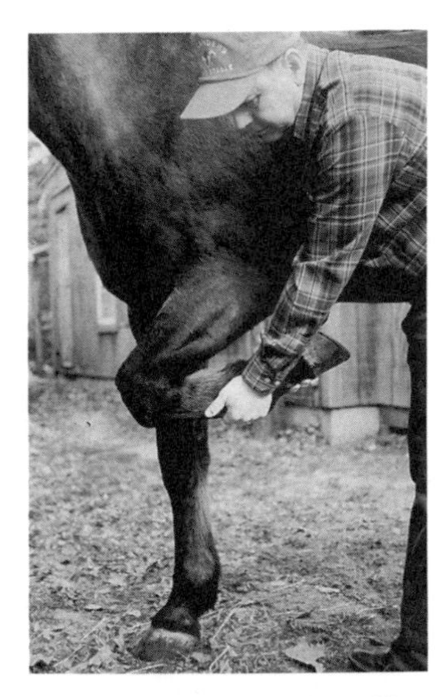

Fig. 15–17. With one leg off the ground, the horse is unable to kick.

Holding the Tail

Holding the horse's tail can be a very effective method of restraining the horse from kicking out with the hind legs. Taking a firm hold of the tail and pulling downward forces the horse to keep its weight on its hind legs. This allows you or an assistant to safely work on the hind legs. *(See Figure 15–18.)*

Rolling the Shoulder Flesh

Rolling the shoulder flesh as a method of restraint is very effective on any horse. It is particularly effective in preventing the horse from striking with its front feet. The person applying restraint in this fashion must stand facing the front of the horse on the near side with both hands gripping the flesh on the shoulder. The flesh should then be "rolled" forward toward the front of the horse. *(See Figure 15–19.)*

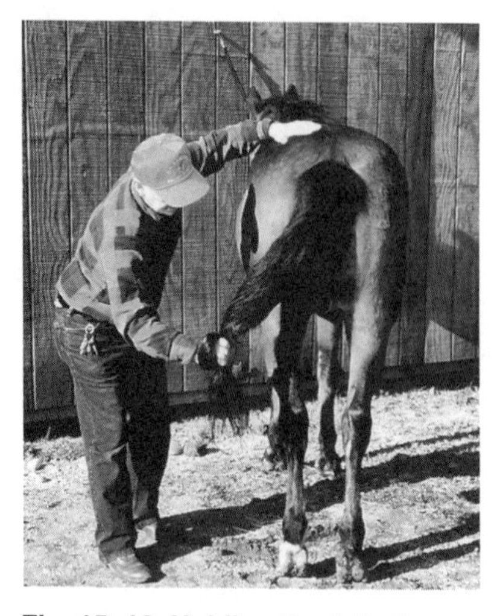

Fig. 15–18. Holding the tail allows you to work safely on the hind legs.

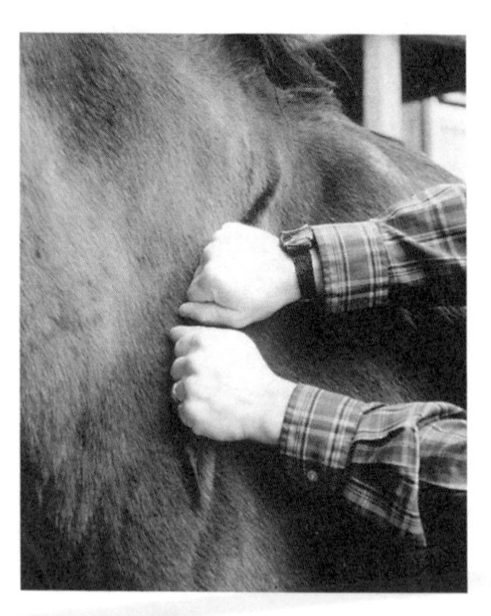

Fig. 15–19. Rolling the shoulder flesh discourages a horse from striking with its front legs.

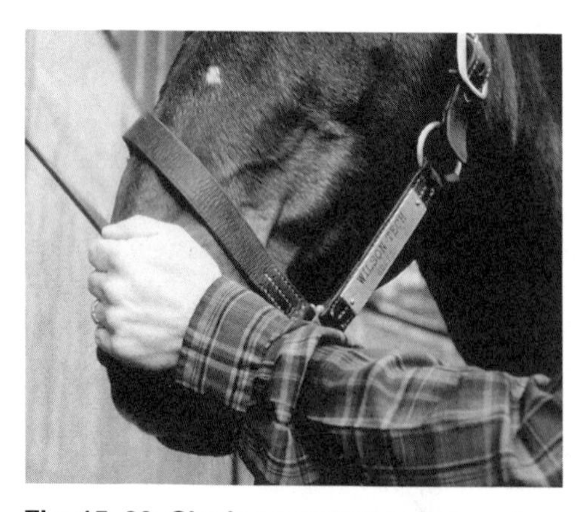

Fig. 15–20. Closing a nostril reduces air intake, causing the horse to stand quietly.

Closing the Nostril

This action reduces the horse's amount of air intake which causes it to stand quietly. With your right hand, hold the horse's halter. Place your left hand on the bridge of the horse's nose and squeeze *one* nostril closed.

SUMMARY

There are several effective methods to humanely restrain a horse. Take the time to master them all so that no matter what situation arises, you are prepared with the appropriate restraining method.

Chapter
16

Miscellaneous Stable Duties

The responsibilities that were outlined in previous chapters usually do not take up the groom's entire day. However, there are always a number of miscellaneous tasks that must be done to keep a stable operating effectively. A dedicated groom will take the initiative and do some miscellaneous jobs around the stable when time allows.

This chapter discusses some of the more common extra duties a caretaker will be faced with, including blanketing a horse, doing the daily laundry, preparing a horse

Fig. 16–1.

for shipping, and light stable maintenance. The caretaker may be expected to carry out miscellaneous duties on a daily basis, or only occasionally.

BLANKETING

There are three basic reasons for blanketing a horse:

- to keep the horse warm and dry in inclement weather
- to prevent a chill and to absorb moisture after bathing
- to keep the muscles warm and supple

Blanketing the racehorse is generally the groom's responsibility, so it is important to be familiar with the types of sheets and blankets, as well as the situations in which each should be used.

Types of Sheets and Blankets

Night or Stable Sheet

This type of blanket prevents the horse from getting a chill while confined in its stall during cold weather. The night or stable sheet may be made of several different materials such as wool, nylon, etc. The night or stable sheet is placed over the body from the withers to the top of the tail. (Never pull the blanket toward the horse's head, against the direction of hair growth. Always put it on higher than it needs to be, then slide it backward.) The night or stable sheet covers the horse's sides and chest, and is secured to the body by a strap and buckle across the chest and two surcingles which are brought underneath the body and fasten on the horse's near side. Make sure the blanket is not folded or wrinkled under the surcingles.

Fig. 16–2. The stable sheet is used at night when the weather is cool.

Woolen Cooler

A woolen cooler is used mainly in colder weather. It prevents the horse from getting a chill after being washed or sponged down after a race or workout, and aids in the drying process by absorbing moisture. The woolen cooler also helps the horse's body temperature gradually return to normal while being walked after a race or workout.

Fig. 16–3. The woolen cooler is used after a race or workout in cold weather.

If the cooler was folded properly, you can unfold it until only one fold remains. Drape it evenly, with the center fold over the horse's withers. Then unfold the top layer of the cooler over the horse's neck. It will be necessary to slide it back into position. A browband is pulled gently over the ears and rests on the horse's brow. The two sets of tie straps are tied in bows in front of the neck and chest. The tailpiece is placed under the tail.

Scrim or Anti-Sweat Sheet

The Scrim or anti-sweat sheet is used primarily in warm weather. Made of a loosely woven mesh, it is designed to allow air to come in direct contact with the horse's body and cool it down, while at the same time keeping the horse free from chills. The anti-sweat sheet fastens with one buckle at the chest and two straps that encircle the body and fasten on the near side.

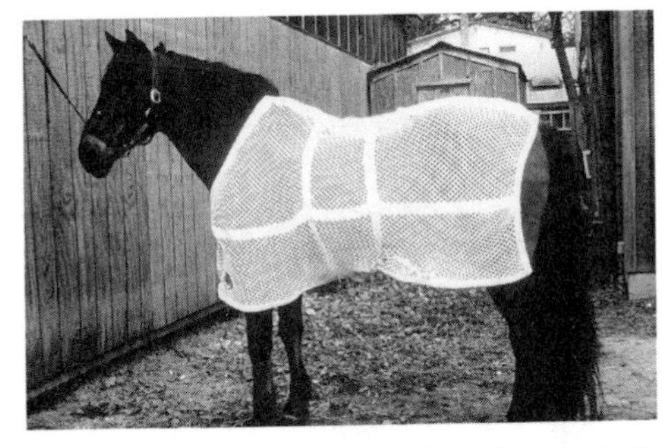

Fig. 16–4. The anti-sweat sheet is used after a race or workout in warm weather.

Fly Net

Used primarily in warm weather, the fly net is similar to the anti-sweat sheet in that it is made of a mesh material. However, the mesh on a fly net is woven much smaller, providing protection from the irritation of insects. The mesh material of the fly net allows air to come into contact with the horse's body. For this reason, using a fly net after bathing the horse usually helps the horse dry and cool out more quickly. This sheet is placed on the horse's body just like the anti-sweat sheet.

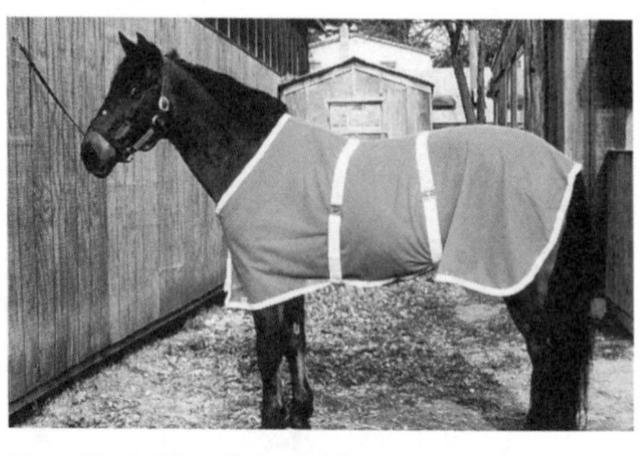

Fig. 16–5. The fly net is used in the summer to protect the horse from insects.

Rain Sheet

The rain sheet is placed on the racehorse when going to and from a race during inclement weather to keep the horse as dry as possible. This sheet is usually made of a rubberized nylon that repels water.

It is placed over the body and neck, and is secured by two ties under the neck and chest. The browband rests on the brow in front of the ears. It is also secured to the body by an elastic surcingle encircling the body over the withers and just behind the front legs. Buckle it snugly, or the rain sheet may slip backward and hurt the horse's ears. A pommel pad may be used under the surcingle to protect it from rubbing against the withers.

Fig. 16–6. The rain sheet is used to keep the horse dry in wet weather.

Paddock Sheet

The paddock sheet is similar to the night or stable sheet but is used only when the racehorse goes to the saddling paddock before a race.

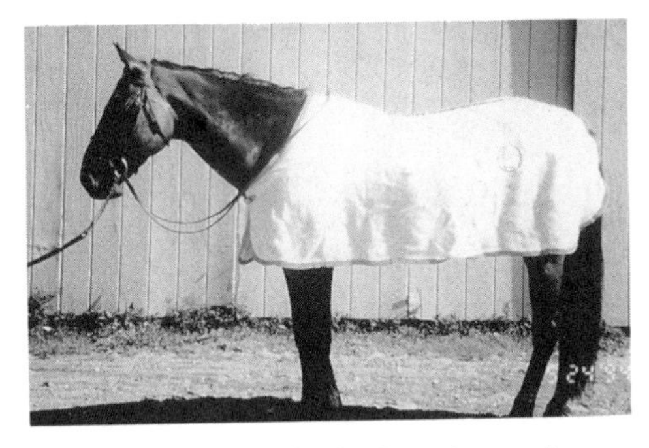

Fig. 16–7. The paddock sheet is used when the racehorse goes to the saddling paddock before a race.

It is usually fastened to the body by a buckle and strap at the chest and only one surcingle which buckles on the near side underneath the sheet. It extends from the withers to the tail and it keeps the horse warm and comfortable before and after a race. The stable name and colors are usually displayed on the sides of the paddock sheet.

Shoulder Piece

This type of blanket is designed to cover only the horse's chest and shoulders. It may be used on a racehorse when going to the saddling paddock. Some trainers rub liniment on the horse's shoulders before a race to "loosen up" the shoulder muscles. The shoulder piece is then placed on the horse (under the paddock sheet) to keep the shoulders warm until it is time to put the saddle on.

The chest piece is solid (with no buckle), and therefore must be pulled over the horse's head. The shoulder piece is fastened by a single surcingle which is buckled on the near side *(see Figure 16–8).*

Jowl Hood

During cold weather, the jowl hood may be used to keep the head warm. Quarter Horse owners and trainers sometimes use the jowl hood to "sweat" the throatlatch to reduce the amount of fat there. (Quarter Horses tend to have a bit more fat in the throatlatch area.) The jowl hood is made of heavy cotton and is placed over the throatlatch. It is secured to the head by a leather strap and buckle *(see Figure 16–9).*

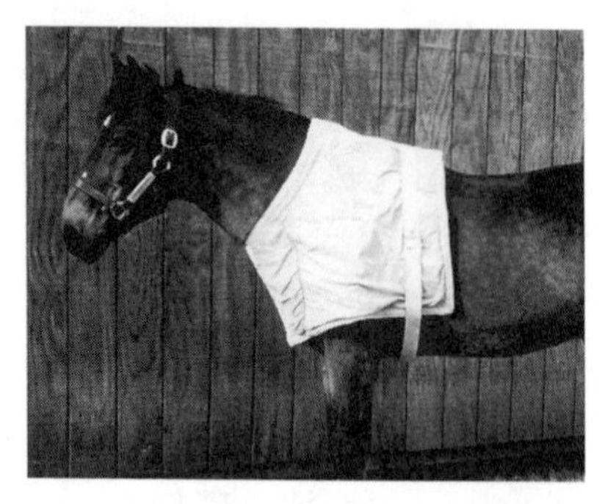

Fig. 16–8. A shoulder piece keeps the shoulders warm before a race.

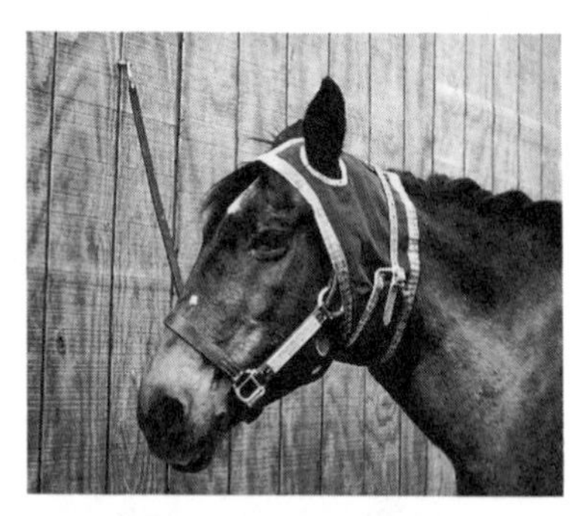

Fig. 16–9. A jowl hood keeps the head and throatlatch warm.

Blanketing Tips

When blanketing horses for any purpose, keep in mind the following safety factors:

- The surcingle should be tight enough to secure the blanket to the horse's body and prevent it from slipping. It cannot be so tight that the horse is restricted when it lays down. And it cannot be so loose that the horse gets a leg caught up in it when lying down or standing up.
- Some blankets have straps that fasten around the back legs. These straps should be crisscrossed so they do not rub against the inside of the horse's legs.
- The chest strap is always fastened first and unfastened last. If the chest strap is unfastened first, and the horse jumps forward, the blanket can slide back and entrap the horse's hind legs because the straps underneath the belly are still fastened.

Properly Folding a Sheet or Blanket

It is important to keep the horses' blankets clean and neat. A blanket that is folded and stored properly lends a neat appearance to the stable. Also, a correctly folded blanket or sheet is much easier to put on a horse. To do this, simply perform the following steps:

1. Lay the blanket out flat on a clean surface.
2. Fold the entire blanket in half lengthwise, keeping the ties on your left side.

3. Fold the left side (with the ties) ⅓ of the way under, in toward the center. This allows the ties to rest in the center of the blanket. Then fold the right side over the left side, "burying" the ties within the folds of the blanket.
4. Fold the entire blanket in half again lengthwise, keeping the ends even.

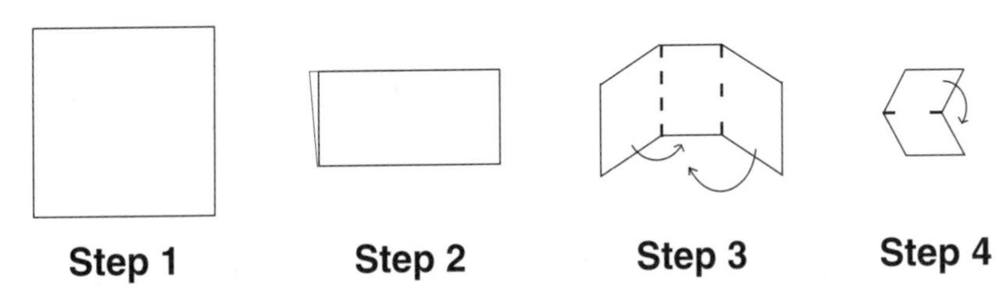

Fig. 16–10. Properly folding a blanket or sheet.

After properly folding the blanket, drape it over the bar of the blanket rack as shown in Figure 16–11.

DAILY LAUNDRY

It is the groom's responsibility to wash the following items daily:

- saddle cloths
- girth covers
- bandages
- rub rags

After the morning training hours, place all the dirty items in a bucket or tub of hot water. Add a mild detergent to the water.

Use a manual agitator to swish the items around in the bucket for a few minutes. After agitation, wring out the water from each item by squeezing or twisting the material until no more soapy water drips out, then dump out the soapy water.

Fig. 16–11. Properly folded blankets lend a neat appearance to the stable.

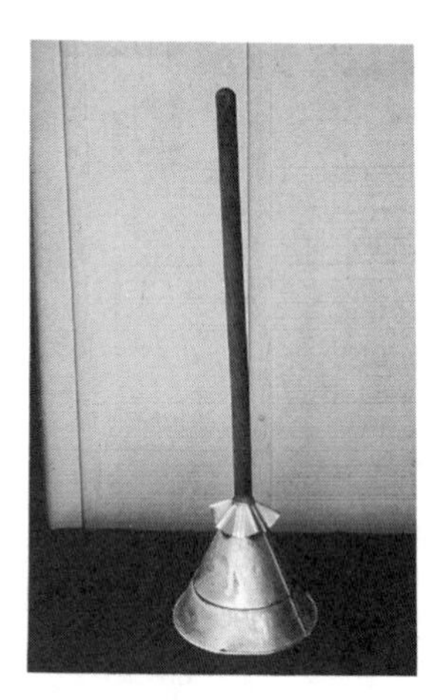

Fig. 16–12. Use a manual agitator to clean the items.

Fig. 16–13. Wring out the soapy water before rinsing.

Next, fill another bucket with clean, warm water, submerge the laundry into this bucket, and rinse the laundry completely. (If the items are not rinsed adequately, the soap residue can irritate the horse's skin.) The laundry should be wrung out a second time and hung on a line to dry. In the case of flannel bandages, it is a normal practice to roll the bandages while wet, wring out the water, unroll them, and then hang them on a line to dry.

If your trainer asks you to disinfect the laundry, you must add a third cycle to the cleaning process. After washing and rinsing the item(s), soak them in a third bucket containing a mixture of water and chlorine bleach (10 parts water to 1 part chlorine). A separate bucket is necessary because adding disinfectant to soapy wash water is inadequate; soap causes bleach to lose its effectiveness. When bleach is mixed with soap it will still whiten the laundry, but it may not disinfect it.

Follow the directions on the product's label regarding the time needed for disinfecting certain items. After the laundry has soaked for the allotted time, it can be removed, rinsed, and hung to dry. (If the laundry is not too soiled, it is equally effective to disinfect the laundry first, before the washing process.)

PREPARING FOR SHIPPING

Shipping horses from racetrack to racetrack, and from farm to farm happens daily all over the world. The common modes of shipping horses include by airplane, van, or trailer.

Horses are prone to injury during shipping, particularly when they are being loaded and unloaded from the transport vehicle. Long trips (over five or six hours) can also make horses vulnerable to illness. It is the groom's duty to properly prepare horses for shipping so as to avoid injury and illness. The trainer may require some or all of the following equipment:

- shipping boots or bandages
- head bumper
- halter wrap
- shipping muzzle
- tail wrap
- blanket
- hay net

Shipping Boots

Shipping boots are designed to protect the shins, tendons, and fetlocks (some shipping boots extend down over the coronets) of the horse while it is being transported. These boots should be used on all four legs. The front leg boots are usually about 12 inches tall and the hind leg boots are about 14 inches tall. The outside of the boot is vinyl, and the inside is lined with either a thick poly-foam or synthetic fleece. Fastened by Velcro or zippers, these foam shipping boots are easy to put on the horse and are typically used alone.

Fig. 16–14. Shipping boots.

Shipping Bandages

Some trainers prefer shipping bandages instead of full shipping boots to support and protect the legs during shipping. Sometimes hock boots and bell boots are used with shipping bandages for increased protection of the hocks and coronets. If either of these boots are used, they should be put on first, and the cotton should overlap them. Like shipping boots, shipping bandages should be used on all four legs.

Applying the Shipping Bandage

The materials used for the shipping bandage are as follows:

- cotton (double thickness 10 – 12 sheets)
- flannel bandage
- two saddler pins
- adhesive tape

The following steps for applying a shipping bandage are illustrated in Figure 16–15.

1. Roll the cotton smoothly on the leg from below the knee or hock down to and including the coronet and heels. (If bell boots are used, they should be put on first, and the cotton should overlap them.)
2. Tuck the end of the flannel bandage just above the ankle and anchor the bandage to the leg with the first wrap. Wrap the flannel downward over the fetlock, pastern, coronet, and heel. After covering the upper part of the foot, continue wrapping the flannel up the leg to the knee. Be sure to leave one inch of the last wrap showing.
3. End the flannel on the outside of the leg. Leave ½ inch of cotton showing at both the top and bottom of the bandage. Insert the two saddler pins in the same manner as for a standing bandage. *(See Chapter 10 for instructions on applying a standing bandage.)*
4. Wrap adhesive tape over the pins to ensure the wraps remain secure throughout the horse's journey. (Some trainers do not require this extra safety precaution.)

Applying a Shipping Bandage (Fig. 16–15.)

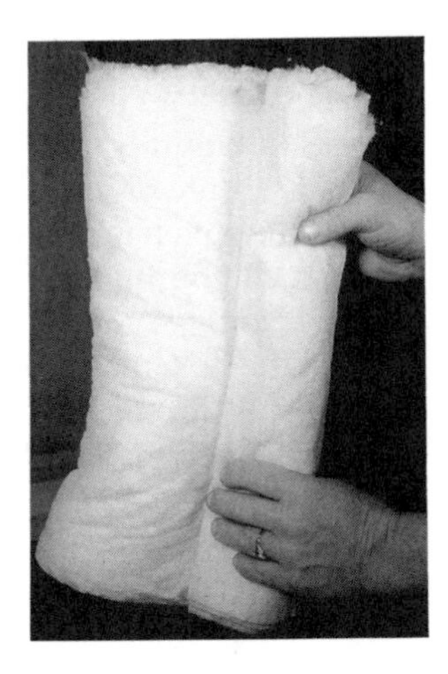

Step 1

Roll the cotton on the leg from below the knee down to and including the coronet and heels.

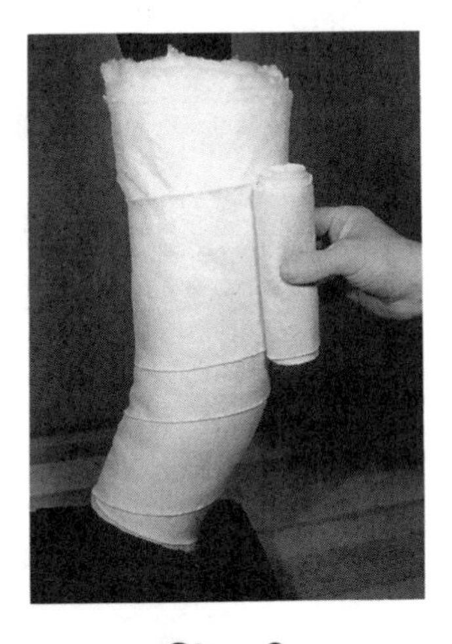

Step 2

Wrap the flannel down over the foot and back up to the knee.

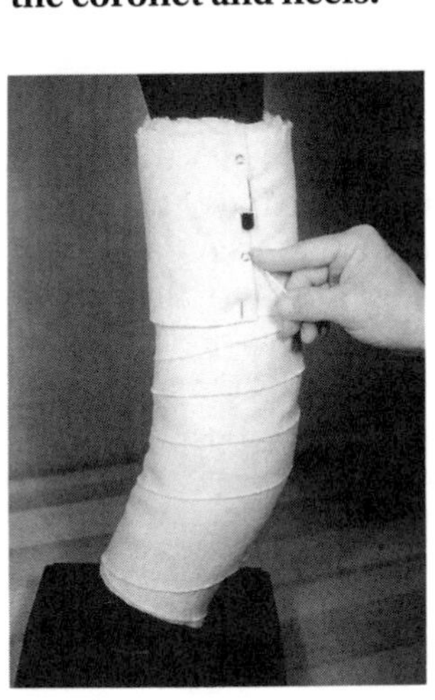

Step 3

End the flannel on the outside of the leg. Insert two saddler pins.

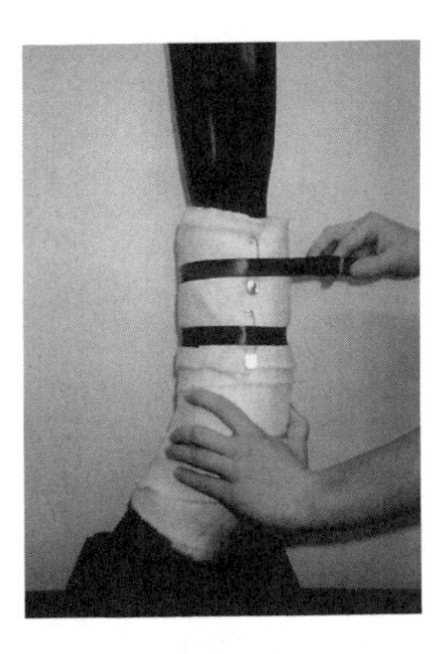

Step 4

Optional: Wrap adhesive tape over the pins as an extra safety precaution.

Fig. 16–16. Bell boots protect the coronet and heel areas during shipping.

Bell Boots

Bell boots protect the coronet and heel bulbs and are used for extra protection during shipping. They are often used with shipping bandages, but should be applied first so the bandage covers the top of the bell boot. *(See Chapter 10 for more information on applying bell boots.)*

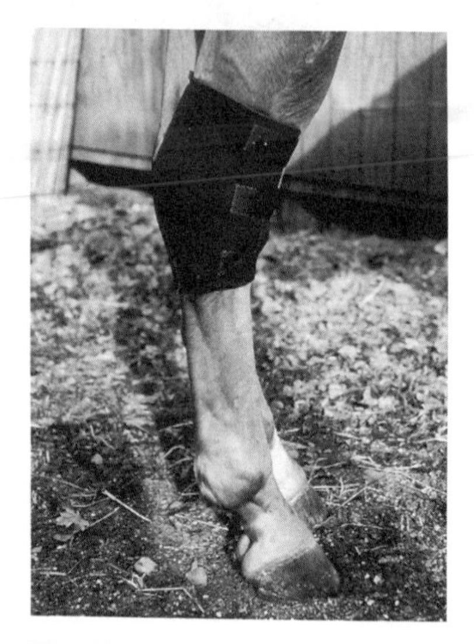

Fig. 16–17. Hock boots protect the hocks.

Hock Boots

Hock boots protect the horse's hocks during shipping. As with the shipping boots, they are made of vinyl and lined with either polyfoam or synthetic fleece. They are fastened around the hock by either Velcro or a leather strap and buckle.

When used alone, hock boots may have a tendency to slip off the hock and slide down the horse's legs—a dangerous predicament for a horse in transport. Because of this problem, it is preferable to use hock boots with shipping bandages, as the bandages help hold them in place.

Head Bumper

A head bumper protects the horse's poll (between the ears) in the event that the horse throws its head up and hits the roof of the trailer or van during shipping. The poll is sensitive and is very susceptible to injury. A heavy blow to the poll is enough to cause instant death. Although it may take a few minutes to place the head bumper on the horse, this precaution is well worth the effort.

Sometimes referred to as a skull cap, poll guard, or shipping cap, the head bumper is made of leather with a soft felt lining and has two openings for the ears. There are three leather loops to allow the crownpiece of the halter to pass through and keep it in place on the head.

To put a head bumper on a horse, first thread the crownpiece of the halter through the loops of the head bumper. Then place the noseband on the horse's nose, gently insert the ears through the two openings, and fasten the halter on the near side as normal.

Fig. 16–18. A head bumper protects the poll, and wrapping the halter with flannel protects the horse's face during shipping.

Halter Wrap

The nosepiece and cheekpieces of the halter should be wrapped with flannel to protect the horse's nose and the side of the face from chafing and rubbing during shipping. There are also commercial fleece coverings which can be purchased and used in the same way as the flannel.

Shipping Muzzle

The shipping muzzle is not necessary for all horses. It is used mostly on horses that have a tendency to bite other horses during transit. It is attached to the lower near and off rings of the halter by metal snaps.

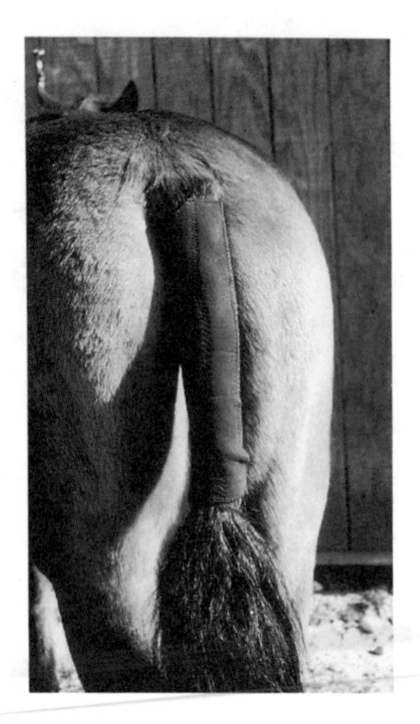

Fig. 16–19. A tail wrap keeps the tail hairs from breaking.

Tail Wrap

The tail wrap keeps the horse from rubbing its tail and breaking tail hairs during shipment. There are two types of tail wraps. The traditional tail wrap involves using a flannel bandage and starting at the base of the tail, working down the tail to the base of the tailbone, then back up to the top. The tail wrap is fastened like a leg wrap, with the tie or pin on the top side of the tail, possibly wrapped with adhesive tape. Be careful not to wrap the bandage so tightly that it cuts off circulation to the tail.

The newer type of tail wrap is a long, narrow strip of synthetic material that is simply folded around the top of the tail and fastened shut by a Velcro strip.

Blanket

It is important to blanket the horse in cold weather to keep it warm during transit. While good ventilation is important during shipping, the blanket helps to avoid cold drafts which can make the horse ill.

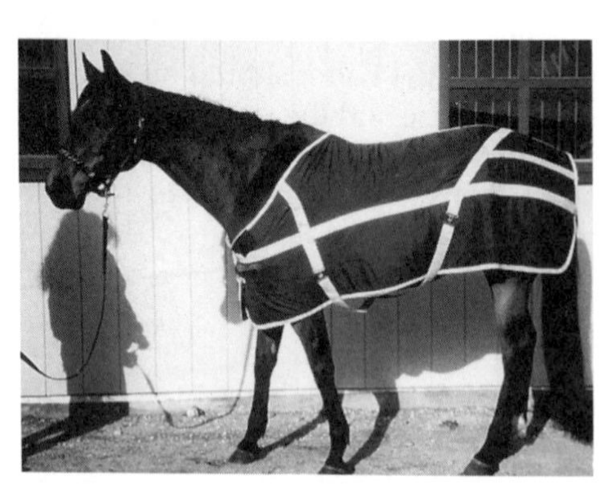

Fig. 16–20. A blanket helps to avoid drafts during transit in cold weather.

When blanketing the horse for shipping, make sure the blanket is fitted properly and all straps are fastened securely. A loose or dangling strap in a trailer can be very dangerous for the horse.

Hay Net

Fig. 16–21. A hay net makes the trip less stressful by keeping the horse occupied.

The hay net is hung inside the shipping vehicle to allow the horse to eat hay during transit. It should be noted that hay nets are excellent for shipping, especially for short distances.

Allowing a horse to eat in a van or trailer takes its mind off the stress of shipping. As in the case of the stall entrance, the hay net should be removed when the horse is either being loaded or unloaded on the van or trailer to avoid a possible hazard.

Some of the equipment used to ship racehorses is optional. It is up to the trainer to decide what equipment is to be used on a particular horse. However, for each journey, the groom must fit the horse with the proper equipment selected by the trainer.

Fig. 16–22. This horse is ready to be loaded in a trailer.

GENERAL MAINTENANCE

In a sense, the groom sometimes functions as a "Jack of all trades" in the stable area. Responsibilities are not limited to just caring for the horse, although that is the most important part of the job. The following sections discuss a few other duties the groom may encounter, as well as the tools that are required to complete these duties. (These tools may be provided by the trainer.)

The Groom's Toolbox

Leather Hole Punch

A leather hole punch is needed to adjust any piece of leather tack or equipment. Extra holes may be needed in halters, stirrup leathers, saddle billets, etc. in order for them to fit properly.

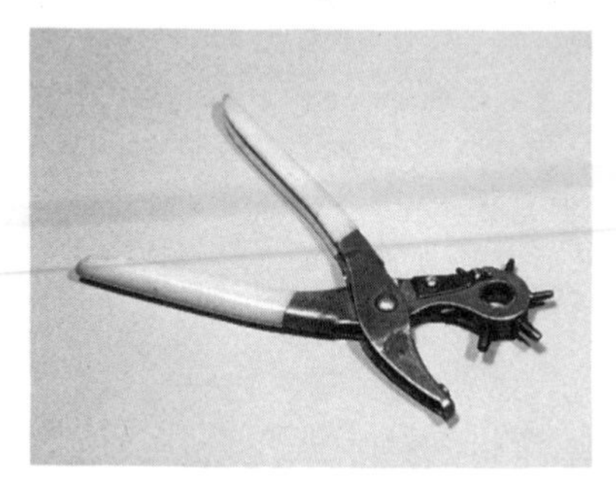

Fig. 16–23. A leather hole punch is used to make extra holes in leather bridles, halters, etc.

The leather hole punch is usually made of steel and has a revolving head with six tubes that make different sized holes. A hole is made by slipping the leather piece under the tube punch and squeezing the two handles together. This tool is very useful around any stable.

Insect Fogger

The insect fogger is used in the stable during the summer months to keep down the fly and mosquito population. This is a task that need only be performed twice per month.

Fig. 16–24. An insect fogger is used during fly season.

Using an insect fogger, a commercial fly spray is usually sprayed inside the stalls as well as around the manure pit. It is recommended to spray the stalls when the horses are not occupying them to prevent direct contact with the insecticide. Likewise, be sure to wear protective gloves, goggles, and a face mask when using an insect fogger to prevent contact with or inhalation of any harmful chemicals.

There are a number of non-toxic insecticides on the market, and a veterinarian should be consulted as to which brand is best and its proper use.

Fig. 16–25. Lay the crossbeam of the measuring stick over the horse's withers to get an accurate height measurement.

Measuring Stick

A measuring stick is used to measure the height of the horse. For instance, if a trainer wishes to determine if a young horse is growing, the groom may be asked to measure the horse's height.

Most measuring sticks for horses consist of a moveable crossbeam (which is placed on the horse's withers) on an upright stick (placed on level ground by the horse's near shoulder). The upright stick is marked in "hands." (A hand is equal to four inches.) The crossbeam also has a bubble level which allows an accurate reading. The horse's height should be read directly *below* the crossbeam.

For a horse that spooks when its groom uses the measuring stick, it is helpful to have someone cup a hand behind the left eye to prevent the horse from seeing the stick.

Screw-Eyes and Double-End Snaps

Metal screw-eyes and double-end snaps are used extensively in most stables. The screw-eyes are screwed into wooden walls or panels. The double-end snaps, which vary in length, hook onto the screw-eyes. They are used together throughout the stable to hang up water buckets, feed tubs, hay racks, tie chains, stall webbings, stall screens, and other stable equipment.

Water Bucket

Metal screw-eyes and double-end snaps are used to hang water buckets on the stall wall. First, drive a screw-eye into the wall. (Remember, a bucket should be hung high enough so that a horse cannot defecate into it.) A double-end snap is then attached to the screw-eye and the handle of the water bucket is attached at the other end of the snap.

Feed Tub

To mount a feed tub in the corner of the stall, first install three screw-eyes into the corner wall. The feed tub is attached to the wall by the use of three double-end snaps: one end of a snap is attached to a screw-eye in the wall and the other end is attached to the horse's feed tub.

Hay Net

A screw-eye is needed to hang a hay net on the stall wall. To do this, drive a screw-eye high on the wall. Then thread the end of the hay net string through the screw-eye. *(The sequence required to properly hang a hay net is described in Chapter 3.)*

Hoof Pick

Many a groom has prevented a lost hoof pick by securing it to his or her belt loop with a double-end snap. One end of the double-end snap is attached to the hoof pick and the other end of the snap is attached to the belt loop. This practice not only prevents lost tools, but makes the hoof pick easily accessible.

Tie Chain

Installing a tie chain is accomplished by inserting a screw-eye high on the stall wall. The tie chain (rubber or metal) is then attached to the screw-eye by means of a double-end snap. At the other end of the

Fig. 16–26. Metal screw-eyes and double-end snaps are used for many things around a stable.

tie chain, another double-end snap is used to hook on to the center ring of the horse's halter.

Stall Webbing and Safety Chain

Many stables hang stall webbing and a safety chain across the stall doorway as a barrier, rather than close the stall door. To hang stall webbing, install three screw-eyes on each side of the stall doorway, at about a horse's chest level, for a total of six screw-eyes. The stall webbing is equipped with three metal snaps on each side to attach to the screw-eyes.

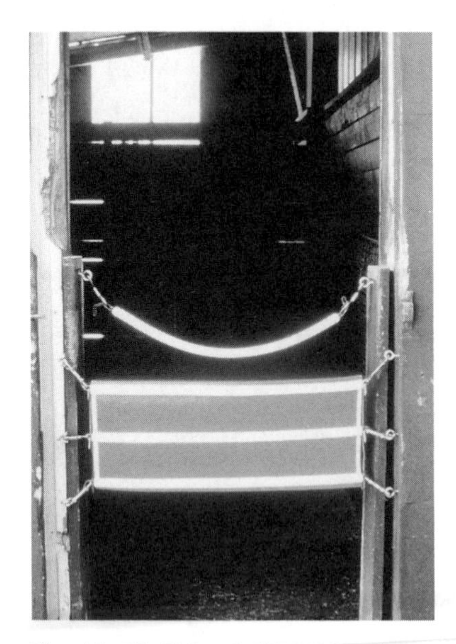

Fig. 16–27. This stall has a safety chain and stall webbing.

Fig. 16–28. A stall screen allows good ventilation.

By installing one screw-eye on each side of the doorway (a total of two screw-eyes), the safety chain can be attached above the stall webbing in the same manner. Both the stall webbing and the safety chain are designed to allow more air to get in the stall, and make the horse feel less boxed in. At the same time, they keep the horse from getting out of the stall.

Stall Screen

Instead of a door or stall webbing, some stables affix a stall screen to the doorway of a stall. A stall screen is installed using three screw-eyes on one side of the door frame and one screw-eye with a double-end snap on the other side. The stall screen allows for the maximum amount of ventilation during hot weather and at the same time serves as a barrier to keep the horse confined to the stall.

SUMMARY

As this chapter indicates, there is much more involved in the care of racehorses than just "rubbing." Trainers will attest to the fact that each racehorse is an individual and requires "customized" care. While the information in this book provides a strong base to work from, some things must be left to time and experience. Good grooms who stand the test of time are valuable assets to trainers and owners, and will be rewarded. But the best reward, as any caretaker will admit, is walking into the barn each morning and being greeted by enthusiastic nickers, and leaving at night to the sound of horses peacefully munching their hay.

Fig. 16–29.

Appendix

Parts of the Horse

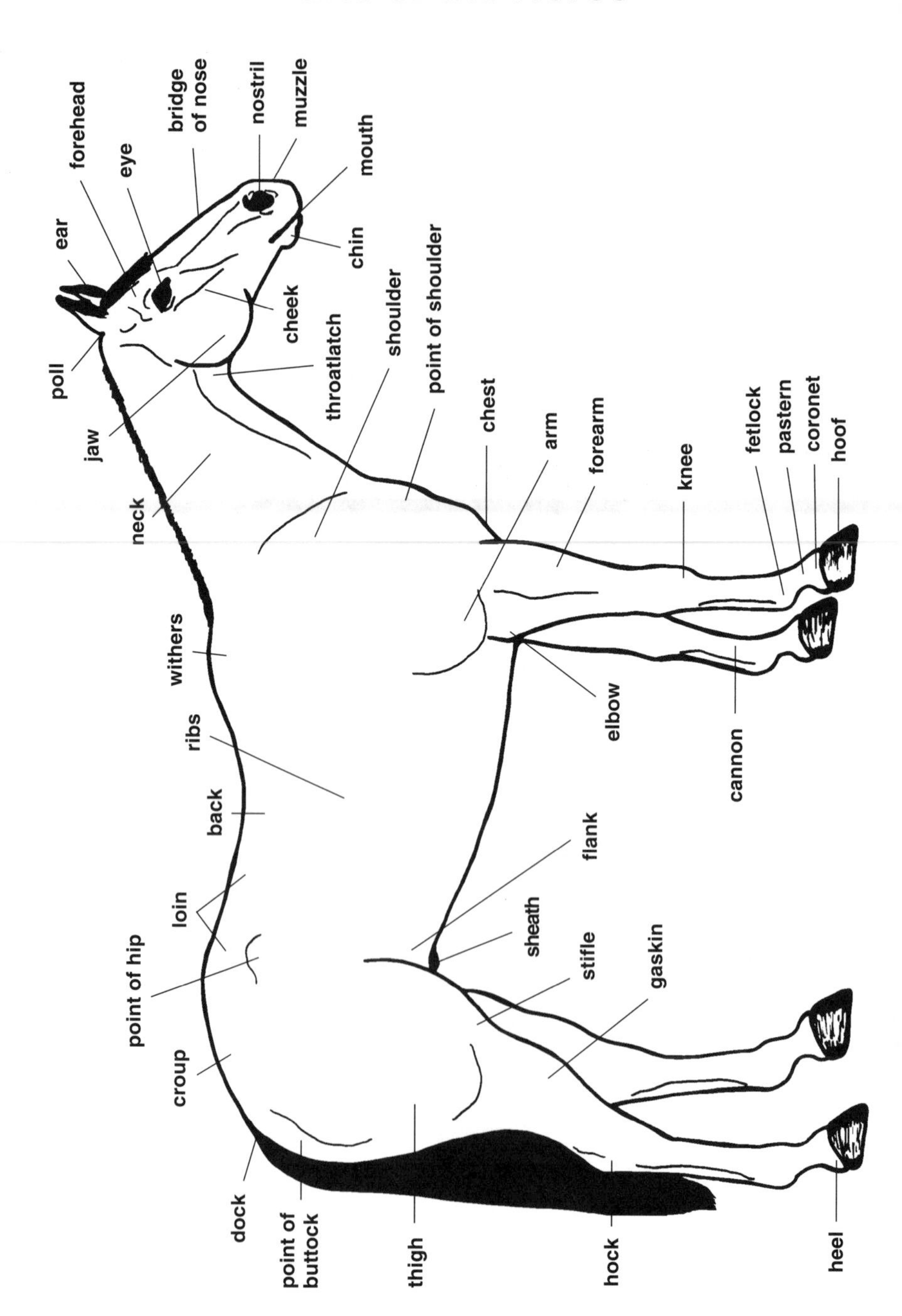

Bones of the Horse

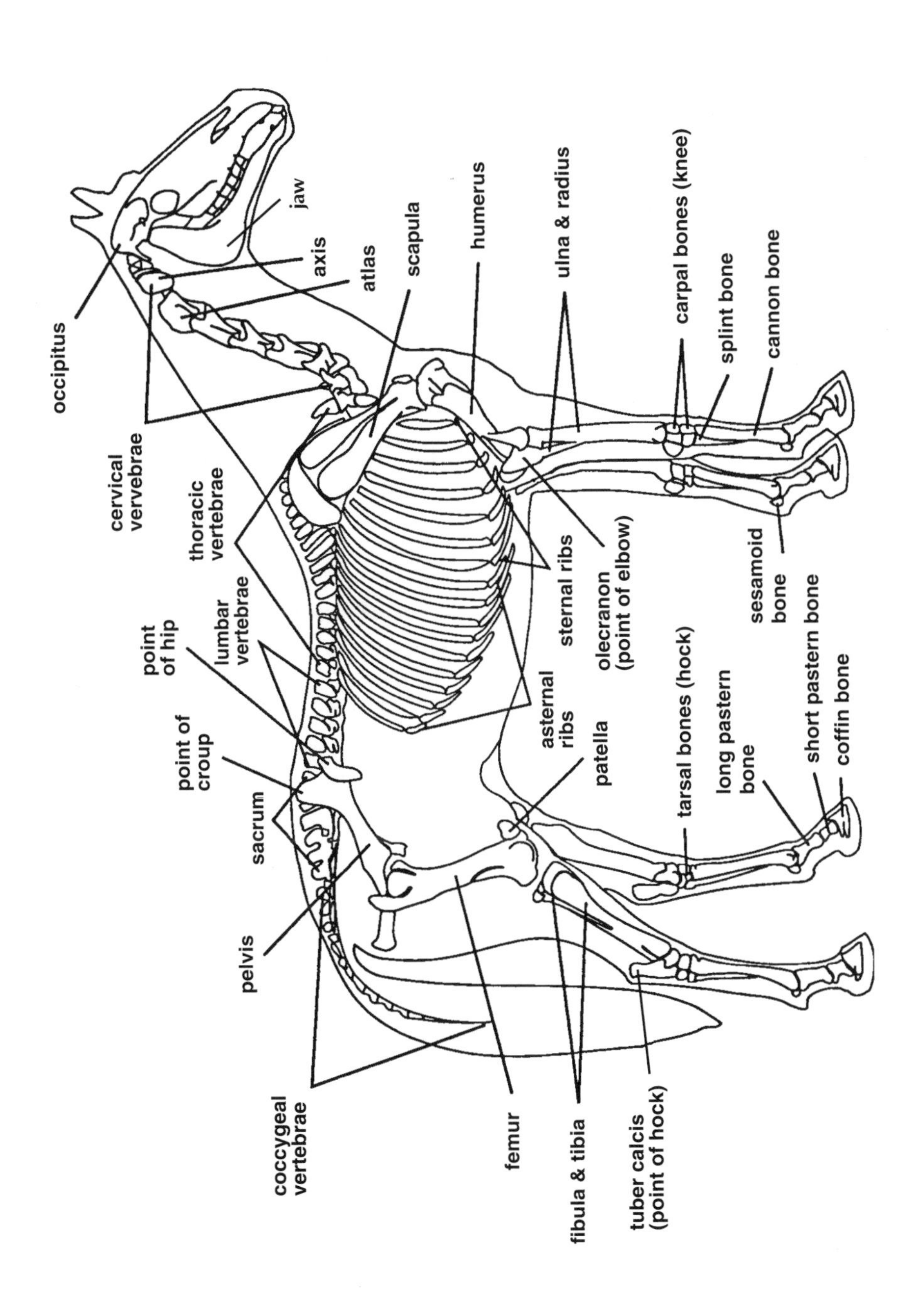

Bibliography

Adams, O.R. *Lameness in Horses.* Third Edition. Lea and Febiger, Philadelphia, PA, 1989.

Ensminger, M.E. *Horses and Horsemanship.* Interstate Publishers, Inc., Danville, IL, 1990.

Equine Research, Inc. *Feeding to Win II.* Grand Prairie, TX, 1992.

Equine Research, Inc. *The Illustrated Veterinary Encyclopedia for Horsemen.* Grand Prairie, TX, 1977.

Equine Research, Inc. *Veterinary Treatments & Medications for Horsemen.* Grand Prairie, TX, 1977.

Evans, J. Warren. *The Horse.* W. H. Freeman and Company, New York, NY, 1990.

Harrison, James C. et al. *Care & Training of the Trotter & Pacer.* The United States Trotting Association, Columbus, OH, 1968.

Hayes, Karen E.N. "Language of Pain." *Horse & Rider,* May 1995, pp. 44 – 46, 106.

Hayes, Karen E.N. "Your Horse is Cut—What Should You Do?" *Horse & Rider,* May 1995, pp. 22.

Kays, John M. *The Horse.* Arco Publishing Company, Inc., New York, NY, 1969.

Loving, Nancy. *Veterinary Manual for the Performance Horse.* Equine Research, Inc., Grand Prairie, TX, 1993.

Morris, Desmond. *Horse Watching.* Crown Publishers, Inc., New York, NY, 1988.

Rose, Mary. *The Horsemaster's Notebook.* Billing & Sons Ltd., Worcester, Great Britain, 1988.

Figure Credits

All photos are by **Joel Silva** except the following:

Figure 1–3. Barbara Ann Giove.
Figure 1–4. Barbara Ann Giove.
Figure 1–7. Cathy Nelson.
Figure 1–8. Barbara Ann Giove.
Figure 1–9. Barbara Ann Giove.
Figure 1–10. Courtesy of *The Racing Journal.*
Figure 1–11. Cappy Jackson.
Figure 1–12. Courtesy of *Quarter Week.*
Figure 1–13. Courtesy of The United States Trotting Association.
Figure 1–14. Courtesy of The United States Trotting Association.
Figure 2–2. Equine Research, Inc. file photo.
Figure 2–3. Barbara Ann Giove.
Figure 2–4. Cappy Jackson.
Figure 2–5. Equine Research, Inc. file photo.
Figure 2–6. Barbara Ann Giove.
Figure 2–7. Equine Research, Inc. file photo.
Figure 2–8. Equine Research, Inc. file photo.
Figure 2–9. Equine Research, Inc. file photo.
Figure 2–10. *Racing Northeast.* Brandy Adams Mitroff photo.
Figure 2–12. Equine Research, Inc. file photo.
Figure 2–13. Equine Research, Inc. file photo.
Figure 2–15. Equine Research, Inc. file photo.
Figure 2–20. Equine Research, Inc. file photo.
Figure 2–24. Barbara Ann Giove.
Figure 2–25. Barbara Ann Giove.
Figure 3–1. Barbara Ann Giove.
Figure 3–2. Equine Research, Inc. file photo.
Figure 3–3. Equine Research, Inc. file photo.
Figure 3–19. Barbara Ann Giove.
Figure 4–1. Barbara Ann Giove.
Figure 4–2. Barbara Ann Giove.
Figure 4–3. Monica Thors.
Figure 4–28. Equine Research, Inc. file photo.
Figure 4–29. Equine Research, Inc. file photo.
Figure 4–30. Equine Research, Inc. file photo.
Figure 5–1. *Racing Northeast.* Brandy Adams Mitroff photo.
Figure 5–29. Courtesy of The United States Trotting Association.
Figure 5–30. Courtesy of Thoro'bred Racing Plate Co.
Figure 5–32. Equine Research, Inc. file photo.

Figure 5–33. Courtesy of *Quarter Week.*
Figure 5–34. Equine Research, Inc. file photo.
Figure 5–36. Courtesy of The American Quarter Horse Association.
Figure 6–1. Barbara Ann Giove.
Figure 7–1. *Racing Northeast.* Brandy Adams Mitroff photo.
Figure 7–2. Equine Research, Inc. file photo.
Figure 7–21. Monica Thors.
Figure 7–22. Monica Thors.
Figure 7–23. Monica Thors.
Figure 7–24. Monica Thors.
Figure 7–25. Monica Thors.
Figure 7–26. Equine Research, Inc. file photo.
Figure 7–27. Equine Research, Inc. file photo.
Figure 7–28. Monica Thors.
Figure 7–29. Monica Thors.
Figure 7–30. Monica Thors.
Figure 7–31. Monica Thors.
Figure 7–32. Monica Thors.
Figure 7–33. Monica Thors.
Figure 8–1. Barbara Ann Giove.
Figure 8–5. Monica Thors.
Figure 8–11. Courtesy of The United States Trotting Association.
Figure 8–12. Monica Thors.
Figure 8–14. Monica Thors.
Figure 8–20. Monica Thors.
Figure 8–21. Monica Thors.
Figure 8–23. Monica Thors.
Figure 8–24. Monica Thors.
Figure 8–25. Monica Thors.
Figure 8–33. Monica Thors.
Figure 9–1. Courtesy of The United States Trotting Association.
Figure 9–4. Monica Thors.
Figure 9–5. Barbara Ann Giove.
Figure 9–6. Courtesy of The United States Trotting Association.
Figure 9–7. Courtesy of The American Quarter Horse Association.
Figure 9–8. Cathy Nelson.
Figure 9–9. Equine Research, Inc. file photo.
Figure 10–1. Equine Research, Inc. file photo.
Figure 10–3. Equine Research, Inc. file photo.
Figure 10–5. Monica Thors.
Figure 10–6. Monica Thors.
Figure 10–8. Monica Thors.
Figure 10–9. Monica Thors.

Figure 10–10, right. Monica Thors.
Figure 10–11. Monica Thors.
Figure 10–15. Monica Thors.
Figure 10–16. Monica Thors.
Figure 10–17. Equine Research, Inc. file photo.
Figure 10–19. Equine Research, Inc. file photo.
Figure 11–1. Cappy Jackson.
Figure 11–7. Courtesy of *Quarter Week.*
Figure 11–9. Equine Research, Inc. file photo.
Figure 11–10. Equine Research, Inc. file photo.
Figure 11–11. Equine Research, Inc. file photo.
Figure 11–12. Barbara Ann Giove.
Figure 11–13. Equine Research, Inc. file photo.
Figure 12–7. Courtesy of The United States Trotting Association.
Figure 12–8. Courtesy of *Quarter Week.*
Figure 12–14. Equine Research, Inc. file photo.
Figure 12–25. Cathy Nelson.
Figure 13–1. Barbara Ann Giove.
Figure 13–3. Equine Research, Inc. file photo.
Figure 13–4. Equine Research, Inc. file photo.
Figure 13–5. Courtesy of *Quarter Week.*
Figure 13–7. Barbara Ann Giove.
Figure 13–8. Monica Thors.
Figure 13–10. Cathy Nelson.
Figure 14–1. Barbara Ann Giove.
Figure 14–3. Equine Research, Inc. file photo.
Figure 14–17. Equine Research, Inc. file photo.
Figure 15–1. Barbara Ann Giove.
Figure 16–1. Barbara Ann Giove.
Figure 16–16. Equine Research, Inc. file photo.
Figure 16–26, bottom left. Barbara Ann Giove.
Figure 16–26, top left, top right, bottom right. Equine Research, Inc. file photos.
Figure 16–27. Equine Research, Inc. file photo.
Figure 16–28. Equine Research, Inc. file photo.

Glossary

across the board A combination pari-mutuel race ticket on a horse. Pari-mutuel means that you collect something if your horse runs first, second, or third.

allowance race Race where horses must meet certain conditions set by the Racing Secretary to be eligible for entry. Allowance races are a step above claiming races.

also-eligible A horse that is officially entered in a race but is not permitted to start unless a spot is made available by another starter being scratched.

also-ran Any horse that runs "out of the money," or worse than a third place finish.

analgesic A pain-reliever that does not cause loss of consciousness.

antibiotic A chemical substance produced by a microorganism that inhibits the growth of, or kills other microorganisms.

antiseptic An agent that prevents the decay of tissue by inhibiting the growth and development of microorganisms.

apprentice A young jockey just starting out who is given from three to seven pounds weight allowance.

asterick 1. Used in front of a horse's name on a pedigree, an asterick (*) indicates that the horse is imported (from another country). 2. Used in front of a jockey's name in a race program, an asterick indicates the jockey is an apprentice rider.

astringent An agent, usually applied to the skin, that causes contraction of blood vessels, stops discharge, and dries the skin.

backstretch 1. That area where the horses are stabled, which is restricted to trainers, owners, grooms, and other horseracing officials and professionals. 2. The straightaway part of the actual track, which is opposite the grandstand.

ball A large pill used to administer various types of solid horse medicines.

balling gun A veterinary tool used to administer solid medication by mouth.

bang To cut the hair of the tail in a straight line midway between the hocks and ankles.

bars 1. Area of the mouth where there are no teeth. 2. Parts of the hoof on either side of the lateral grooves that support the wall and keep the sole from coming in contact with the ground.

bat A short whip used by jockeys or exercise riders in a race or workout.

bleeder A horse that bleeds from the lungs (it may appear to be bleeding from the nose) due to a stressful race or workout. A horse can, in some states, be classified as a bleeder upon endoscopic examination by a veterinarian.

blemish A defect that may diminish the appearance of the horse, but does not compromise the horse's usefulness.

blinds Attachments to the bridle that restrict what the horse can see from the sides and rear.

blinkers Attachments to the blinker hood that restrict what the horse can see from the sides and rear.

blowing Heavy breathing when a horse is tired after a race or workout.

blowout A short, fast final workout one or two days before a race.

bolt 1. When a horse suddenly takes off or ducks to the outside rail during a race or workout. 2. When a horse eats its feed rapidly, seemingly without chewing.

breaking Breaking gait, usually from a trot or pace into a gallop. When a horse breaks gait in a race, the driver must immediately pull the horse back to the proper gait.

breaking out When a horse suddenly begins to sweat. This condition can be caused by high temperatures, nervousness, or stress.

breeze Workout where the horse is traveling at a brisk gallop, but not running all out.

bridle path The area of mane behind the horse's ears (where the crownpiece of the bridle rests) that is kept clipped.

broken-down A horse that is unable to race due to an injury.

broodmare A mature female horse used for breeding.

brushing A general term for light interference.

bug boy Slang term for an apprentice jockey.

bursa A sac or sac-like cavity filled with fluid and situated between body structures where friction would otherwise develop.

calcification When the body deposits calcium in strained or injured soft tissues (such as muscles or ligaments) in an attempt to strengthen these areas. Calcification may cause a hardening of soft tissues, but does not create new bone growth.

cartilage The tissue that covers the ends of bones within joints to protect the bones from compression and concussion.

cast When a horse rolls in its stall and its legs become pinned against the stall wall, making it impossible for the horse to rise.

catch driver A person who is engaged by the trainer to drive the harness horse in a race.

calks Projections on the bottom of horseshoes which are thought to allow the horse better traction on the racetrack surface.

chaff The dust or debris left from hay or straw.
check rein The strap that runs from the saddle to the crownpiece of the bridle on a harness horse.
chestnuts Horny growths found on the inside of all four legs, which may be used as a means of identification of horses stabled at a racetrack. Like fingernails, chestnuts (also called "night eyes") are unique to each individual.
chronic Condition that persists for a long time.
chute Straightaway extensions of either the backstretch or homestretch.
claiming race A race where all horses entered may be claimed (bought) for the amount stated. Claiming races make up a large percentage of the program at most tracks.
clean legs A horse's legs that are free from any sign of lameness or blemishes.
cleaned up When a horse has finished eating and no feed is left in the feed tub.
clockers People who record the workout times of racehorses, and who may supply that information to a racing publication, such as *the Daily Racing Form.*
clubhouse turn The first turn on a racetrack after passing the grandstand and clubhouse area.
Coggins test A blood test used to determine if a horse has Equine Infectious Anemia (also called Swamp Fever). Most racetracks require a negative Coggins test result dated within the last year before allowing any horse to enter the grounds.
colic Any disorder characterized by severe abdominal pain. The two main types are spasmodic and impaction colic.
colt A male horse, uncastrated, under five years old.
condition book A book issued by the Racing Secretary that lists all the conditions of future races such as weight, distance, purse, etc. It is usually set for a two-week period. For a horse to be eligible for a particular race, it must meet the conditions of that race.
condition of track The surface condition of a racetrack, which includes fast, off, sloppy, muddy, heavy, slow, and good.
congenital Physical characteristics of a horse that have been inherited, rather than acquired.
contagious Disease which is able to be spread from one horse to another by direct contact (two horses touching noses) or indirect contact (brushing both horses with the same grooming tool).
cooling out Washing and walking a horse after a race or workout to restore normal body temperature, pulse, and respiration rate.

cording Wrapping a bandage so tightly or awkwardly around the horse's lower leg that it causes a tendon to bow.

counterirritant A substance that is applied to the skin to irritate an injury with the goal being to increase blood flow to that area and thus speed up the healing process.

corrective shoeing Any type of shoeing done to correct a conformational defect, minimize a lameness, or generally improve the way a horse travels.

cross-firing Interference which is generally confined to pacers where the inside of the hind foot hits the inside quarter of the diagonal forefoot.

cross-ties Two ties attached to opposite sides of the stall (or aisle) which meet in the middle. One tie fastens on the near ring of the horse's halter and the other tie fastens on the off ring.

crupper That part of the harness that encircles the tail and is attached to the saddle by the back strap to keep the saddle from slipping forward.

cushion The surface layer of dirt on a racetrack.

dash A harness race decided in a single trial.

dead heat When two horses finish a race at exactly the same time.

dead weight Lead weights carried by a jockey who cannot meet the weight requirements of a race. These "dead" weights are carried in a special saddle pad in whatever quantity is necessary to meet the specified weight requirement.

deworming A periodic "flushing out" of parasites to keep the horse healthy and encourage more efficient feed usage.

disinfectant A chemical or physical agent used to kill bacteria on inanimate objects. Disinfectants are usually too toxic for use on animals.

dogs Wooden horses or rubber traffic cones set up on the track just outside the inside rail to keep horses off the rail in morning workouts. These "dogs" preserve that part of the track for the afternoon races.

doing up Placing medication on the legs and covering them with standing bandages.

double-gaited A horse that can trot and pace competitively.

draw To remove all hay from a horse's stall, usually before a race.

driving 1. A horse traveling at full speed and all out at the end of a workout or race. 2. Controlling a horse from the rear, usually with a harness, some type of cart, and long driving lines attached to the bit.

EIPH (Exercise-Induced Pulmonary Hemorrhage) When a horse

bleeds from the lungs as a result of intense exercise.
elbow-hitting When the horse hits its elbow with the shoe of the hind foot on the same side. It rarely occurs except in those horses with weighted shoes.
entry Usually two or more horses entered in a race that are owned or trained by the same person. A bettor gets to bet on all the horses in an entry for the price of one.
farrier A professional horseshoer.
fatigue When a horse can no longer physically maintain its speed or level of exertion. A fatigued horse should be rested in order to replenish body fluids and avoid injury.
filly A female horse, under five years old, not used for breeding.
flake A single square of hay. (There are 10 – 12 flakes in an average bale of hay.)
flat race Any horserace that is not over jumps.
flehmen response When a horse lifts its head and curls its upper lip back, usually in response to a strong smell.
forage The leaves and stalks of plants, such as the grasses and legumes used as hay.
forging When the toe of the hind foot hits the sole of the forefoot on the same side.
form The past performance of a racehorse; often a table giving details relating to a horse's past performance.
founder Inflammation of the laminae within the foot; also called laminitis.
free-legged pacer A pacer that races without hobbles.
front side The area of the racing facility that racegoers see, including the grandstand, the track, and the paddock area.
freshening Resting a horse that has become stale or sour from racing or training. Most racehorses are freshened at a farm.
furlong A unit of measurement used on the racetrack to designate 220 yards or ⅛ mile. (There are eight furlongs in a mile.)
futurity A type of race, usually a stake race, strictly for two-year-old colts and/or fillies.
gait A specific, repeated sequence of limb movements. Examples of gaits are walk, trot, pace, and gallop.
gap An opening on the outer wall of the racetrack to allow horses to enter and exit the stable area.
gelding A male horse which has been castrated (both testicles removed surgically).
good doer A horse that cleans up all of its feed at each feeding, keeps weight on easily, and maintains good overall condition.

grab When a horse strikes a front leg with its hind foot.
grease heel Chronic bacterial infection of the skin on the fetlocks and pasterns (also known as "moist eczema").
hand A unit of measurement of a horse's height. One hand is equal to four inches.
hand ride When a jockey or exercise rider urges the horse to perform its best by using the hands and not the whip.
handicap race Race where the horses carry weight based on their past performances; the better the horse, the more weight it will be assigned.
handily Racing or working with little effort or urging.
hay The cured materials cut from grasses, grains, or legumes.
head-shy When a horse throws its head up because it is afraid of being hit in the face or head.
headstall The entire bridle, excluding the bit and reins.
heart A personality trait of a horse that exhibits extreme courage and ambition in competition.
heat A single trial in a race that is decided by winning two or more trials.
heaves A respiratory disorder characterized by forced (noisy) expiration of breath, frequently caused by allergies or dust.
hogging down When a harness horse pulls its head down against the action of the overcheck.
homestretch The last straight stretch of track before the finish wire, directly in front of the grandstand.
hobbles Leather, nylon, or plastic straps with semicircular loops which are placed around a horse's gaskins and forearms, connecting the fore and hind legs on the same side. They are used used to keep a harness horse on gait.
horsing When a filly or mare exhibits signs of being in heat.
hotwalker The person who walks a horse after a race or workout to cool the horse.
hotwalking machine A machine that horses are attached to after a race or workout that leads them in a circle (at a walk) to cool them out. This machine may also be used to walk the horse for exercise.
hung When a horse becomes tired and slows down at the end of a race.
hydrotherapy Using water to reduce injury-related pain, heat, and inflammation.
impost The amount of weight that is carried by a horse in a race.
in the money Those horses finishing first, second, or third in a race.

interfering A defect in a horse's way of going, whereby a front foot hits the opposite leg, or a hind foot hits the opposite leg anywhere between the coronet and cannon. This condition is predisposed in horses with base narrow, toe wide, or splay footed conformation.

irons Another term for stirrups.

Jockey Club The Jockey Club is the organization in charge of the American Stud Book, registry of Thoroughbred horses.

jog A slow trot or pace, typically used as a warm-up or exercise gait with the horse traveling the wrong way of the track.

jog cart A cart used only for exercising the trotter or pacer.

joint The point where two or more bones of the skeleton come together.

knee-hitting High interference, where the horse hits a knee with the opposite front foot.

lameness A defect that compromises the horse's form and/or function temporarily or permanently.

laminitis See founder.

lead pony A horse used to lead a racehorse to the starting gate to keep the racehorse quiet and calm. A lead pony is also used in the morning workouts to accompany a racehorse to and from the racetrack.

length A unit of measurement of distance between horses in a race. A length is the number of feet from the nose to the tail (about eight feet).

ligament Band of fibrous tissue that connects bones and cartilage and supports the joints.

lugging in To suddenly bear in toward the inside rail during a race or workout.

lugging out To suddenly bear out toward the outside rail during a race or workout.

maiden A horse that has never won a race.

mare A mature female horse five years old or over.

microorganism Microscopic organisms such as bacteria, viruses, and fungi.

morning glory A horse that trains well in morning workouts but fails to perform well in the afternoon.

mucking The process of cleaning a stall.

mudder A horse that runs unusually well on a sloppy, muddy, or heavy racetrack.

near side The left side of a horse.

nerving Cutting a nerve in the lower leg of a lame horse to eliminate feeling in that area. The procedure is called a neurectomy.

nutrient Any feed element or group of feed elements that is essential for life, namely proteins, carbohydrates, fats, minerals, vitamins, and fiber.

off side The right side of the horse.

off track A racetrack that is listed as muddy or sloppy.

oral dose syringe A device used to administer liquid or paste medication by mouth.

ossification The natural development of bone in a young animal or the production of bone in an animal at any age as a result of stress or injury (e.g. sidebone).

overcheck That part of a harness that attaches to the top of the saddle, runs over the crownpiece of the bridle, and extends down the face to attach to the overcheck bit. The overcheck holds the Standardbred's head up and in position.

over-extension of the fetlock See running down.

overnight A race for which entries close as late as the day preceding the race. Also refers to the rough draft of the next day's schedule of entries.

over-reaching The toe of the hind foot catches the heel of the forefoot on the same side. The hind foot advances more quickly than with forging (stepping on the heel of the forefoot). The toe of the hind foot may step on the heel of the shoe of the forefoot on the same side and cause shoe pulling.

pace Lateral two-beat gait.

paddock That part of the racetrack where the horses are saddled or harnessed before a race.

palatable Of appealing taste.

periosteum Membrane covering all bones of the body that also has potential bone-forming elements.

place To finish second in a horserace.

pocket When a horse is surrounded by other horses during a race. This situation usually causes the horse to lose concentration.

poles Markers along the rail of a racetrack that designate distance, such as the ¼ mile pole, or the ⅛ mile pole. Poles are usually marked in different colors for quick identification.

poor doer A horse that does not clean up its feed as well as it should and is difficult to keep at the proper weight and condition.

post The starting gate.

post position The number of the box in the starting gate from which the horse breaks in a race.

post time The designated time when all horses entered in a race are expected to arrive at the post, or starting gate.

poultice A paste applied to the skin to pull swelling from an inflamed area.
prop See refuser.
proud flesh Excessive granulated and protruding tissue in a healing wound.
puller A horse that leans heavily on the bit and is difficult to slow down.
pulling up The gradual slowing down of a horse either during or after a workout or race.
purse The total money that is divided among the owners of those horses placing first, second, and third in a race.
quick hitch The method of hooking jog carts and sulkies to the harness by snapping them into hooks on each side of the saddle.
rate When a rider holds a horse back to "save" the horse's speed for the end of the race or until the rider is ready to make a bid to overtake the leaders. Often, the leaders attempt to rate their horses to save some speed for fending off challengers.
ration The amount of feed allowed a horse within 24 hours.
refuser A horse that refuses to break from the starting gate.
receiving barn Barn where horses coming from different racetracks are isolated just before their race.
record The fastest time made by a horse in a heat or dash that it wins, or in a performance against time.
roughage Any feed that is high in fiber.
rubbing down Rubbing a racehorse after a difficult race or workout (in cold weather) to help the horse dry off and cool out faster.
running down When a racehorse's ankle gives or flexes so much during a race or workout that the back of the ankle comes in contact with the track (also called over-extension of the fetlock). This contact is abrasive and can cause the ankle to become raw, open, and sore.
saddling paddock See paddock.
savage The act of (a horse) attempting to bite another horse during a race or workout.
set A group of horses from a particular stable that have their morning workouts scheduled at the same time. When one set of horses returns to the barn, another set is usually sent out.
scalping When the toe of the forefoot catches the hairline at the top of the hind foot (on the same side) as it breaks over. In more serious cases, the forefoot may hit the front of the pastern or cannon. Scalping is most commonly seen in the trotting horse.
schooling Getting a horse used to breaking from the gate, tacking

up in the paddock, and relaxing around crowds.

scoring Preliminary warm-up before the start of a harness race; the horses are turned away briefly and trotted as if in a race.

scratch To withdraw a horse from a race in which it was formally entered.

sheath The skin surrounding the penis of a male horse.

show To finish third in a horserace.

smegma A combination of dirt and glandular secretions that collects on the sheath or udder of a horse. Smegma should be cleaned off regularly.

speedy cutting Interference in the fast gait. It may be the same as cross-firing, or it may mean that the outside wall of the hind foot strikes the back of the front leg (in the middle) on the same side.

spit box Slang term used to describe the barn where saliva, urine, and blood samples are taken by veterinarians before and after a race. Horses finishing first, second, or third as well as any other horse the veterinarians wish to spot test, must go to the spit box after a race.

spooked When a horse suddenly becomes frightened.

spot test A random drug test for racehorses.

stable The horses, equipment, and employees under the supervision of a racehorse trainer; not the barn that houses horses. For instance, there may be three "stables" in one barn.

stake money Bonus money sometimes given to grooms when one of the groom's horses or one of the stable's horses wins a stake race.

stake race A race that has paid entry fees (by the racehorse owners) added to the purse. Stake races are considered the most prestigious class of races.

stallion An uncastrated male horse five years old or over that can be used for breeding.

stale A horse with a poor mental and physical condition as a result of overtraining and over-racing.

state racing commission The state organization that establishes and enforces rules and regulations governing racing in that state.

stick See bat.

stocking up An obvious swelling in the lower legs (front and/or back) often due to poor circulation or lack of exercise.

stride The distance and/or time from when a particular foot hits the ground until the same foot hits the ground again.

sulky A cart that is used strictly for racing Standardbreds.

supplement A feed or feed mixture that is richer in a specific nutrient than the ration to which it is being added.

surcingle A leather or fabric band that encircles the horse behind the front legs and over the back (just behind the withers). Surcingles are commonly used over blankets to hold them in place.

sutures Stitches closing up a wound.

tack Equipment used on a horse during a race or workout including saddle, bridle, harness, noseband, martingale, breast collar, blinkers, etc.

tailing off A horse that is gradually losing its peak form.

tendons Bands of strong fibrous tissue that connect muscles to bones.

terrets The rings on the harness's saddle through which the driving reins are run.

testing barn See spit box.

thrush A bacterial infection in the medial and lateral grooves of the frog, typically caused by poor foot maintenance.

tie chain A chain hung by a screw-eye that is used to tie Thoroughbreds and Quarter Horses to their stall walls.

tongue tie A strip of cloth used to tie the horse's tongue down in its mouth to prevent the horse from swallowing the tongue during a race or workout.

topping off Adding water to a horse's water bucket so that it is filled to the top.

toxemia A toxic substance produced by bacteria that damages cells and tissues, including blood vessels.

trot Diagonal two-beat gait.

two-minute lick A slow exercise session whereby a horse runs the distance of one mile in two minutes. The horse will run an average of 15 seconds for each eighth of a mile.

tying-up A condition whereby a horse's hind end muscles begin to cramp and the horse takes short steps and may eventually fall down.

udder The mammary gland and surrounding skin (including the teats) of a female horse.

under wraps When a rider restrains a horse from performing at its peak during exercise.

United States Trotting Association (USTA) Organization that registers Standardbred horses and regulates harness racing.

vaccine A dosage of weakened or killed microorganisms administered to a horse that stimulates the immune system and protects the horse from the disease caused by that microorganism.

valet Person assigned to care for a jockey's equipment, assist the trainer in saddling the horse, help untack the horse after a race, and carry the saddle back to the jockey's room.

vital signs Temperature, pulse, respiration rate, etc.

washy When a horse breaks out in a sweat due to heat and humidity or nervousness before a race.

watered off When a horse has been sufficiently cooled out after a race or workout so that it no longer desires a drink of water when it is offered.

wheel When a horse turns sharply to the left or right immediately after leaving the starting gate.

works Training sessions which are normally scheduled early in the morning every day at a racetrack.

weight allowance The amount of weight "given" to a horse and/or rider to make them level with the competition. For example, a filly running in the Kentucky Derby only has to carry 121 pounds, whereas a colt must carry 126 pounds. In this case, the filly is given a weight allowance of 5 pounds.

winner The horse whose nose reaches the finish wire first.

winner's circle That area on the front side where the winning horse, jockey/catch-driver, trainer, owner, and groom go immediately after a race for awards and/or photographs.

Index

A

B

D

E

H

I

J

K

L

M

N